Revolutionizing Aircraft Materials and Processes

Spiros Pantelakis • Konstantinos Tserpes
Editors

Revolutionizing Aircraft Materials and Processes

Springer

Editors
Spiros Pantelakis
Department of Mechanical Engineering
and Aeronautics
University of Patras
Patras, Greece

Konstantinos Tserpes
Department of Mechanical Engineering
and Aeronautics
University of Patras
Patras, Greece

ISBN 978-3-030-35345-2 ISBN 978-3-030-35346-9 (eBook)
https://doi.org/10.1007/978-3-030-35346-9

This Springer imprint is published by the registered company Springer Nature Switzerland AG
The registered company address is: Gewerbestrasse 11, 6330 Cham, Switzerland

Preface

The increase of international competition in aeronautics and the need for modern aircrafts to cope with the demanding environmental goals ask for challenging technological solutions and the development of breakthrough technologies and concepts. The progress on advancing existing materials and the ability to develop novel materials offering less weight, enhanced mechanical properties, more functionalities, manufacturing flexibility, recyclability, etc. are the key for responding to the emerging needs of increasing efficiency, increasing safety, reducing costs, and decrease in the environmental footprint. In this book, current advances and emerging needs on aeronautical materials and manufacturing processes are presented.

First, an overview of the historical evolution of aeronautical materials is provided along with the technological significance of the advancements achieved through the years for the development of aircraft structures. The progress on the development of aircraft aluminum alloys is presented, and the strengths and limitations of this significant class of materials are discussed. Composites have been proved to be excellent aircraft structural materials due to their desirable properties. To them belong their high specific tensile properties, improved resistance to corrosion, and possibilities for manufacturing integral structures. Special emphasis is given on the impact of composite materials to progress from differential to integral aircraft structures. Furthermore, challenges and problems related to joining of composite structures by bolted joints and adhesives are discussed. On the other side, wide use of composites is related to the urgent need to face issues like fatigue problems of parts subjected to compression, environmental attack (e.g., humidity, UV aging), lack of electrical conductivity, lack of experience in repair, problems with recycling of thermosettings, etc. The development of novel aeronautical materials, such as multifunctional materials, represents a promising way to revolutionize aerostructures. Yet, as the level of technological readiness of these materials remains still low its advancement is currently subject of intensive research efforts, worldwide. Achievements made with these materials as well as the challenges, which still need to be faced, are presented. In this context, special focus is given to the development of nanocomposites and biocomposites. Considering a prospective widespread exploitation of the aforementioned novel materials, the interactive development of novel materials, disruptive design concepts, and suitable manufacturing techniques are discussed. In this context, the example of the integral metallic structures designed by disruptive concepts, including topology optimization and manufactured by additive manufacturing will be highlighted.

This book contains 13 interesting chapters. The chapter by Spiros Pantelakis presents the historical development of aeronautical materials from Icarus till today. Alexis Kermanidis gives an overview of the evolution of aircraft aluminum alloys from the original Al-Cu alloys to modern nano-crystalline and hybrid aluminum alloys. The chapter of Falzon and Pierce presents the evolution, current status, and future challenges of the thermosetting composite materials in aerostructures. Barile and co-authors discuss the most recent advancements and applications of thermoplastic composites, focusing on the reasons why they are increasingly representing a more viable manufacturing solution for structural components. The first chapter of George Lampeas presents the recent developments on additive manufacturing by focusing on the design by topology optimization, materials, and processes. In the next chapter, he presents the specific characteristics of cellular materials that act as cores in aeronautical sandwich structures. The chapter of Sachse and co-authors introduce highly promising concepts on integral, disruptive, and multifunctional aircraft structures. Tserpes and Pantelakis discuss the status and prospective of the nano-enabled multifunctional materials developed for aircraft applications by focusing on the mechanical behavior and multi-scale modeling. In the next chapter, Yi and co-authors present the recent developments of biopolymers and biocomposites by focusing on different materials, their manufacturing processes, properties, and prospective applications. In the next chapter, Guadagno and co-authors present the current shortcomings of self-healing materials for their applications in aeronautics. The chapter of Peer Woizeschke deals with laser-based processes for joining light metals and lightweight structures. Konstantinos Tserpes presents the evolution of adhesive bonding technology in aircraft structures by focusing on materials, processes, NDT methods, design concepts, and certification aspects. Finally, the book closes with the chapter of Kanderakis and co-authors on the contemporary repair challenges and the latest innovations in equipment and methodologies on bonded repair for composite structures.

The editors of this special issue would like to thank the authors for their high-quality contributions and for making this book a success. Additionally, the editors would like to thank the Springer editorial office for their useful support.

Patras, Greece

Spiros Pantelakis
Konstantinos Tserpes

Contents

Chapter 1
Historical Development of Aeronautical Materials

Spiros Pantelakis

1 The First Man in the Air

The dream of flying is not new. The myth of Icarus, who has been the first “flying man,” refers to the time before 1500 BC. The flying dove, that has introduced a sort of propulsion principle, has been constructed and probably also demonstrated by Archytas the Tarentine at 405 BC. About 35 centuries have been needed up to Otto Lilienthal who became at 1895 the first real “flying man.” To design and construct his flying vehicle Otto Lilienthal studied the wings of the stork. December 17, 1903 is widely recognized as the birthday of aviation; it refers to the first flight of the aircraft of the Wright brothers. This aircraft that was made 100% by wood and covering fabric was patented on May 22, 1906. Already at this very early stage of aviation it has been obvious that materials and structures are an essential key for any progress in aeronautics.

2 The First Metallic Aircrafts

A full metallic aircraft, the Junker J1, was presented in December 1915; its material has been 100% steel. This heavy airplane was retired 1 year later to introduce in 1917 an airplane made 100% by an aluminum alloy, namely the duralumin (Fig. 1.1).

S. Pantelakis (✉)
Laboratory of Technology and Strength of Materials, Department of Mechanical Engineering and Aeronautics, University of Patras, Patras, Greece
e-mail: pantelak@mech.upatras.gr

S. Pantelakis, K. Tserpes (eds.), *Revolutionizing Aircraft Materials and Processes*, https://doi.org/10.1007/978-3-030-35346-9_1

Fig. 1.1 Junker J7 (100% duralumin) (Hugo Junkers—Ein Leben für die Technik 2019)

Fig. 1.2 Fuselage fragment of G-ALYP at the Science Museum in London (En.wikipedia.org 2019)

The development of the Al–Cu alloy duralumin represents a revolutionary step in aviation. The Al–Cu alloys, although recently driven back by the polymer base composites, remain up to our days the dominant material for primary aircraft structures of the international flying aircraft fleet. A further step change in aviation represents the use of the jets; it was introduced in July 1949 with the De Havilland Comet aircraft. The aircraft material remained 100% aluminum alloys. The new propulsion concept required and at same time allowed for new design concepts.

Yet, on March 3, 1953 a Canadian Pacific Airlines Comet crashed during takeoff at Karachi, Pakistan. Three fatal Comet crashes due to structural problems, specifically on May 2, 1953; on January 10, 1954; and on April 8, 1954 led to the grounding of the entire Comet fleet. High stress concentration at the quadratic windows of the aircraft as well as extensive corrosion damage of the structural aluminum alloys have been recognized as essential causes of the structural failures observed (Fig. 1.2). After several design modifications were implemented, the Comet services resumed in 1958. In a very painful way we have been confronted with the fact that higher technologies are often associated to higher risks.

3 The Era of Composite Materials

In the 1970s a new era began concerning the aircraft structural materials: the era of composites. Already in 1969 Boeing used 1% composites on the 747 aircraft. The Airbus A300 would feature in 1972 the first use of 4% composite materials of any European passenger large aircraft. The introduction of composites and the progress achieved on developing rivetless joining techniques for "non weldable" aluminum alloys paved the way for a significant step change in the evolution of the aircraft structures: the introduction of the integral structures. It means that riveted structures like the one in Fig. 1.3 can be replaced by rivetless structures of and in addition metals can be replaced by composites (Fig. 1.4). Yet, the evolution of aircraft structures and the replacement of metals by composites have not been so fast as it has been several times anticipated. In 1987 the weight of the Airbus A320 consisted of 68% aluminum alloys, only 15% composites, 9% steel, 6% titanium alloys, and 2% other materials. A number of reasons have caused this slow penetration of composites in aircraft manufacturing. Expensive manufacturing (mainly autoclave), expensive

Fig. 1.3 Rivets in the Boeing 747 front section (Airliners.net 2019)

Fig. 1.4 The full composite forward fuselage of the Airbus A350 XWB (Aero.jaxa.jp 2019)

material, lack of sufficient knowledge concerning the long-term behavior of this class of materials (fatigue, environmental attack by humidity, UAV radiation, etc.), the need to join parts by using bolts, insufficient repair technologies, the need to rethink damage-tolerant design and also some psychological barriers "to trust this black metal" have been some of the causes of the delay on expanding the use of composites in aircraft structural components.

4 The H2020 Vision

It is trivial to notice that in our modern economies in most cases technological and scientific progress are driven by the needs of the economy. When the needs of the economy coincide with the needs of the society the conditions for boosting the technologies that are required for satisfying these needs are fulfilled. In January 2001 the Advisory Council for Aeronautics Research in Europe published its Vision for the year 2020 (Acare4europe.org 2019). The goals of these Visions are to appreciably decrease the environmental impact, increase safety, and boost the competitiveness of the European Aviation sector. The quantitative targets set have been demanding: reduction of fuel consumption and CO_2 emissions by 50%, reduction of NOx emissions by 80%, reduction of perceived external noise by 50%,

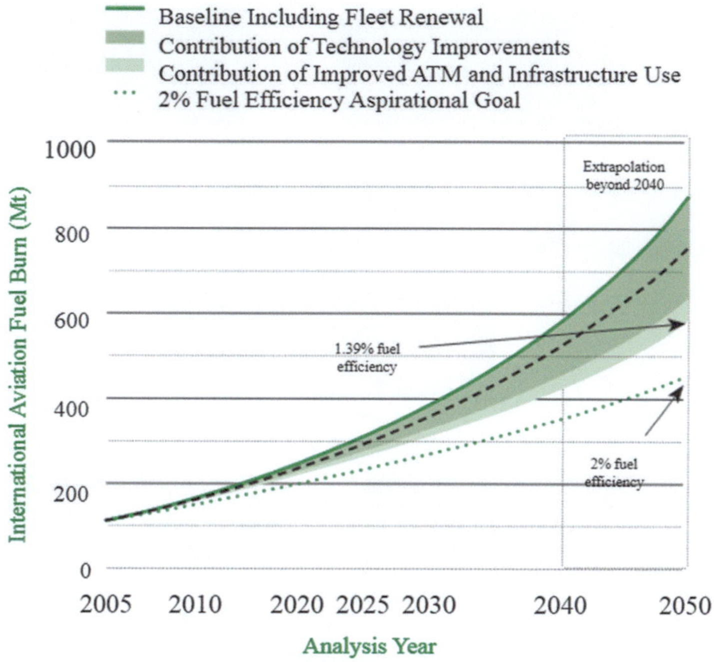

Fig. 1.5 Results for global full-flight fuel burn for international aviation from 2005 to 2040, and then extrapolated to 2050 (Cfapp.icao.int 2019)

substantial progress in reducing the environmental impact of the manufacture, maintenance and disposal of aircraft, reducing accidents by 80%, etc. Displayed in Fig. 1.5 are the results for global full-flight fuel burn for international aviation from 2005 to 2040, and then extrapolated to 2050, considering the contribution of aircraft technology, improved air traffic management and infrastructure use, i.e., operational improvements, to reduce fuel consumption. The figure also illustrates the fuel burn that would be expected if ICAO's 2% annual fuel efficiency aspirational goal were achieved (Cfapp.icao.int 2019).

The advancements achieved in aviation over the years and particularly the advancements achieved on aerostructures make evident that material development, design concepts as well as manufacturing and assembly techniques represent rigs in the same chain. The progress on and the consequences for the evolution of aircraft structures is the result of their interaction. In other words, to exploit the full potential of any essential progress in material development, new design concepts and proper manufacturing techniques may be required. On the other hand, the application of new disruptive design concepts may require for new materials and manufacturing techniques. Figure 1.6 underlines the significance of material development, disruptive design concepts, and advanced manufacturing techniques for achieving emission reductions. The progress achieved on designing, manufacturing and joining aerostructures and the resulting increased use of integral aircraft structures have been essential keys for gradually approaching the ACARE goals.

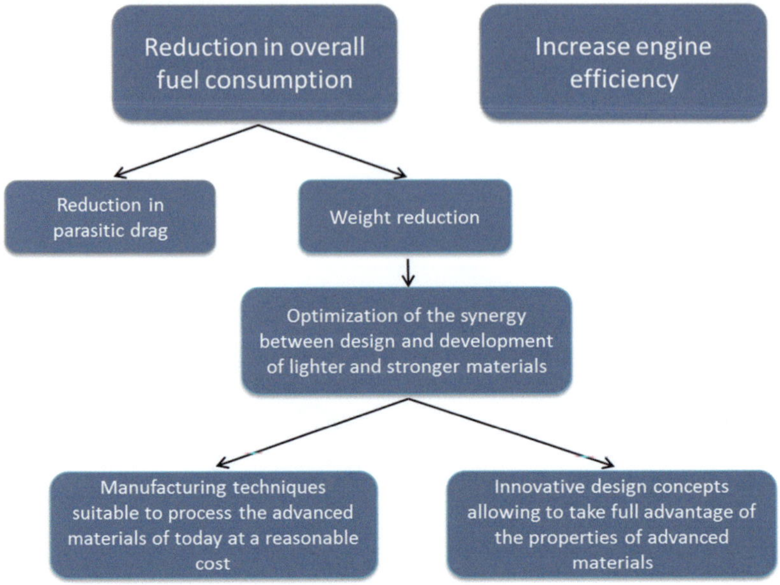

Fig. 1.6 The synergistic effect of material development, disruptive design concepts, and advanced manufacturing techniques for achieving emission reductions

5 Toward Composite Aircrafts

In 2005 the Airbus A380 made its first flight. Its weight consisted of 61% aluminum, 22% composites, 5% steel, 5% titanium, and 7% other materials. This aircraft had up to the time of its first flight the lowest percentage of aluminum by weight of all flying Airbus models. In this aircraft 20 different alloys and tempers had been used compared to the six utilized on the A320/330 aircraft. The A380 also featured the application of a new material for fuselage skins, the aluminum matrix composite material GLARE, that shows improved fatigue and impact properties at a lower density than existing metallic materials. It is worth noticing that up to the introduction of the Boeing 787 and the Airbus A350 airplanes one could summarize the use of structural aircraft materials as following: movable structural parts were made by Fiber Reinforced Plastics, not movable parts, like the fuselage, by aluminum alloys, the pylons for the engines were made by titanium alloys and some specific steels, the landing gear was made by steels. The increased use of CFRPs as well as the use of GLARE in A380 had led to weight reduction of 15 tons compared to what would be if metallic materials were used. Displayed in Table 1.1 are the aluminum alloys used on A380 along with their application. The use of the various types of materials on the A380 may be seen in Fig. 1.7.

Table 1.1 Aluminum alloys used on A380 and their application

Alcan aerospace advanced alloys and their main application on the Airbus A380-800 and A380-800F aircraft			
Form	Alloy/temper	A380 application	Comments
Plates	7056-T7951	Upper wing panels	A380-800F
	7449-T7951	Upper wing	
	2024A-T351	Lower wing reinforcement	
	2050-T84	Lower wing reinforcement	
	2027-T351	Lower outer wing panel	A380-800F
	7010-T7651	Upper outer wing panel, heavier gauge wing ribs	Integrally machined
	7040-T7451	Fuselage main frames, cockpit window frames, beams	
	7449-T7651	Lower gauge wing ribs	
	7040-T7651	Wing spars	Inner front & inner center
Heavy sections	7449-T79511	Upper wing stringers	
	2027-T3511	Lower wing stringers	
	2196-T8511	Floor beams	
Small sections	7349-T6511	Seat rails, stiffeners of center wing box	
	7349-T76511	Fuselage stiffeners	
	2024HS-T432	Fuselage frames	
	6056-T78	Fuselage stiffeners	Associated with 6056-T78 sheet
	6056-T6	Fuselage stiffeners	Associated with 6156CI-T6 sheet
	2196-T8511	Floor structure, fuselage stiffeners	
Sheet	6056-T78	Pressure bulkhead below cockpit floor	
	6156CI-T6	Fuselage panels	

Only 4 years after the first flight of A380, namely in 2011, a dramatic change on the material use in aircraft structures was introduced: the use of composites for the fuselage as well as for most of the components of the wing. Figure 1.8 illustrates the use of materials on the Boeing 787 body. The use of materials on the wide body A350, released in 2015, is nearly the same, namely 52% composites, only 20% aluminum alloys, 14% titanium, 7% steel, and 7% other materials. The evolution in the use of materials on the aircraft body that with the Boeing 787 and Airbus A350 led to a dramatic step change is illustrated in Table 1.2; the table shows the usage of composites in the commercial aircraft industry over the years. In parallel, a significant evolution on the available aluminum alloys also took place. The development of aerospace aluminum alloys, the approximate year of first application as well as the types of product and application are displayed in Fig. 1.9.

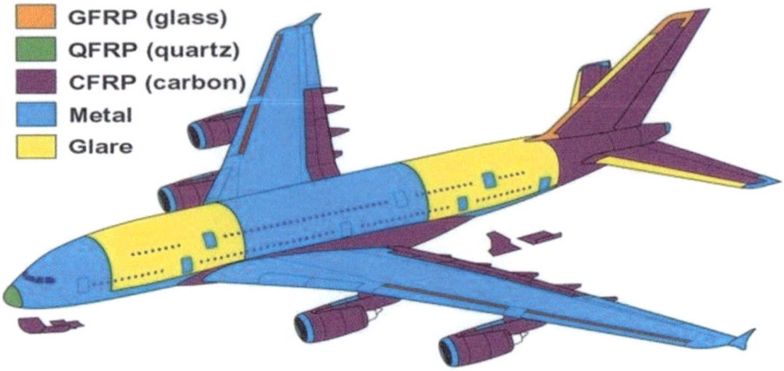

Fig. 1.7 Types of materials on the A380 (Mrazova 2013)

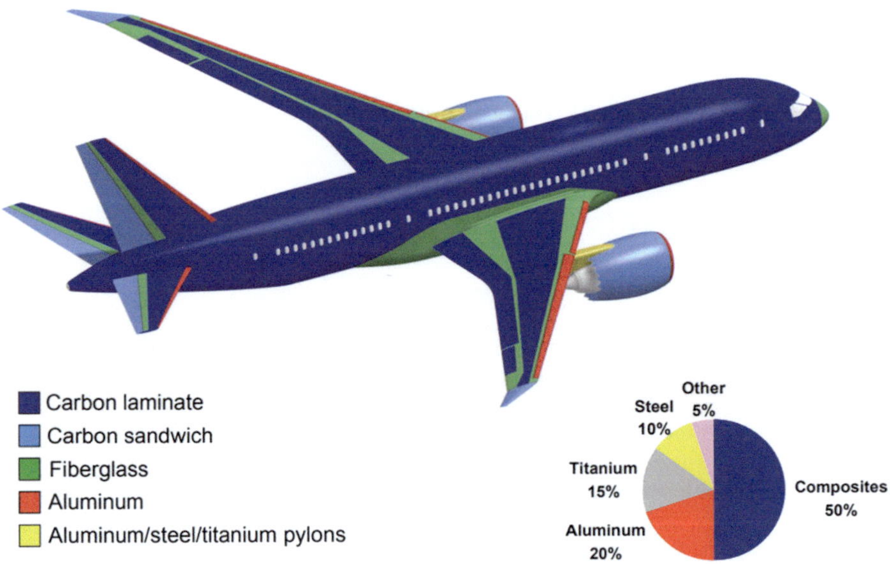

Fig. 1.8 Materials used in Boeing 787 body (Ata-divisions.org 2019)

Table 1.2 Composite usage over the years in commercial aircraft industry

1960–1970	1970–1980	1980–1990	1990–2000	2000–2010	2010–2020	2020–2030
Boron Filaments		First primary structure		A380/787		
0-5%	5-8%	8-12%	12-30%	30-50%	50-70%	70-80%

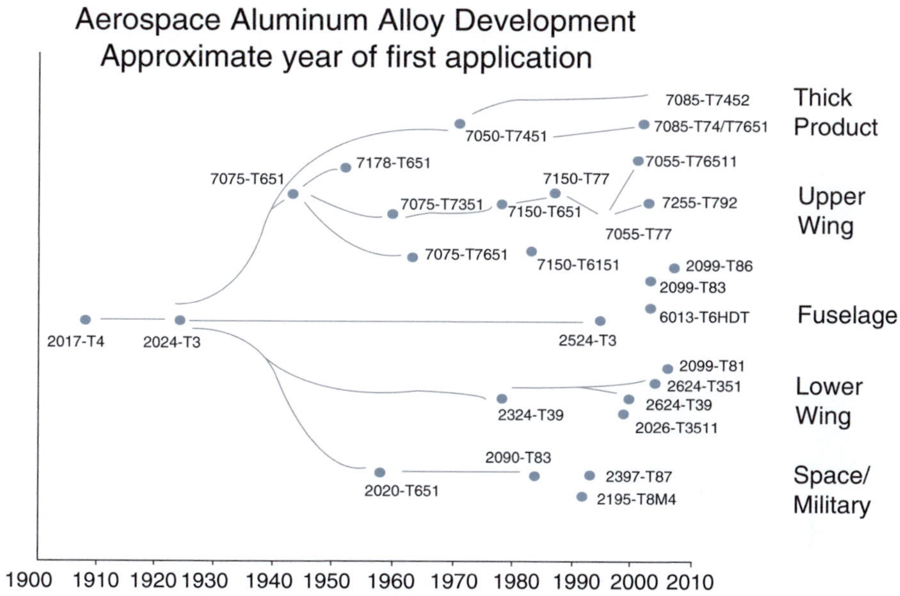

Fig. 1.9 Development of aerospace aluminum alloys, approximate year of first application, and types of product and application (Sae.org 2019)

For completeness, some typical titanium and nickel-based alloys used in modern aircrafts as well as their chemical composition and main mechanical properties are given in Table 1.3, respectively.

The main mechanical properties of the most widely used aluminum alloys used on the aircraft body are compared to the respective properties of composites in Table 1.4.

The desire to replace metals by composites is understandable and well justified when comparing the properties of these categories of materials and particularly the specific properties that are of concern for light weight structures (Table 1.5). A more global comparison between metals and composites is made in Table 1.6. The progress achieved on materials motivated the development of new design concepts and

Table 1.3 Titanium-based alloys and nickel-based alloys

Alloy	Main alloying elements (% wt)	Typical grain size (μm)	Ultimate tensile strength (MPa)	Young modulus (GPa)	Density (g/cm^2)	Elongation at fracture (%)
Titanium-based alloys						
Ti-6A1-4V Grade 5	89.4% Ti–6% Al–4% V–0.4% Fe–0.2% O	40	950	114	4.43	14
Ti-5A1-2.5Sn Grade 6	90.6% Ti–5% Al–3% Sn–0.1% C–0.2% O–0.05% Fe–0.05% N–0.02% H	35	861	110	4.48	16
Nickel-based alloys						
INCONEL X-750	70% Ni–15% Cr–7.5% Fe–2.5% Ti–1% Nb–1% Co–1% Mn–0.5% Cu–0.5% Si–1% Al	35	1250	214	8.28	30
NIMONIC 80A	69% Ni–21% Cr–2% Ti–2% Co–2% Fe–1% Mn–1% Si–1.8% Al–0.2% Cu	40	1250	222	8.19	30

Table 1.4 Mechanical properties of composite materials and aluminum alloys

	Ultimate tensile strength (MPa)	Young modulus (GPa)	Density (g/cm^3)
Composites			
Carbon fiber reinforced epoxy resin (unidirectional)	1550	137.8	1.55
Glass fiber reinforced epoxy resin (unidirectional)	965	39.3	1.85
Kevlar fiber reinforced epoxy resin (unidirectional)	1378	75.8	1.38
Boron fiber reinforced Al 6061-matrix	1109	220	1.55
Aluminum alloys			
AA2024	510	70.3	2.83
AA7075	570	71	2.80
AA7050	470	72.4	2.77

advanced manufacturing and assembly techniques. It has resulted to a significant evolution of the aircraft structures. Characteristic examples of structural advancements made possible through the evolution of aircraft materials are given in Fig. 1.10; they refer to the Airbus A380 aircraft.

Table 1.5 Comparison of the specific properties of aluminum and composites

Aeronautical structural materials		Density (g/cm³)	Specific young modulus (MPa∗cm³/g)	Specific strength (MPa∗cm³/g)
Metals	2024 T3	2.78	26,290	161.15
	7075	2.81	25,510	203.55
	6063	2.70	25,518	89.25
	AZ31C-F	1.77	25,420	146.89
	Ti-6Al-4V grade 5	4.43	25,730	214.44
	INCONEL X-750	8.28	25,845	150.96
Composites	Kevlar fiber reinforced epoxy resin	1.38	46,086	927.53
	Carbon fiber reinforced epoxy resin	1.55	89,873	1443
	Carbon fiber reinforced thermoplastic (CFRTP)	1.57	85,350	915.50

Table 1.6 Advantages and disadvantages of composite materials and metals

	Composite materials	Metals
Advantages	High strength-to-weight ratio	Complex shapes
	Tailored mechanical properties in various directions	High ductility
	Corrosion resistance	Damage resistance
	Impact resistance	Conductive
	Design flexibility	Easy to fabricate
	Fatigue resistance	Joinable by various methods
		Cost effective
Disadvantages	Reduced electrical conductivity	Heavy structures (low strength-to-weight ratio)
	Inferior fire resistance	Prone to corrosion damage
	Brittleness	Inferior creep resistance
	Unpredictability	
	Lack of recyclability (thermosettings)	

6 The Flightpath 2050

The efforts made and the advancements outlined above make no doubt that our airplanes are by far quicker, less polluting, safer, more cost efficient. Yet, at same time new challenges lie ahead. In 2011 the Flightpath 2050 was published by the European commission. Its goals are particularly demanding: 75% reduction in CO_2 emissions per passenger kilometer, a 90% reduction in NOx emissions and a reduction of the perceived noise emission of the flying aircraft by 65%, all compared to the year 2000, emission-free aircraft movements when taxiing, air vehicles that are

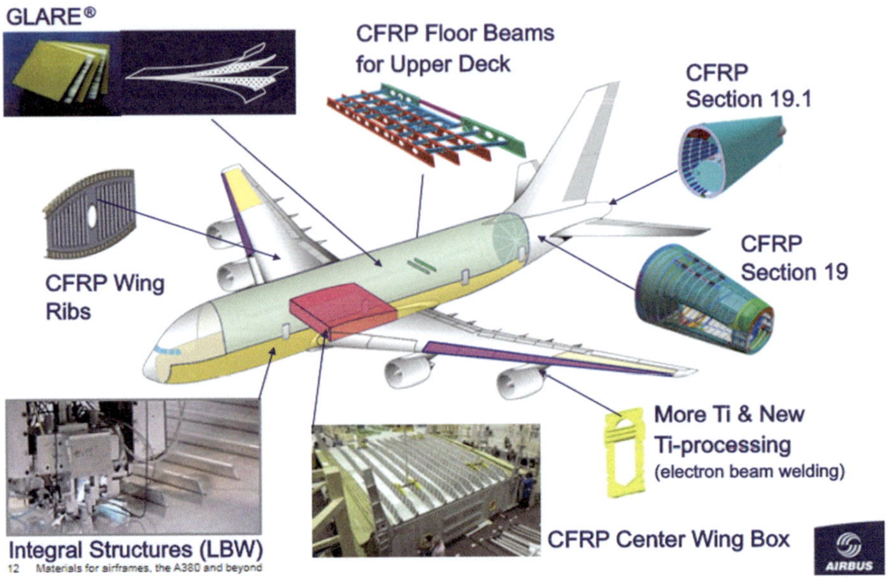

Fig. 1.10 Structural advancements in A380 (Slideplayer.com 2019)

designed and manufactured to be recyclable, extensive use of alternative fuels, less than one accident per ten million commercial aircraft flights are some of the goals set in the Flightpath 2050.

Furthermore it has been widely recognized that global warming represents a major threat to our planet. To face this major problem urgent political decisions and drastic actions are required. The Paris Agreement that was negotiated by representatives of 196 state parties in Le Bourget, Paris and adopted by consensus on December 12, 2015, is a political agreement setting challenging technological goals. One of the main aims of this agreement is to hold the global average temperature increase to well below 2 °C above preindustrial levels and to pursue efforts to limit the temperature increase to 1.5 °C above preindustrial levels. It is obvious that the above very challenging but at same time indispensable objectives cannot be achieved by following the evolutionary approaches that are typical for the aeronautics sector. To meet these challenges new revolutionary concepts will be needed. Hybrid propulsion concepts for large aircrafts, bio-fuels, new batteries, disruptive aircraft design and flight concepts, large thermoplastic structural components such as the fuselage are only some of the research topics of the next day.

7 Materials of the Future

Therefore, it is not surprising that the development of the aircraft materials of the future is subject of extensive efforts. Some of these efforts are following a rather evolutionary approach and have already reached a higher level of technological readiness. To this category belong alloys lattice materials, hybrid laminates consisting of layers of composite and aluminum to provide high-impact strength as well as directional strength, new advanced thermoplastic composites suitable for out-of-autoclave processes, new more cost-efficient titanium alloys, advanced magnesium structural alloys with improved corrosion and fatigue behavior, etc. Other approaches are aiming rather to a step change. Yet they remain up to now at a low level of technological maturity and find so far only limited use. A characteristic example of this category of materials is advanced composites reinforced by carbon nanotubes. Carbon nanotubes having a diameter of few nanometer and length from a few nanometers up to several micron offer a Young's Modulus of the order of 1 TPa and a tensile strength of the order of 60 GPa. As a comparison the Young's Modulus of the aluminum alloy 2024 is about 73 GPa and its strength of the order of 480 MPa. In addition carbon nanotubes offer an electrical conductivity of 109 A/cm^2 that is higher than the electrical conductivity of copper (106 A/cm^2). By using as reinforcement graphene that provides a tensile strength of 1100 GPa a further strength increase could be expected. In 2016 at Texas' Rice University scientists have developed a new graphene-based coating that continuously melts ice by conducting an electrical current. This was achieved by incorporating carbon nanotube graphene nanoribbons in an epoxy/graphene composite. In laboratory tests the scientists were able to heat a rotor blade over 93 °C which in turn resulted in melting away 1 cm layer of ice. Also in 2016 at Farnborough International Air Show the first model aircraft incorporating a graphene skinned wing developed by the University of Manchester, the University of Central Lancashire and several small and medium enterprises, including Haydale Composite Solutions made a successful flight. The development of composites reinforced with carbon nanotubes paved the way to develop multifunctional materials. Multifunctional structural materials possess properties beyond the basic strength and stiffness that typically drive structural design. Multifunctional structural materials can be designed to have integrated mechanical, electrical, thermal, magnetic, optical, power generative, and possibly other functionalities. A subcategory of this type of materials refers to technologies inspired by biology, as for example, the self-healing materials which are inspired by the human skin. Self-healing materials have the ability to autonomically heal cracks. For the self-healing materials a number of concepts are proposed and investigated. They include self-healing via molecular interdiffusion as well as self-healing by activating the polymerization of a healing agent when a propagating crack breaks a hollow fiber or a microcapsule that entails the healing agent. The polymerization is triggered by the contact of the healing agent with an embedded catalyst (Fig. 1.11). As shown in Fig. 1.12 the concept works (Polydoropoulou et al. 2018). Yet, the implementation of this promising category of materials in aircraft structural applications implies the need to advance production techniques for high-quality multifunctional materials at affordable effort and cost, to assess the effect of the additives on the mechanical behavior of the materials and on a possible degradation of certain mechanical properties, such as the fatigue strength, due to the additives or the self-healing mechanism involved,

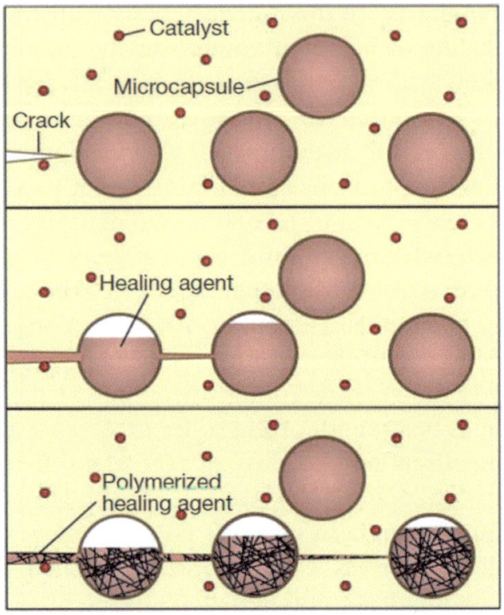

Fig. 1.11 Self-healing mechanism with embedded microcapsules (White et al. 2001)

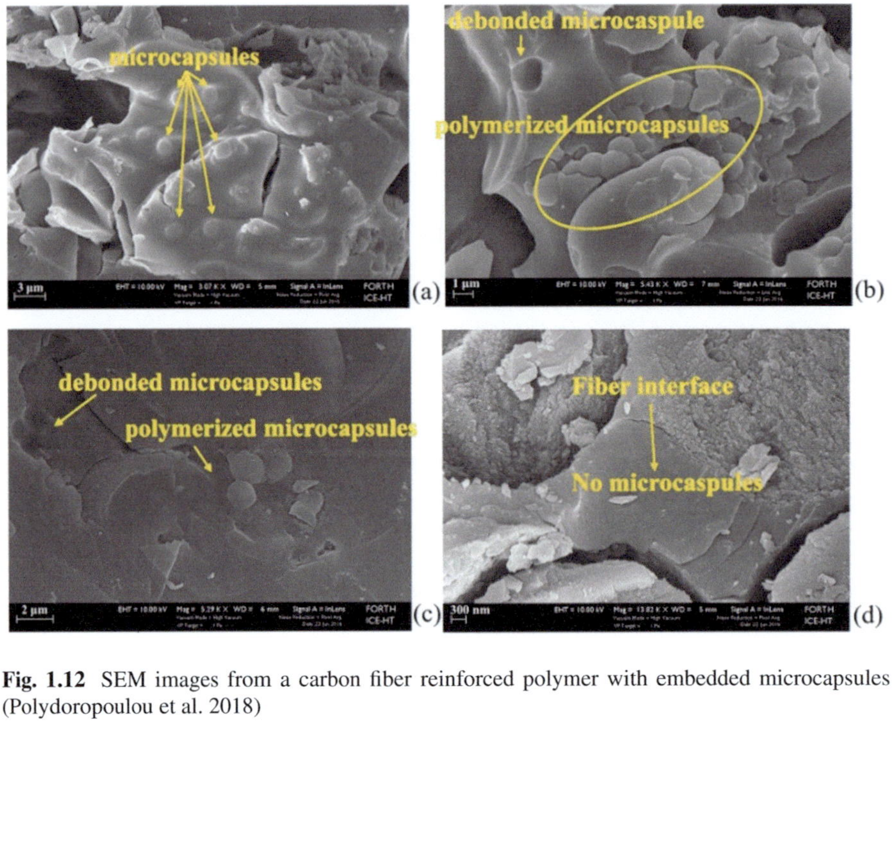

Fig. 1.12 SEM images from a carbon fiber reinforced polymer with embedded microcapsules (Polydoropoulou et al. 2018)

the need to ensure constantly high material quality by avoiding internal defects, the advancement of multiscale models capable to predict the mechanical behavior of potential structures using these materials, etc.

Bio-composites, i.e., composites produced by using, e.g., bio-epoxys and natural fibers as reinforcement, e.g., bamboo fibers, represent already alternative materials for secondary aircraft structures. Several efforts are in progress to further improve the mechanical and electrical properties of this class of composites as well as to advance their manufacturing techniques so as to make the use of bio-composites for load carrying primary structures also possible.

Technologically less mature but very challenging and attractive are efforts to develop new classes of metallic alloys that would be competitive to composites. An example of these efforts represents the nanocrystalline alloys, i.e., metallic alloys with grain dimensions of the order of 100 nm or less. By assuming the validity of the Hall–Petch equation one may expect a huge increase on their yield strength values. By considering also other advantages of metals against composites (e.g., repairability and recyclability) achieving the anticipated yield strength values would make these metallic alloys competitive to composites. Yet, the validity of this equation under a certain value of the nanograin is very doubtful. Furthermore, the grain boundaries are producing an increase in the Gibbs free energy of the system. The consequence of the driving force to reducing free energy is grain growth. Moreover, ductility of this type of alloys is usually very low and information concerning their fatigue behavior is practically missing.

8 Novel Manufacturing Techniques and Simulation-Driven Design

In parallel to the efforts to advance aeronautical materials essential progress could be achieved on developing innovative aircraft structure manufacturing techniques. Additive manufacturing is the flagship among these techniques. The ASTM International Committee F42 on Additive Manufacturing Technologies (Astm.org 2019) defines additive manufacturing as the process of joining materials to make objects from 3D model data, usually layer upon layer, as opposed to subtractive manufacturing methods. Synonyms include additive fabrication, additive processes, additive techniques, additive layer manufacturing, and freedom fabrication. Additive Manufacturing is the general term for a number of technologies. ASTM International Committee F42 on Additive Manufacturing Technologies has categorized AM variations to material extrusion, material jetting, binder jetting, sheet lamination, vat photopolymerization, powder bed fusion, and directed energy deposition. The materials used for AM are both metals and polymers. Hybrid techniques using different categories of materials, e.g., manufacturing a metallic part coated by a polymer have been also developed. Additive Manufacture offers significant advantages. It allows manufacture com-

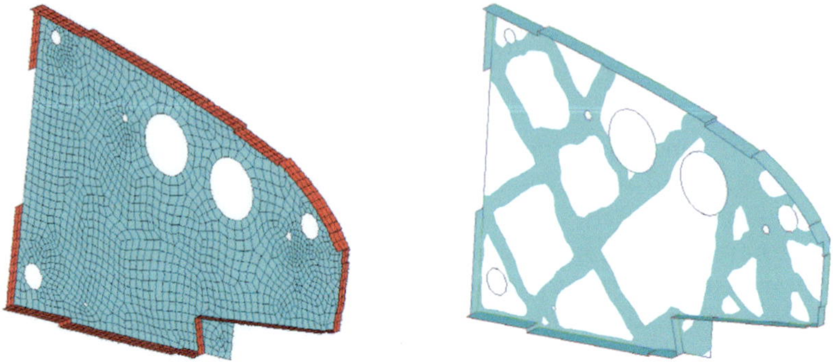

Fig. 1.13 The "famous" airbus A380 LE rib, conventional design (left) and AM future design (right) (Altair HyperWorks Insider 2019)

plex and demanding aircraft components offering optimum material topology, i.e., optimum performance and maximum functionality, load transfer and mechanical behavior. Figure 1.13 is the "classical" figure showing the Airbus A380 LE rib in its conventional (left) as well as in its AM future design (right) (Altair HyperWorks Insider 2019). Therefore, it is not surprising that AM is already exploited to produce aircraft parts. Yet, for a wide spread implementation of AM to produce topology optimized aircraft components a number of issues like powder material systems with appropriate properties, mechanical behavior (part anisotropy, distortion, fatigue behavior, coupling geometrical and structural topology optimization by accounting for failure), production time and cost, appropriate NDT methods, etc., need to be faced.Essential progress could be also achieved on developing out-of-autoclave manufacturing techniques for the use of advanced thermoplastic composites to produce aircraft components. Resin Transfer Molding (RTM) and its variation Vacuum Pressure Resin Transfer Molding VARTM are the most implemented from this category of techniques in aeronautical applications. Industrial potential has been demonstrated also for a number of further techniques. To them belong the Automated Fiber Placement and the Automated Tape Placement as well as the Diaphragm Forming Technique that uses infrared, microwave, or induction for heating; it makes this technique fast and low cost. To mention is also the Continuous Compression Molding which is a proper technique for continuous production of simple profiles.

9 Adhesive Bonding Technology

Motivated by the advancements on materials and manufacturing techniques disruptive structural design concepts have been introduced. Displayed in Fig. 1.14 is a typical design concept for future use. As shown in the figure the semimonocoque

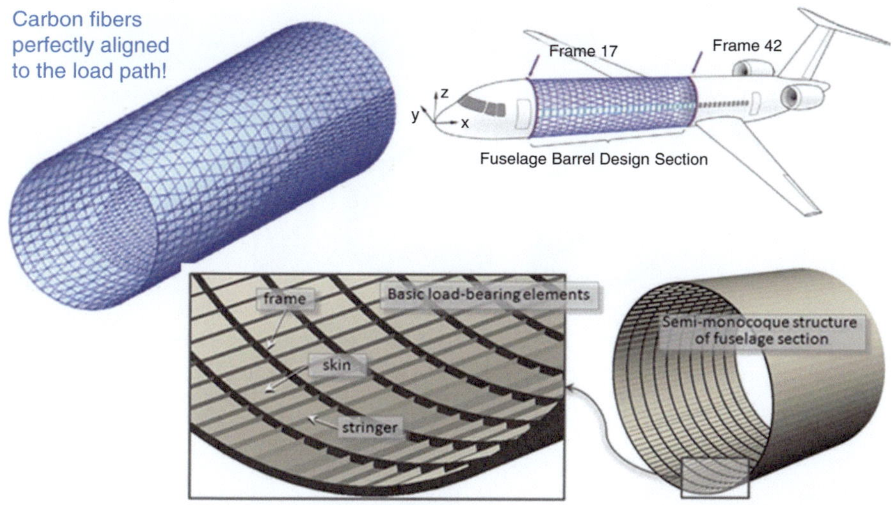

Fig. 1.14 Composite lattice fuselage design instead of semimonocoque fuselage (Cordis.europa.eu 2019)

structure of a fuselage section is replaced by a composite lattice fuselage section (Cordis.europa.eu 2019). It allows to perfectly align carbon fibers to the load path. Not so far from a wide spread application are like the lattice fuselage are adhesives. Adhesive bonding is increasingly replacing bolted joints. Its widespread use requires for varying the joining design concepts (e.g., Fig. 1.15) and facing drawbacks like the environmental ageing, the quality of the bonding the lack of reliable nondestructive techniques for identifying various types of contamination at the bonded surfaces, the low resistance to delamination, etc. Yet, the significant advantages of adhesive bonding like the high joint efficiency index (relative strength/weight of loading joint region), the low part count, the absence of cutouts and the low cost potential along with the progress achieved on developing better adhesives, hybrid NDT techniques to identify contaminations at the bonded surfaces and assess the quality of the bond, improved design and modeling concepts of the joints, etc., increase confidence that adhesive bonding of primary aircraft structures lies not far.

The above short outline of research trends that are expected to appreciably influence the development of the aerostructures of the future is obviously not complete. The progress on the aerostructures of the future will be definitely influenced by the progress on achieving virtual testing in every scale up to full scale, i.e., validated simulations using testing results at lower length scales and virtual testing of very complex aerostructures up to full scale thus reducing development time and cost, the progress on digitalization, the 4.0 Industry, the factory of the future, etc.

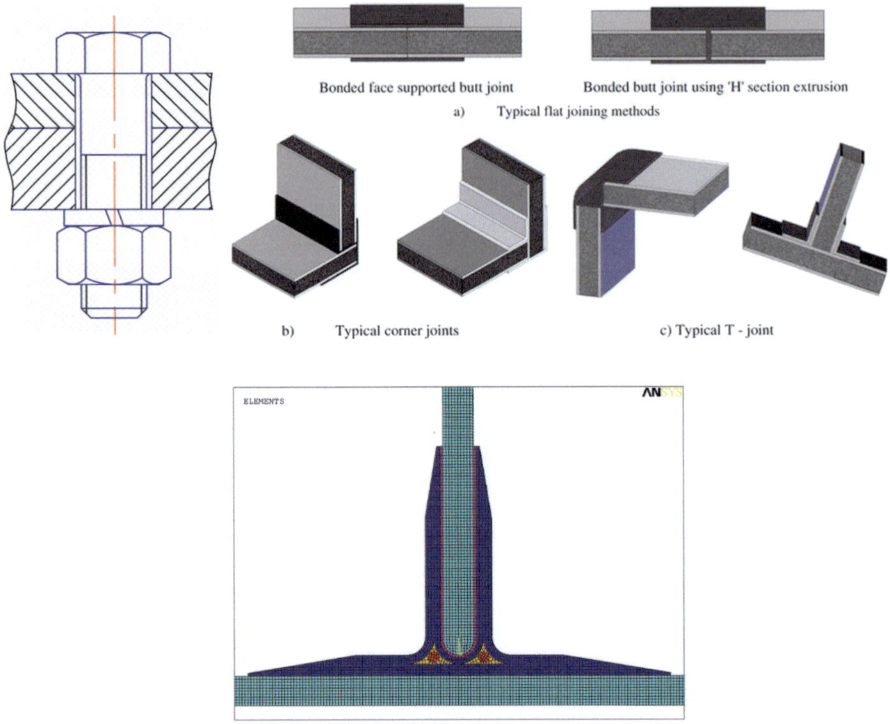

Fig. 1.15 From bolted joints to adhesive bonding

Our future aerostructures will be morphing, i.e., capable of adopting their shapes to the aerodynamic loads, will include health-monitoring abilities at real time, will be self-healing, will be fully recyclable and biodegradable, will be ultralight. The challenges to achieve these objectives are high. They require for evolutionary progress as well as breakthrough and revolutionary steps. Like always in the history of aviation very much will depend on the development of aeronautical materials. Yet, their amazing story will be the subject of future amazing books.

References

Acare4europe.org (2019) [online]. https://www.acare4europe.org/sites/acare4europe.org/files/document/Vision%202020_0.pdf

Aero.jaxa.jp (2019) Structural/composite materials technology|Science & Basic Tech.—Aeronautical Science and Basic Technology Research| Aeronautical Technology Directorate. [online]. http://www.aero.jaxa.jp/eng/research/basic/structure-composite/

Airliners.net (2019) Aviation photo #1096104: Boeing 747-2J9F/SCD—Northwest airlines cargo. [online]. http://www.airliners.net/photo/Northwest-Airlines-Cargo/Boeing-747-2J9F-SCD/1096104

Altair HyperWorks Insider (2019) INNOVATION, SIMULATION, OPTIMIZATION applied to the Architecture, Engineering and Construction (AEC) market. [online]. https://insider.altairhyperworks.com/intuitive-technology-foraec/

Astm.org (2019) ASTM F2792—12a standard terminology for additive manufacturing technologies, (2015). [online]. https://www.astm.org/Standards/F2792.htm

Ata-divisions.org (2019) [online]. http://www.ata-divisions.org/S_TD/pdf/other/IntroducingtheB-787.pdf

Cfapp.icao.int (2019) ICAO 2013 environmental report. [online]. https://cfapp.icao.int/environmental-report-2013/files/assets/basic-html/page35.html

Cordis.europa.eu (2019) [online]. https://cordis.europa.eu/docs/results/265881/final1-2014-03-06-alasca-final-report-submitted.pdf

En.wikipedia.org (2019) BOAC flight 781. [online]. https://en.wikipedia.org/wiki/BOAC_Flight_781

Hugo Junkers—Ein Leben für die Technik (2019) Junkers J7. [online]. https://www.junkers.de/flugzeuge/junkers-j-7

Mrazova M (2013) Advanced composite materials of the future in aerospace industry. INCAS Bulletin 5(3):139–150

Polydoropoulou PV, Katsiropoulos CV, Loukopoulos A, Pantelakis SG (2018) Mechanical behavior of aeronautical composites containing self-healing microcapsules. International Journal of Structural Integrity 9(6): 753–767. https://doi.org/10.1108/IJSI-12-2017-0075

Sae.org (2019) [online]. https://www.sae.org/magazines/pdf/13AERD0201.pdf

Slideplayer.com (2019) Materials for airframes, the A380 and beyond—ppt video online download. [online]. https://slideplayer.com/slide/1554816/

White S, Sottos N, Geubelle P, Moore J, Kessler M, Sriram S, Brown E, Viswanathan S (2001) Autonomic healing of polymer composites. Nature 409(6822):794–797

Chapter 2
Aircraft Aluminum Alloys: Applications and Future Trends

Alexis T. Kermanidis

1 Introduction

1.1 Brief Historical Overview

Aluminum alloys are materials resulting from the addition of alloying elements in pure aluminum, one of the most widely distributed metals on Earth in order to enhance its properties. While its existence has been known for more than 100 years, it was only after the first quarter of the nineteenth century when the first scientists succeeded in the production of aluminum in pure metallic form. Pioneer contributions in the isolation of the metal were in 1825 the Danish Worker, H.C. Oersted, who succeeded in preparing aluminum powder by mixing anhydrous aluminum chloride with potassium amalgam and several years later F. Wohler by replacing the amalgam by potassium and reporting in the years between 1827 and 1847 the main chemical and physical properties. In 1855 French scientist Henri Sainte-Claire Deville improved Wohler's method of preparation by developing an industrial method including a reduction process replacing potassium by sodium, and by using the double chloride of sodium and aluminum as his source of the metal. The world's first industrial production of aluminum was established in 1888, when the electrolytic production of aluminum from alumina and molten cryolite (AIF3 NaF) was independently developed by French engineer Paul Héroult and American engineer Charles Martin Hall, which is now known as the Hall–Héroult process.

Modern production of aluminum is based on the electrolytic process proposed by Bayer who advanced the Hall–Heroult process for making aluminum oxide from mineral bauxite, which contains approximately 25% of aluminum.

A. T. Kermanidis (✉)
Laboratory of Mechanics and Strength of Materials, Department of Mechanical Engineering, University of Thessaly, Volos, Greece
e-mail: akermanidis@mie.uth.gr

S. Pantelakis, K. Tserpes (eds.), *Revolutionizing Aircraft Materials and Processes*, https://doi.org/10.1007/978-3-030-35346-9_2

With the Bayer process bauxite is converted to alumina by digestion with a solution of sodium hydroxide under pressure, the purified alumina produced is added to a molten mixture of cryolite and fluorspar, and is electrolyzed in a cell with carbon anodes. The pure aluminum produced may be cast into ingots, be submitted to addition of alloying elements to produce alloys, and perform specific processes to adjust its mechanical properties (heat treatment) or shape it into a desired form (forming process). Aluminum alloys are used in modern industrial applications having different requirements regarding scale size and design. Typical applications involve their use as electrical conductors, as structural and support materials in the sectors of air and inland transport, building and architecture, in packaging of goods as well as other miscellaneous applications of everyday life.

In air transport aluminum alloys have been used as primary airframe materials since the 1920s. Their lightweight metal structure, which is receptive to heat treatment for tailoring the desired mechanical properties has a density of 2.7 g/cm^2, that allows to produce lightweight materials with high specific properties. Additionally, their relatively low cost and flexible manufacturing, forming and joining capabilities, have extended over the years their applications to mass production of thin and large aircraft structural parts. In the more recent years they have faced significant competition in the lightweight advanced materials sector from composite materials and their use has been today limited from 60% to 20% in modern aircrafts, in parallel to a respective increase of Carbon and Glass Fiber Reinforced Polymers (CFRP and GFRP). Still, the enormous knowledge gained over the years about their metallurgical structure, behavior, damage tolerance and maintenance, in combination with potential for new aluminum alloy development to meet specific existing disadvantages, makes their sustainability for future application as aircraft materials relevant for the upcoming years.

2 Aircraft Aluminum Alloys

2.1 Classification and Composition

Depending on the method of manufacturing and the involved changes in microstructure and properties, there are two major classifications for aluminum alloys, the wrought and cast materials. Cast aluminum alloys contain alloying elements in large amounts and are manufactured in molds in liquid form and made to solidify into the desired shape. They can only be machined further, but cannot be subjected to any metal working operations like forging, etc. Wrought aluminum alloys are amenable to mechanical working operations like forging, rolling, extrusion, deep drawing processes etc. and contain alloying elements in low percentages compared to cast alloys.

Table 2.1 Main alloying elements in the wrought aluminum alloy designation system

Alloy	Main alloying elements
1xxx	Mostly pure aluminum; no major alloying additions
2xxx	Copper
3xxx	Manganese
4xxx	Silicon
5xxx	Magnesium
6xxx	Magnesium and silicon
7xxx	Zinc
8xxx	Other elements (e.g., iron and silicon)
9xxx	Unassigned

2.1.1 Wrought Aluminum Alloys

Aluminum alloys are classified based on a four-digit numerical system (Gilbert and Kaufman 2000). For wrought alloys the first digit indicates the alloy group or major alloy addition, the second digit indicates modifications of the original alloy or impurity limits and the last two digits identify the specific aluminum alloy (Table 2.1).

2.1.2 Cast Aluminum Alloys

For cast alloys the first digit refers to the major alloying element, and the second and third identify a particular composition. The zero after the decimal point identifies the product. The main alloying elements are for each series the following: 1XX.O aluminum 99.00% minimum, 2XX.O copper, 3XX.O silicon, with added Cu and/or Mg 4XX.O silicon, 5XX.O magnesium, 6XX.O unused, 7XX.0 zinc, 8XX.0 tin, 9XX.0 others. Other numerals are used to designate ingots. The other designations are similar to those used to identify wrought alloys. Often, a letter prefix is used to denote either an impurity level or the presence of a secondary alloying element. These letters are assigned in alphabetical sequence starting with A but omitting I, O, Q and X. X is reserved for experimental alloys. For example, A201.0 and A357.0 have higher purity than the original 201.0 and 357.0.

2.1.3 Heat-Treatable (Precipitation-Hardenable) Aluminum Alloys

The initial strength of aluminum alloys is produced by the addition of alloying elements, which exhibit increasing solid solubility in aluminum as the temperature increases. In solution heat treatment, the material is typically heated to temperatures of 480–560 °C, depending upon the alloy. This causes the alloying elements within the material to go into solid solution, procedure which is followed by rapid quenching in water, to trap the alloying elements in solution. Precipitation heat treatment (artificial aging) is used after solution heat treatment, which includes heating the material for a controlled time at a lower temperature (around 120–205 °C). This process, used after solution heat treatment, both increases strength and stabilizes the material. Precipitation strengthening of supersaturated solid solutions involves the formation of finely dispersed precipitates during aging, which impede the dislocation movement.

Heat-treatable aluminum alloys are preferred to non-heat-treatable alloys as aircraft materials mainly due to the significant additional strengthening obtained by the artificial aging process in order to meet the design requirements of structural mechanical strength. With alloy development a systematic improvement of mechanical properties over the years has been achieved as a result of the increasing understanding of the correlation between composition, microstructure, processing, and properties. This has been achieved through modifications in chemical composition, the use of multiple stage heat treatments, and appropriate manufacturing processing. In that regard, heat-treatable aerospace aluminum alloys have been developed with advanced mechanical properties for use in aircraft structures. Wrought, heat-treatable aluminum alloys include the 2XXX, 6XXX, 7XXX, and some of the 8XXX alloys. Because they develop the highest specific strength (strength-to-weight ratio), the heat-treatable alloys have been used for airframes. The predominant aircraft alloys have been the 2XXX when damage tolerance is the primary requirement and 7XXX when strength is the primary requirement. Recently, 6XXX and 8XXX alloys have seen limited use.

2.2 *Alloying Element Concentration*

The effects of main alloying elements on the mechanical properties of aluminum alloys cannot be independently examined but in combination with other factors like type and concentrations of other chemical constituents, heat treatment, etc. Some basic effects of major alloying elements shown in the alloy series of Fig. 2.1 that have been reported may be summarized in the following.

Silicon (Si) Silicon is an important single alloying element used in majority of aluminum casting alloys. Machinability is poor with addition of silicon in aluminum (Aluminum-silicon alloys 1999–2010). In wrought alloys, silicon is used with magnesium at levels up to 1.5% to produce hardening phase Mg_2Si in the 6*xxx* series of heat-treatable alloys (ASM 1990).

Copper (Cu) Copper provides relatively high strength because it provides solution strengthening and the ability for precipitation hardening. The highest strengthening may be achieved when Cu concentration is between 4% and 6% Cu, depending upon the influence of other constituents. Depending on the concentration of copper the resistance to atmospheric corrosion is reduced and Stress Corrosion cracking

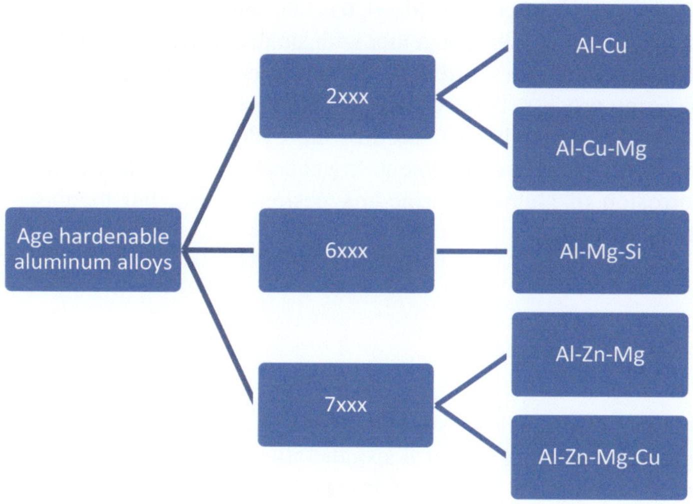

Fig. 2.1 Classification of heat-treatable wrought aluminum alloys

(SCC) susceptibility becomes an issue in specific tempers (T3 and T4). The weldability reduces as the copper content increases (Yeomans 1990).

Copper–Magnesium (Cu–Mg) An increased strength following solution heat treatment and quenching may be obtained by adding magnesium to aluminum–copper alloys. In certain wrought alloys an increase in strength accompanied by high ductility occurs by natural aging, while by artificial aging the further increase in strength is accompanied by reduction of ductility.

Magnesium (Mg) Magnesium provides substantial strengthening, improvement of the work hardening characteristics of aluminum. Corrosion resistance and weldability are good (Davis 1999). Addition of manganese (Mn) in Al–Mg system provides higher strength in the work-hardened condition, high resistance to corrosion, and good welding characteristics.

Magnesium–Silicon (Mg–Si) Wrought alloys of the 6xxx group contain up to 1.5% each of magnesium and silicon in the approximate ratio to form Mg_2Si. The maximum solubility of Mg_2Si is 1.85%, which decreases with increasing temperature. Precipitation hardening occurs by formation of Guinier–Preston bands and fine precipitate dispersion resulting in an increase in strength, though not as great as in the case of the 2xxx or the 7xxx alloys.

Zinc–Magnesium (Zn–Mg) The addition of magnesium to the aluminum–zinc alloys enhances the strength of this system, especially in the range of 3–7.5% Zn. The strength of the wrought aluminum–zinc alloys is substantially improved due to the formation of $MgZn_2$ phase via heat treatment.

Zinc–Magnesium–Copper (Zn–Mg–Cu) The addition of copper to the aluminum–zinc–magnesium system, together with small amounts of chromium and manganese, results in the highest-strength aluminum-base alloys. Zinc and magnesium control the aging process, copper increases the aging rate by increasing the degree of supersaturation, through nucleation of the $CuMgAl_2$ phase and also increases quench sensitivity upon heat treatment. In general, copper reduces the resistance to general corrosion of aluminum–zinc–magnesium alloys, but increases the resistance to stress corrosion.

2.3 *Temper Designation*

A substantial improvement in strength properties is achieved by heat treatment depending on the temper used for a specific alloy system. Temper designation for aerospace alloys supplied in the as-fabricated or annealed condition uses the suffixes F and O, respectively. Alloys supplied in the solution heat-treated condition that have not attained a substantially stable condition by aging are designated by the letter W, and those supplied in the solution heat-treated and aged condition by the letter T. Digits following T designate the type of aging treatment. The aging treatments used are (Rambabu et al. 2017):

T1—Cooled from an elevated-temperature shaping process and naturally aged to a substantially stable condition
T2—Cooled from an elevated-temperature shaping process, cold-worked, and naturally aged to a substantially stable condition
T3—Solution heat treated, cold-worked, and naturally aged to a substantially stable condition
T4—Solution heat treated and naturally aged to a substantially stable condition
T5—Cooled from an elevated-temperature shaping process and artificially aged
T7—Solution heat treated and overaged or stabilized
T8—Solution heat treated, cold-worked, and artificially aged
T9—Solution heat treated, artificially aged, and cold-worked
T10—Cooled from an elevated-temperature shaping process, cold-worked, and artificially aged

3 Performance, Development, and Applications

3.1 *Material Performance*

Aluminum alloy selection for structural aircraft design is based on material performance, which is related to factors associated with structural safety, cost, and environmental friendliness. Extensive knowledge of microstructure–property relationships, experimental observation of mechanical behavior, good understanding of involved damage mechanisms combined with advanced manufacturing, processing and joining capabilities are prerequisites for good material performance. Cost is another critical factor, which includes manufacturing, processing as well as operating costs. End-of-life management (maintenance and recycling) and environmental aspects related to the recyclability potential of the materials also contribute to the decision-making for the use of aluminum alloys.

Respective properties–parameters, which are relevant for the design criteria described above and are used in material selection for aircraft structural applications are given in Table 2.2. Depending on which degree a material type may combine different mechanical, physical or chemical, economical and environmental characteristics, it can achieve a certain performance level based on the schematic representation of Fig. 2.2, thus making it a candidate material for a specific aircraft application. Aluminum alloys from the 2XXX, 7XXX, and 6XXX series may offer a high-performance level based on the schematic diagram of Fig. 2.2 and have been the primary materials during the last decades for the manufacturing of aircraft fuselage skin and wing structures, as well as other secondary applications of military and commercial aircraft. Their high performance may be associated with advantages such as good specific properties, excellent damage tolerance, corrosion resistance, good machining and formability as well as affordable cost in comparison with modern composite materials.

Currently, the 7XXX series Al–Zn alloys are being used for strength; 2XXX series Al–Cu alloys are used for fatigue and damage tolerance critical applications, 6XXX series are medium strength, corrosion resistance, and have good machinability, while Al–Li alloys are chosen where high stiffness and lower densities are required. In Fig. 2.3 a comparison in terms of strength is made highlighting the benefits of 7XXX series alloys with regard to yield and tensile strength with regard to other series aluminum alloys.

Typical drawbacks of aluminum alloys include the limitations in stiffness and specific properties, poor corrosion behavior specifically in the high-strength 2XXX and 7XXX series, and certain limit in mechanical strength potential. In Figs. 2.4 and 2.5 specific elastic moduli and yield strength values of aluminum alloys are presented, showing good performance compared to other technological materials but at

Table 2.2 Design criteria for material selection in aircraft structural applications

Design criteria	Relevant property/parameter	
Properties	Mechanical	Yield strength, tensile strength Fatigue limit, compressive strength, Young's Modulus, elongation at fracture, fracture resistance
	Physical	Density
	Thermal	Thermal coefficient
Formability	Forming energy	
Cost	Price	
Environmental aspects	CO_2 emissions, recycling energy	

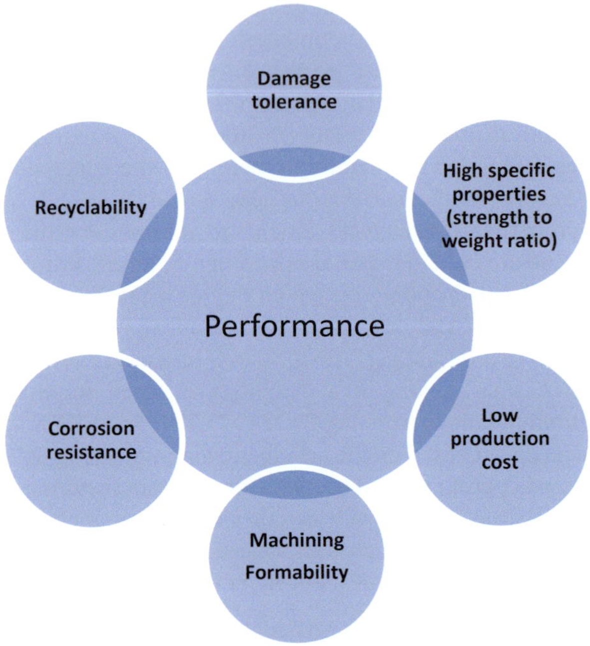

Fig. 2.2 Performance of aluminum alloys for aircraft structural applications

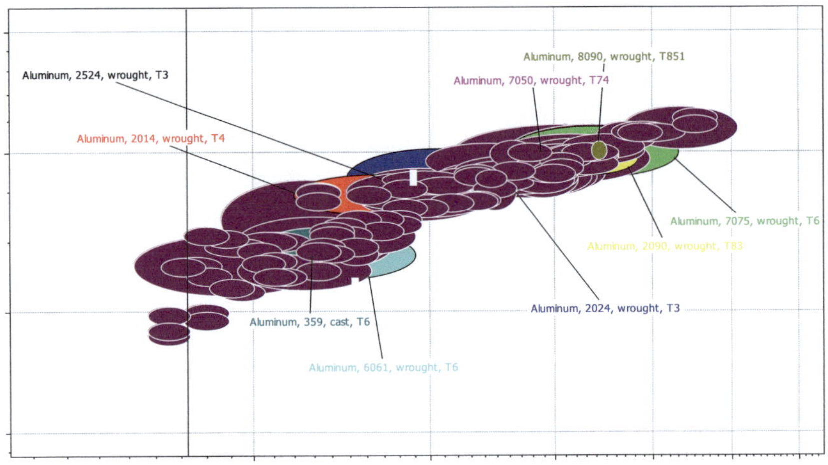

Fig. 2.3 Strength properties of heat-treatable aircraft aluminum alloys. (Figure drawn with CES 2009 EduPack (Charts/data/etc 2009))

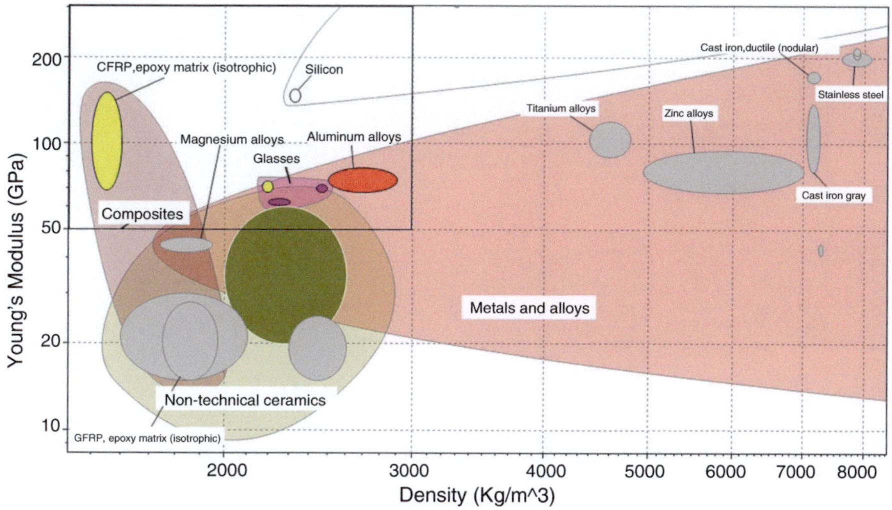

Fig. 2.4 Specific elastic moduli of aluminum alloys (in red) with regard to other technological materials. (Figure drawn with CES 2009 EduPack (Charts/data/etc 2009))

the same time indicating the existing limitations against modern CFRP composites. The new advanced 7XXX series alloys and next-generation Al–Li alloys aim to improve this aspect. On the other hand, in Fig. 2.6 the benefits of aluminum alloys in criteria such as fracture resistance and material cost with regard to modern composite materials are highlighted.

Aluminum alloys are excellent damage-tolerant materials (specifically the 2XXX series alloys) and therefore are used in damage critical airframe parts. The understanding of the property—microstructure relationship is of great importance in aluminum alloy development for the design of microstructures with superior damage tolerance performance (Kermanidis and Pantelakis 2011; Tzamtzis and Kermanidis 2014). An example for better understanding is provided here. With increasing strength, ductility and fracture toughness usually deteriorate. An increased resistance to fracture requires a microstructure, which can accommodate significant plastic deformation, but this generally works at an expense for strength. The balance between strength and toughness is greatly affected by a variety of processing parameters, such as amount of grain boundary precipitates and constituent phases, grain structure and degree of recrystallization (Starke and Staley 1996). In Fig. 2.7 the relationship between toughness and strength for heat-treatable aluminum alloys is presented, showing that the 2024 material of the 2XXX series exhibits superior damage tolerance (combination of high fracture toughness and lower yield strength) compared to alloys 7050 or 7075 of the 7XXX series. The above examples indicate that with regard to specific requirements (e.g., strength or damage tolerance), the performance level of different aluminum alloys (e.g., 7XXX or 2XXX series) may vary, making the use of different material types for different applications a necessity.

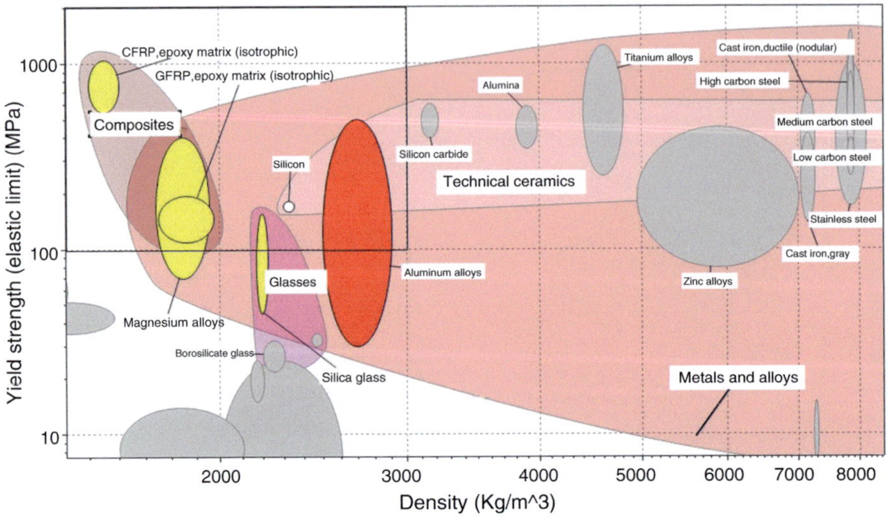

Fig. 2.5 Specific yield strength values of aluminum alloys (in red) with regard to other technological materials. (Figure drawn with CES 2009 EduPack (Charts/data/etc 2009))

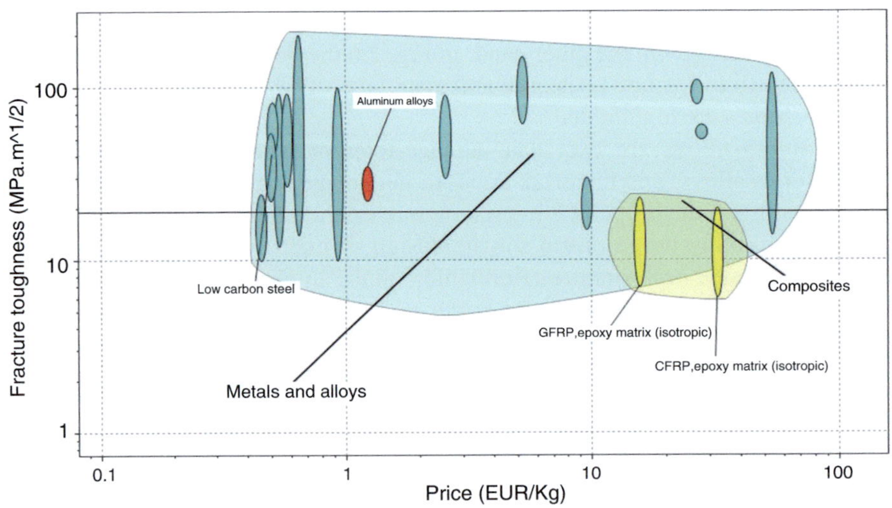

Fig. 2.6 Strength properties of heat-treatable aircraft aluminum alloys. (Figure drawn with CES 2009 EduPack (Charts/data/etc 2009))

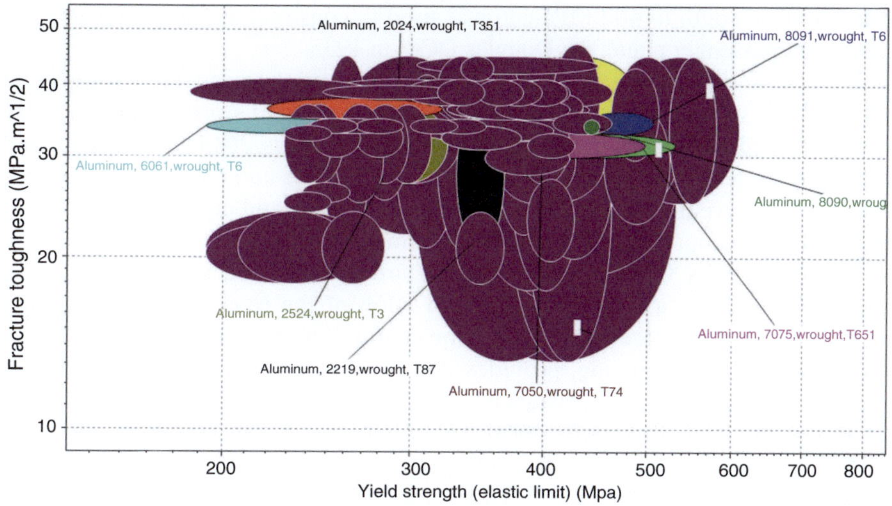

Fig. 2.7 Fracture toughness vs yield strength of heat-treatable aircraft aluminum alloys. (Figure drawn with CES 2009 EduPack (Charts/data/etc 2009))

3.2 Aluminum Alloy Development for Aircraft Applications

3.2.1 Wrought Aluminum Alloys

The aim of the incremental development of heat-treatable aluminum alloys since the beginning of the twentieth century has been to produce lightweight materials, with advanced properties to meet the evolving aircraft structure requirements for low density combined with superior strength, fatigue and damage tolerance, corrosion resistance, machinability and formability at a reasonable manufacturing cost. Based on this concept, various improvements in composition, processing, and study of mechanical behavior have led to the last-generation aluminum alloys from the 2XXX, 6XXX, and 7XXX used today in aircraft applications.

2XXX Series Alloys

The first Al–Cu alloy called Duralumin was discovered in 1903–1911 by Alfred Wilm, with approximately 4% Cu, was amenable to specific heat treatment and had an average mechanical tensile strength of 340 MPa (Merica et al. 1919). Based on Duralumin, Alcoa Company fabricated in 1930 an alloy with much higher strength, the 24s, which is the ancestor of 2024 alloy. The 24s alloy used an increased concentration of Mg of 1.5% rather than 0.5%, moderate cold working and aging to increase its strength to 400 MPa and was used to construct the first commercially successful passenger plane, the Douglas DC-3 in 1935.

The 2024 alloy, which was based on the 24s, used with specific tempers like T3 is a high-strength alloy, which offers very good yield strength and ductility values exceeding the strength of similar primary alloys 2014 (T6 temper) and 2017 (T4 temper) and has excellent fracture toughness and fatigue crack propagation characteristics. However, 2024 does not have good corrosion resistance. In fact, having copper as the main alloy element means that 2024 is susceptible to corrosion. 2XXX alloys in sheet form are usually clad with a high-purity aluminum (alclad), a magnesium–silicon alloy of the 6*xxx* series, or an alloy containing 1% Zn protection coating. The coating, usually from 2% to 5% of the total thickness on each side, provides galvanic protection of the core material and thus greatly increases resistance to corrosion. In alclad form, the 2024 alloy combines high strength with corrosion resistance of the commercially pure cladding. Also, the 2024 alloy is usually difficult for welding using fusion welding methods (Preston et al. 2004). Today, modifications of 2024 alloy with lower impurity levels, types 2224 and 2324 are used in the lower wings of Boeing 777, while the alloy 2524 with even lower levels of impurities to increase toughness is used in the fuselage section.

6XXX Series Alloys

Although not as strong as most 2XXX and 7XXX alloys, 6XXX series alloys combine good formability, weldability, machinability, and corrosion resistance, with medium strength. Alloys in this heat-treatable group may be formed in the T4 temper (solution heat treated but not precipitation heat treated) and strengthened after forming to full T6 properties by precipitation heat treatment. In 1930 Alcoa developed the precipitation hardened alloy 61s (6061) alloy using as main alloying elements in aluminum magnesium (1%), silicon (0.6%), and copper (0.3%), which offered corrosion resistance and good weldability. The 6061 aluminum alloy, including less zinc than for example the 7075 material, provides superior welding abilities and workability over other alloys, but does not exhibit the same high strength and stress resistance as 7075 offers. In 6013 form it has found applications in fuselage skin and stringer due to its improved strength over the 6061 and toughness, its good stretch formability and weldability (Davis 2001).

7XXX Series Alloys

7XXX are high-strength alloys, which exhibit reduced resistance to stress corrosion cracking and inferior damage tolerance compared to 2XXX series and thus are often utilized in a slightly overaged temper to provide better combinations of strength, corrosion resistance, and fracture toughness.

Historically, research on the addition of zinc and magnesium in aluminum showed that strengths higher than 2XXX series may be achieved, which prepared the ground for 7075 alloy using a concentration of 5.5% zinc. Increasing the concentration of zinc to 6.8% and using artificial aging strength values up to 540 MPa could be achieved in the form of 7178 alloy, which was introduced in 1951 (Hatch 1984). As the aircrafts became larger, thicker airframe sections were required to increase the load bearing capacity. This highlighted on one hand the susceptibility of 7178 alloy to stress corrosion cracking (SCC), promoted by the low Cu content (<1% for thick sections), and on the other its low fracture toughness, partially due to high Mg content (>2.5%). A need for optimized Cu and Mg levels was apparent. Some years later (1960s) the development of overaged tempers (such as T73 and T76) made the alloy more resistant to SCC and exfoliation corrosion. Regarding fracture toughness, the development of alloys 7050 (which replaced Cr and Mn with Zr and lowered the concentration of Mg compared to 7075) and 7475 with lower Fe and Si content was able to improve the toughness. The main successor to 7178 was alloy 7055, which using higher concentrations of zinc 8.0%, copper 2.3%, and lower magnesium 2.0% compared to 7178 was able to achieve strengths up to 620 MPa for specific products. The alloy is presently used at the upper wing skin of Boeing 777 (Starke and Staley 1996), which is a highly stressed aircraft part.

The AA7050-T7351 was developed during the 1970s to fit the need for a material that would develop high strength in thick-section products, good resistance to SCC and exfoliation corrosion, and good fracture toughness and fatigue characteristics. These developments are continuing, with one of the latest alloys being AA7085 with very low quench sensitivity and high strength and damage tolerance property combinations in thick sections, which is used in the wing spar and rib structures of the Airbus A380 plane (Chakrabarti et al. 2004).

3.2.2 Al–Li Alloys

Certain limitations in properties described in Sect. 3.1 and the continuous competition from advanced composites toward a lighter aircraft has triggered the search for new solutions in the development of aluminum alloys. The addition of lithium, which is the lightest metallic element (ASM 1990), has proven to be an attractive potential solution for airframe structural weight reduction, since for each 1% Li added, the density of an aluminum alloy is reduced by 3% (ASM 1990). Lithium contributes also to the mechanical property improvement of aluminum alloys. It is unique amongst the more soluble alloying elements in that it causes a considerable increase in the elastic modulus (5% for each 1% Li added) (Starke et al. 1981). Therefore, aluminum–lithium alloys offer higher specific properties compared to other aluminum alloys as shown in Fig. 2.8. They also exhibit better fatigue performance and resistance to fracture under cyclic loading (Welpmann et al. 1984; Jata and Starke 1986; Lavernia et al. 1990; Venkateswara Rao and Ritchie 1992; Wanhill 1994).

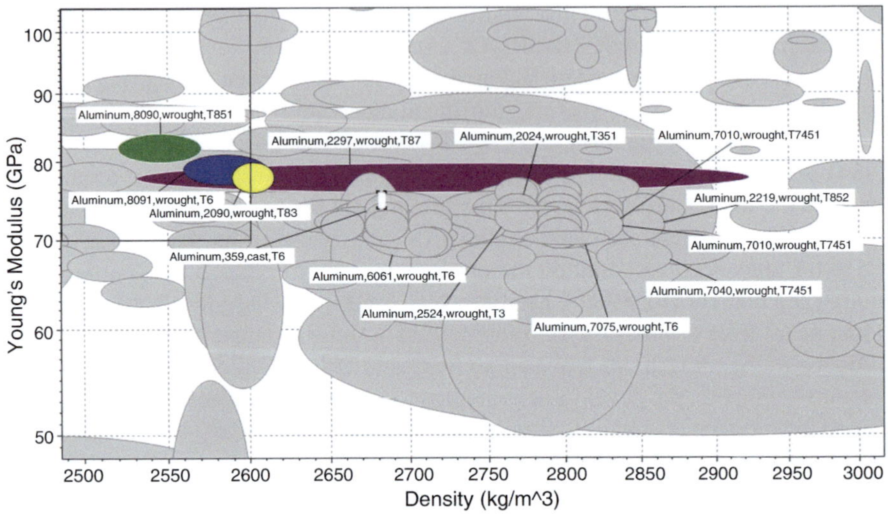

Fig. 2.8 Specific property advantage of Al–Li alloys compared to other aluminum alloys. (Figure drawn with CES 2009 EduPack (Charts/data/etc 2009))

Despite the obvious advantages obtained by the low density and superior cyclic performance, certain disadvantages prohibited until now the extensive use of Al–Li alloys in aircraft. Such disadvantages include low ductility (Lin et al. 1982; Sanders and Starke 1982; Gregson and Flower 1985; Webster 1987), inferior low cycle fatigue resistance (Venkateswara Rao and Ritchie 1992; Sanders and Starke 1982; Eswara Prasad et al. 1996; Eswara Prasad et al. 1997; Eswara Prasad and Rama Rao 2000), inadequate fracture toughness not only in the in-plane but more significantly in the through-thickness directions (Jata and Starke 1986; Gregson and Flower 1985; Suresh et al. 1987; Venkateswara Rao and Ritchie 1989; Lynch 1991; Eswara Prasad et al. 1993a) and high degree of crystallographic texture, which results in anisotropy in the mechanical properties (Peel et al. 1988; Eswara Prasad et al. 1993b; Jata et al. 1998). In addition, poor corrosion resistance and thermal stability at high temperatures resulting in decrease in mechanical properties were factors of concern in the aircraft industry.

The first-generation Al–Li alloy 2020 was developed in 1958 and contained 1.2% Li. In 1984, the second-generation alloys 2090 and 8090 were manufactured containing Li in a concentration higher than 2%, which contributed to 10% lower density and 25% higher specific stiffness than the 2000 and 7000 series alloys.

In the third generation, produced near the beginning of the twentieth century, efforts have been made to improve the performance of Al–Li alloys. The concentration of Li was kept below 2% and the Cu/Li ratio has been increased, while also other alloying elements have been added in order to improve the properties encountered in the previous generations. Characteristic examples are alloys 2090 and 2091

(Dursun and Soutis 2014). In the third generation fracture toughness has been improved, with processes including grain size refinement (Starke and Lin 1982), composition optimization, thermal–mechanical processing, and precipitate microstructure control. Characteristic examples are the new-generation 2199 and 2099 Al–Li alloys, which found applications in the aircraft for fuselage skin-stringer components and lower wing applications and extrusions. In certain cases the addition of Zn and optimization of alloy composition and temper provided improved corrosion resistance (Eswara Prasad et al. 1993b). With the above, alloys 2199 and 2099 presented superior properties compared to conventional alloys as the 2024 alloy in terms of static strength, corrosion resistance, fatigue crack growth performance (Bodily et al. 2012), toughness without accounting for the contribution of lower density, and better weldability (Dursun and Soutis 2014).

Pechinet/Alcan introduced in 2004 another new-generation Al–Cu–Li alloy, the 2050, which was developed for high strength and damage tolerance applications (Lequeu et al. 2010). The alloy was superior compared to 2024-T351 in terms of strength, corrosion resistance, and fatigue and competitive to alloy 7050-T7451 due to improved strength and toughness combination, while at the same time providing 5% lower density and significantly improved stress corrosion resistance.

Al–Li alloy 2198 was developed in 2005 by Alcan to replace modern 2xxx series alloys 2024 and 2524 for damage tolerance and fatigue (in corrosive environment) improvement (Moreto et al. 2011). The newest third-generation Al–Li alloys 2060 and 2055 introduced in 2012 and 2011 by Alcoa (Rajan et al. 2016) showed improved strength/toughness relationship and good thermal stability, excellent corrosion performance compared to that of common aerospace aluminum alloys such as 2024-T3 and 7075-T6. Therefore, these alloys could be alternative materials for fuselage, lower wing and upper wing constructions (Figs. 2.9 and 2.10).

3.2.3 Cast Aluminum Alloys

Heat-treatable casting alloys include the 2*XX*, 3*XX*, and 7*XX* series. Applications of heat-treatable cast alloys in the aircraft sector include alloy A242.0 for diesel and aircraft pistons; air-cooled cylinder heads; aircraft generator housings, alloy 355.0 for aircraft fittings and jet engine compressor cases, alloy 356.0 for aircraft wheels and airframe castings; alloy A356.0 for structural parts, A 360.0 and 518.0 for aircraft castings, alloy 520.0 for aircraft fittings.

The two most widely used aluminum casting alloys for premium quality aircraft castings are alloy A357-T6 and A201-T7. The former exhibits excellent casting characteristics and can produce minimum tensile properties of up to 350 MPa ultimate, 280 MPa yield and 5% elongation in designated areas of sand castings. Non-designated areas can achieve minimum properties from 280 to 315 MPa ultimate, 210 to 245 MPa yield, and 2% to 3% elongation, depending on the casting technique employed. Alloy 201 can provide minimum properties of 420 MPa ultimate, 370 MPa yield, and 5% elongation in designated areas, although this alloy is more difficult to cast.

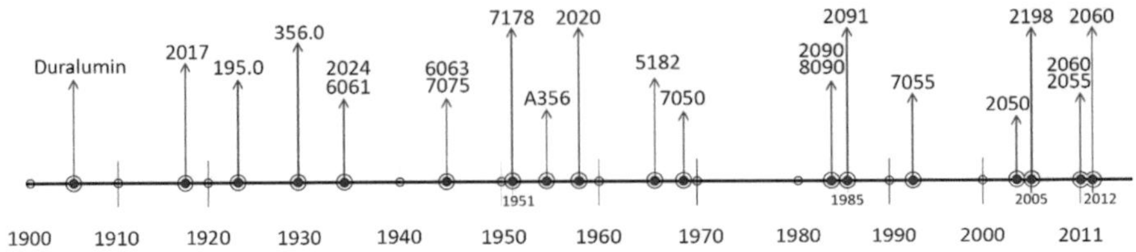

Fig. 2.9 Timeline for development of aircraft aluminum alloys

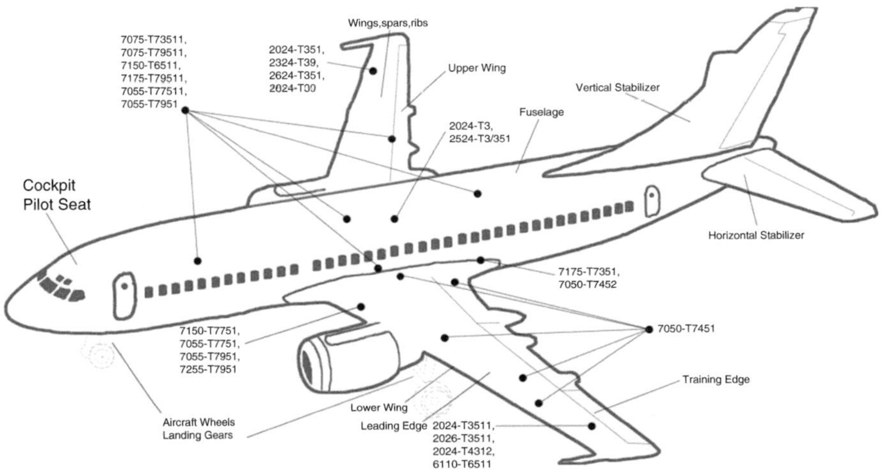

Fig. 2.10 Typical aluminum alloy used in aircraft structural applications

4 Innovation in Processing Technology

The increasing demand of improving the performance level of aluminum alloys by: (1) improving structural integrity, (2) reducing the energy resources and thus increasing the cost efficiency of manufacturing, have led to the continuous search for innovative solutions of fabricating aluminum alloy components as integrated structures with complex shapes and more efficient design.

The aforementioned search has resulted in recent years in the development and implementation of advanced manufacturing and joining processes. The trend of building larger structures with fewer parts has led to demands for thicker and longer plate from which more complex sections can be machined. Alternatively, smaller parts can be joined together with advanced welding techniques (Flower and Soutis 2003). Hence, machining has focused in the processing of thick plates, while the rapid development of forming techniques including cold, warm, and hot forming conditions has contributed to the manufacture of complex-shaped high-strength aluminum alloy panel components (Wu and Yang 2005). In the joining processes, advanced welding techniques of new aluminum alloys offer the potential for replacing differential with integral structures with an associated reduction of aircraft structural weight. Joining methods of aircraft fuselage are developed aiming to replace the use of rivets (Lenczowski 2002).

In the described processes, the requirements involve precision processing with small tolerances, consistency in thickness, flatness, surface appearance, and high-quality control due to the relatively low safety factor used in the design. Also, a specific manufacturing process has to be linked with the specific microstructural modifications or changes to ensure the required properties are obtained. In the next paragraphs an overview of the most up to date, main processes of manufacturing and joining of aluminum alloys is provided.

4.1 Manufacturing

4.1.1 High-Speed Machining from Thick Plate

High-speed machining (HSM) allows economical production of integral parts by providing the opportunity to produce monolithic structural components from thick aluminum plates with reduced machining time, consistent quality, and the ability to machine thin-walled components. With HSM the quality of the machined surface is improved, since the cutting forces are considerably smaller than in conventional machining operations (Perez et al. 2018; Mativenga and Hon 2005). The higher spindle rates are used to make products with lighter, shallower cuts in order to increase the metal removal rates and lower operating costs.

4.1.2 Rolling

Hot rolling is used for producing plate and strip, where cold rolling is used for strip, sheet, and foil. In conventional rolling the technical definitions of hot and cold rolling suggest that metal temperature is high enough to avoid strain hardening (hot rolling) as the metal is deformed. Ductility is enhanced due to the distribution of constituents from grain boundaries, which lessen the flow resistance. The temperature influences the grain morphology with elongated grains resulting from the directionality of rolling and high rolling temperature inducing recrystallization of grains.

In cold rolling the metal is processed at a temperature low enough for strain hardening to occur due to the increase of density of mobile dislocations with increasing plastic deformation, which interact along glide planes. If the hardening effect occurs prematurely and is undesirable in the case of a soft product, annealing is performed to undo the work hardening that has occurred. With cold rolling, which takes more energy as a process, a smoother final surface and different tempers than hot rolling may be obtained (Rolling Aluminum 2008).

Innovations in rolling methods in latest years include: (1) improved methods for monitoring visual quality of the strip, mechanical properties, and flatness; (2) efforts to predict the materials flatness behavior depending on in-going material and cold rolling mill's capabilities; and (3) improvements on fine edge profile control by adapting hot oiling on edges or partially inductive heating applications (Yorulmaz et al. 2016). Rolling may be used in the production with Direct Chill Casting (DC) of ingots or with the Twin Roll Casting (TRC) method. The ingots, which are cast by the conventional DC method are then mainly subjected to preheating and hot rolling operations. With the TRC method, aluminum cast coils are directly produced from liquid aluminum alloys by eliminating a few steps compared to DC. TRC coils are then processed by cold rolling operation in order to achieve desired thickness (Yorulmaz et al. 2016).

4.1.3 Cold/Hot Forming

Cold forming is the process by which a component is formed using a press tool made up of a punch, die, and grip ring after first being put through an approved heat treatment process, which may be used also to relieve stress after forming. Cold forming is much faster than hot forming or superplastic forming, and hence is appropriate for higher volume production. Relatively complex shapes can be formed which include ducting, external engine parts, acoustic panels, and engine cowl skins. Cold forming is lower cost than hot forming or superplastic forming and very much lower cost than machining from solid.

In the field of cold forming methods of particular interest are hydroforming and incremental sheet forming. In sheet hydroforming, instead of the conventional deep drawing with a die cavity, oil or other pressurizing liquid medium are utilized to press the sheet metal tightly onto the punch when it is drawn into the die by the rigid punch. This fluid flow results in a lubrication effect that reduces frictional forces and the quality of the part can be improved (Lang et al. 2004a). With this process, the sheet metal can be drawn to a greater depth and drawing ratio (DR) value of sheet metal can be increased. Sheet hydroforming can also be integrated with other forming processes, such as stamping (Zampaloni et al. 2003) and stretch forming (Lang et al. 2004b). Compared to cold hydroforming, warm hydroforming utilizes the improved formability of aluminum alloys at elevated temperatures, which is effective for high-strength aluminum alloys (Palumbo et al. 2016), such as 6XXX and 7XXX with poor ductility at room temperature. In incremental sheet forming, the single point incremental forming (SPIF) is the most commonly used method, where a blankholder is utilized for clamping and holding the sheet blank, a backing plate to support the sheet. The rotating single point forming tool progressively shapes the sheet to a specific geometry with a CNC machining center (Zheng et al. 2018). Warm/hot incremental sheet forming is an improvement from conventional incremental sheet forming by utilization of a heating device to improve the formability of aluminum alloys (Ji and Park 2008; Fan et al. 2008). Typical forming methods of aluminum alloys are the following.

Stretch Forming

The stretch forming process involves the use of hydraulic rams to force the tool up into the sheet, stretching the whole sheet material to shape (Fig. 2.11a). Stretch forming is a rapid process that can be used for large parts and can be used also with heat treatment for minimizing internal stresses, which reduce the undesired spring-back effect (Gardiner 1957; Pourboghrat; Chandorkar 1992 and Foster et al. 2009). It has lower cost than cold draw forming due to simpler forming tools. The stretch forming process is used for producing large wing leading edges, door panels, and nose sections, due to lower nonrecurring costs and less physical size restrictions on the plant required.

Stamping

Stamping sheet alloys using conventional rigid dies is the most commonly used forming technique, specifically in automotive industry. In warm stamping the sheet blank is heated using an external furnace or hot dies. The process can be either isothermal or non-isothermal depending on the die temperatures. Stamping of high-strength heat-treatable aluminum alloys at high forming temperatures has become feasible and efficient (Gardiner 1957), and is believed to be a leading-edge technique in this area (Bariani et al. 2013; Maeno et al. 2017). It is a hot stamping process integrated with heat treatment to achieve higher strength.

Superplastic Forming (SPF)

Superplastic forming (SPF) shown schematically in Fig. 2.11b is a method especially useful for obtaining complex and precise lightweight metal components (Kappes and Liewald 2011), which is achieved at a particular range of strain rates and temperature. Sheet metal is clamped over a male or female form tool and heated to a high enough temperature to give the material high ductility at low strain rates.

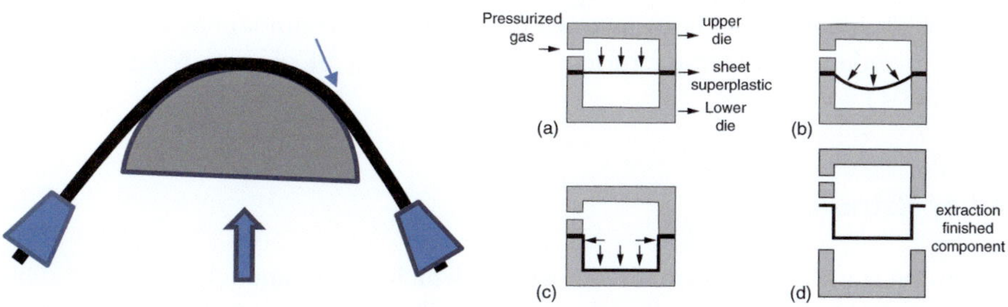

Fig. 2.11 Aluminum alloy forming processes (**a**) stretch forming (**b**) superplastic forming

Pressurized gas (typically argon) on the back face of the sheet forms the material into cavity or over the surface of the tool. One of its advantages is the capacity to produce complex parts in a single operation with a great surface finish, what implies in weight savings as opposed to traditional processes. Moreover, there are slight or no residual stress and "springback" effects, leading to a high-quality structural integrity. In the aerospace industry, the most common applications of SPF use aluminum alloys (2XXX, 5XXX, 6XXX, and 7XXX) for lightly loaded or nonstructural components, such as inlets, wing tips, access doors, and equipment covers (Pereira et al. 2016).

Quick Plastic Forming (QPF)

A main problem with SPF has been the low forming rates for the aerospace materials. Quick plastic forming was developed by General Motors (Krajewski and Schroth 2007) in order to produce less complex-shaped parts and at the same time increase productivity of manufactured parts. The major difference between SPF and QPF is the strain rate and raw material candidate. SPF deforms aluminum alloys at a strain rate exhibiting maximum ductility, while QPF process aims to deform aluminum alloys at strain rates significantly greater than those in SPF to reduce the processing time. In addition, SPF requires raw material that is intrinsically superplastic with fine grains.

Age Forming

Age forming is one of the cost-effective processes for forming skins, which utilizes creep deformation of an aluminum alloy during its heat treatment. In this process, a structural piece is placed on a tool, pressed against the tool by vacuum bagging, and heated under pressure in an autoclave. After aging, the piece is cooled down and released. Springback that occurs because of residual strain at the time of release has to be taken into account in determining the process parameters and tooling so that the final configuration of the piece conforms to the drawing. Investigations on age forming focus on its application to double curvature skins with variable thickness and methods to predict the amount of springback in order to reduce the manufacturing cost and weight of aircraft wings (Adachi et al. 2004).

4.1.4 Extrusion

Extrusion is a plastic deformation process in which an aluminum billet is forced by compression through a smaller die opening. The billet is typically preheated to facilitate the deformation process (Aircraft extrusion 2017). In order to turn the alloying elements into the desired metallurgical structure the extrusions are heat treated to predetermined temperature followed by water quenching to stabilize the mechanical properties. There are many applications for aircraft extrusion. Not just for aerospace use, but material is used where any stress or strength performance is required. Examples of extrusion applications are commercial and military wing and fuselage structures.

4.1.5 Forging

Forging is a manufacturing process where a bar stock is inserted into a die and squeezed with a second closed die. The difference with stamping is that it is used for thicker parts rather than sheets finding broader application in aircraft components. The deformation starts at room temperature (cold forging) and changes the shape and size of the initial part until it has assumed the shape of the die. The hot forging manufacturing process enables the manufacture of parts with complex geometries due to the large amounts of plastic deformation and the consequent increased ductility and is performed at a higher temperature than the recrystallization point of the work metal, thus avoiding strain hardening. A full heat treatment is required on the parts after the forging, to optimize the metallurgical structure and obtain the required mechanical properties. Characteristic applications are frames for aircraft fuselages and precision forged wing components (Aluminum forged products for aircraft applications 2018).

In Fig. 2.12 the usage of major wrought products in a metal aircraft are given as percentage values (Saha 2017), while in Fig. 2.13 the major manufacturing processes of aluminum wrought products according to Boeing such as forging, extrusion, and rolling are displayed schematically.

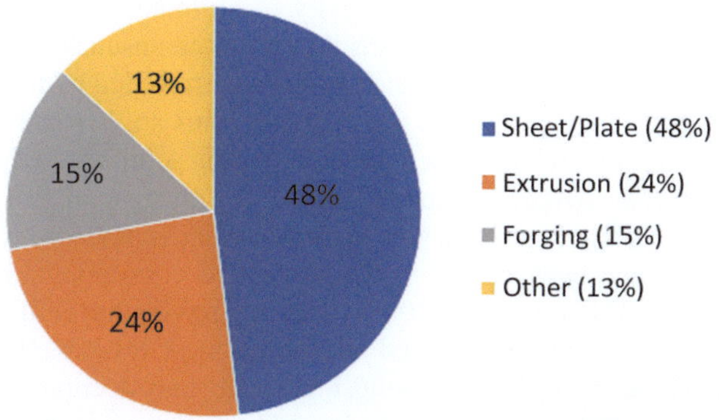

Fig. 2.12 Usage of major wrought products in a metal aircraft. (From Saha 2017)

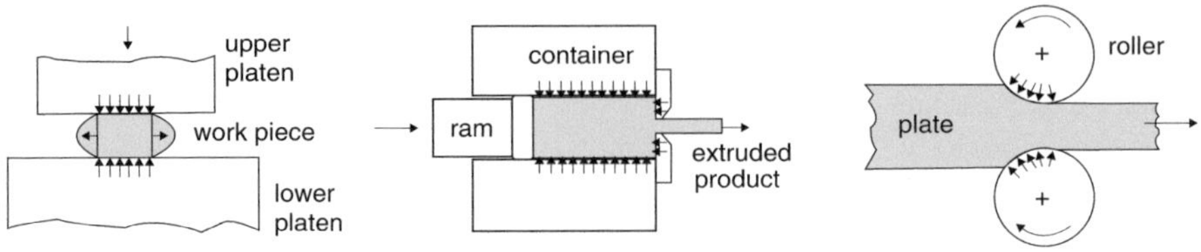

Fig. 2.13 Major manufacturing processes of aluminum wrought products. (From Saha 2017)

4.1.6 Additive Manufacturing

Metal-based additive manufacturing (AM), or three-dimensional (3D) printing is a novel manufacturing methodology, where a single layer of metal is "cast" upon a previous layer allowing for the layer-by-layer building up of a metal component, which is expected to create in the near future new production solutions and opportunities for lightweight aluminum components, complementary to existing machining techniques. With AM, increasing design flexibility, manufacturing of complex geometries, local design optimization and product customization are feasible, expanding the capabilities of aluminum alloy applications in aircraft.

Metallic additive manufacturing systems (Lewandowski and Seifi 2016; Frazier 2014) depend on the heat source (electron beam, laser, or arc), and on how the raw material is supplied. For metal processing, the Selective Laser Melting (SLM) power bed fusion technique (Fig. 2.14b) has received significant attention for production of reliable, defect-free high-strength aluminum alloy microstructures.

The AM method is still not adequate for the majority of commercial high-strength aluminum alloys (Frazier 2014; Herzog et al. 2016). The complex, time-dependent temperature profiles within the part being fabricated resulting from repeated solid state and liquid–solid phase transformations, in association with the poor flowability, high reflectivity and thermal conductivity of aluminum, and the presence of dense and stable oxide layers, are reasons that promote unsatisfactory microstructural characteristics (e.g., large columnar grains) accompanied by the generation of microstructural defects (e.g., cracks, porosity) and thus are barriers for successful processing. Current research is concentrated on finding solutions to the above problems (Kaufmann et al. 2016; Sercombe and Li 2016; Zhang et al. 2016).

Aluminum alloys that have been so far reliably manufactured with SLM are Al–Si alloys, which have been based on the respective conventional casting alloys (Sercombe and Li 2016) and AlSi10Mg (Lewandowski and Seifi 2016; Frazier 2014) alloys. New methods are pursued to produce more tolerable microstructures of high-strength aluminum alloys. Amongst them, the introduction of nanoparticles of nucleants that control solidification during additive manufacturing (Martin et al. 2017), elemental powder mixing, and optimization of processing parameters (Kaufmann et al. 2016; Roberts et al. 2016) have been satisfactorily used for the manufacturing of aluminum alloys 6061 and 7075.

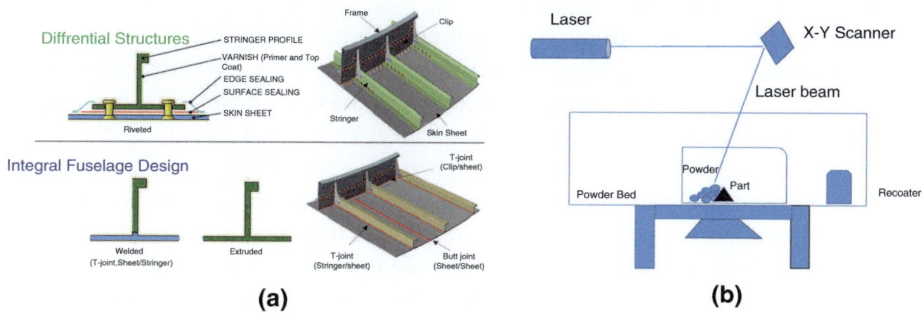

Fig. 2.14 (**a**) Advanced integral structure compared to riveted component; (**b**) power bed fusion process

4.2 *Welding*

Research focus on joining methods of aircraft materials have been directed toward the development of technologies that can reduce the weight of the aircraft, eliminate stress concentrations, reduce heat affected zones, and improve joint efficiency (Fig. 2.14a). Aircraft manufacturers have adopted new welding methods thereby replacing the use of riveted joints, which have certain disadvantages that include stress concentrations promoting fatigue crack initiation and adding a weight burden on the airframe.

4.2.1 Friction Stir Welding (FSW)

Friction stir welding has been considered as the most significant development in metal joining of the past decade and is a solid-state, hot-shear joining process that was developed by The Welding Institute (TWI) in 1991 (Thomas et al. 1991). It is regarded as a green technology because of its energy efficiency, environment friendliness, and versatility. In FSW the welding process begins when the frictional heat developed between the shoulder and the surface of the welded material softens the material, resulting in severe plastic deformation of the material (Nandan et al. 2008; Grujicic et al. 2010). Consequently, the friction stir welding process is both a deformation and a thermal process occurring in a solid state; it utilizes the frictional heat and the deformation heat source for bonding the metal to form a uniform welded joint.

The use of FSW has gained a prominent role in the production of high-integrated solid-phase welds in 2XXX, 6XXX, 7XXX, and Al–Li series, and specifically in materials that are difficult to weld using conventional fusion techniques (Prater 2014) like in the case of 2XXX and 7XXX series alloys (Lequeu et al. 2010). In the precipitation-hardened aluminum alloys (2XXX, 6XXX, and 7XXX), reduction in strength occurs during FSW, in the heat-affected zone due to significant dissolution/coarsening of the precipitates (Liu and Ma 2008; Kermanidis and Tzamtzis 2017).

Investigations toward the improvement of weld mechanical performance have shown that if FSW is performed in specific environment including rapid cooling of the FSW the post-annealing effect, occurring after the FSW process may be remarkably restricted due to cooling, thereby accelerating the refinement of the microstructure and promoting strength characteristics (Liu et al. 2011; Xu et al. 2015).

Variants of the method have been developed to increase weld quality and reduce peak temperatures of the weld cycle. Reverse Dual Rotation Friction Stir Welding (RDR-FSW) supports very low welding loads and improved weld quality. In this process the tool pin and the assisted shoulder are independent and so they can rotate reversely and independently during welding process. This promotes improved weld quality and low welding loads, by adjusting the rotation speeds of the tool pin and the assisted shoulder independently. Experiments (Thomas et al. n.d.) conducted on 7050-T7451 aluminum alloy proved that RDR-FSW significantly reduced the peak temperature reached in the thermal cycle when compared to conventional FSW.

4.2.2 Laser Beam Welding

Laser welding is a crucial joining technology to obtain welds with high depth–width aspect ratios, high quality, high precision, and minimal distortion. LBW uses the radiant energy carried in a very small beam cross section of particularly very high power density, to concentrate on the boundary surfaces of the two parts to be welded together. During the LBW a high-power laser beam is focused onto a metal surface, which melts and vaporizes the metal under the focus creating a weld keyhole eventually generating a weld bead. A laser beam has comparably higher energy density than a typical plasma arc. With the LBW method, welds with high degree of thermal efficiency, deeper penetration as a consequence of metal vaporization in keyhole welding conditions, lower thermal distortion of the weld assemblies, higher welding speeds, narrower HAZ, and better productivity are obtained compared to conventional welding process.

However, the industrial implementation of the system has been perceived as costly in the early days of its introduction due to its very low power conversion rates. Recent developments in laser delivery techniques and resonator technology for CO_2, solid-state fiber, and disk laser configurations have improved the quality of high power laser beams with good conversion efficiencies. CO_2 lasers generally have an electrical to optical conversion efficiencies approaching 20% with very good beam quality, high precision, and high welding speed.

Laser welding of aluminum has great challenges as it involves several physical and chemical processes. For example, aluminum has very low absorption rates due to its high reflectivity, which ranges between 0.86 and 0.90 for pure aluminum at laser wavelengths between 900 and 1000 nm. Therefore, very high specific energy is required in welding of aluminum. LBW is used for specialized operations where minimum heat-input and stress to the weld is required (Ghainia et al. 2009; Sun et al. 2014). Several studies have been performed to understand the behavior of AA2024 welding using different laser power sources (Hu and Richardson 2005;

Alfieri et al. 2011). More focus has been shown toward analyzing the effects welding AA2024 thin sheets that are under 2 mm in thickness using Nd:YAG and CO_2 lasers and satisfactory results have been achieved (Ludovico et al. 2005). Research results have shown that combining laser welding with appropriate heat treatment the corrosion resistance of the aluminum alloy laser weld may be improved (Kermanidis et al. 2010).

4.3 Recycling

Effective recycling when the aircraft structure enters the End of Life (EOL) increases the performance level of aircraft aluminum alloys due to reasons of environmental friendliness and economic aspects. Aluminum manufacturing is an energy-intensive process due to the electrolysis step. On the other hand, when aluminum is directly recovered and reused, it reduces the initial energy by 90% (PAMELA 2008), which in turn also reduces raw material consumption. A typical sequence of activities followed in the recycling procedure is depicted in Fig. 2.15.

Recycling of aircraft aluminum is far from being a fully established process and important challenges encountered today for efficient aircraft aluminum recycling involve:

1. An appropriate selection of recycling methods. Manual methods provide higher quality scrap materials to be recycled, while mechanical methods give the opportunity for reducing the recycling operation costs. Also optimized technologies

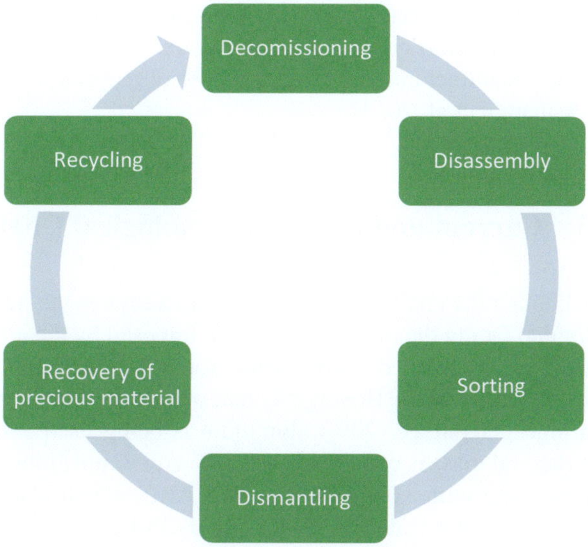

Fig. 2.15 Recycling procedure for aluminum

for the recovery, shredding, sorting, and remelting of aluminum scrap need to be exploited.

2. Applications of recycled materials: (1) Identification of the aluminum alloy composition requirements: In specialized alloys produced today in the aerospace industry with requirements for exceptionally high ductility and toughness, very tight composition controls are allowed. In order to increase the opportunities to reuse recycled aluminum products, there is a need for a more precise identification of composition ranges of current and future recycled metal content, and associate them with candidate alloy composition limits that would most effectively make use of the recycled metal (Merica et al. 1919; The Aluminum Design Manual 2015; Kaufman 2001). (2) Taking advantage of the downcycling potential, which is the reuse of recovered aluminum materials in applications as a lower-value product, such as buildings bicycle frames, vehicles, etc.
3. Correct management and disposal of hazardous materials in aircraft paint primers (e.g., hexavalent chromium).
4. Examination of the recycling of future alloys (Al–Li): The recycling of future aircraft alloys such as aluminum–lithium (Al–Li), aluminum–scandium (Al–Sc) and aluminum–magnesium–lithium (Al–Mg–Li) is another question that remains to be addressed. Currently, the presence of lithium creates an explosion hazard in the aluminum remelting phase and its presence in aluminum scrap is therefore undesired.

An increasing number of companies are becoming active in aircraft dismantling and recycling of aluminum material. Dismantling and recycling demonstrations by individual companies and the formation of the Aircraft Fleet Recycling Association are evidence of aerospace metal recycling efforts (Asmatulu et al. 2013) and helped to set targets for overall aircraft recycling (Eckelman et al. 2014). Within the Airbus program "Process for Advanced Management of End of Life Aircraft" (PAMELA) it was revealed that 85% of the total mass of materials led by aluminum alloys, in obsolete aircraft could potentially be recycled, on the other hand a study of aerospace manufacturers reported that 20% of recyclable materials are actually recycled (Asmatulu et al. 2013).

5 Current and Future Technological Challenges

Although the challenges for aluminum alloys in the next decades are huge, primarily into facing the tough competition imposed by the advanced composite materials, their use is expected to hold strong for years to come as underpinned by modern day applications. In the Boeing 787 half of its structural mass is made of composites and a high percentage (20%) aluminum is being used and in Airbus 350, 40% of its structural mass is composite and 20% aluminum (Jones et al. 2017). Airbus 380 has a fuselage with extensive use of GLARE, which is a combination of carbon fiber and aluminum (Schijve 2009). The F-35 fighter jet has a titanium and aluminum internal structure supporting the exterior composite skin. Moreover the widespread increase in use of composite materials in the manufacturing of aircraft is uncertain since problems related to manufacturing processes, cost as well as damage detection, are factors, which still require specific attention. In any case, the future competitiveness of aluminum alloys as structural materials depends on the improvement of their performance level as described in paragraph 3.1 by exploiting effectively their physical, chemical and foremost mechanical property potential. Research is focused on aspects where aluminum alloys present specific weaknesses compared to composites, especially in microstructural design for the development of new advanced materials with increased strength-to-weight ratio, improved corrosion resistance, and efficient and flexible design.

5.1 Novel Materials Research

5.1.1 Lithium and Scandium

Some good opportunities to enhance performance by increasing the strength-to-weight ratio and corrosion resistance in aluminum alloys come from lithium. Although used in aircraft for more than 20 years, development of the third-generation Al–Li alloys have shown to be very promising materials in that direction. They are cheaper than both CFRP and titanium alloys, provide low density with increased stiffness and improved corrosion resistance as well as damage tolerance compared to traditional alloys of 2XXX and 7XXX series. Still further research is required to address specific manufacturing and joining issues like hole cracking and delamination during drilling in mechanical fastening, poor ductility obtained in laser beam welds, etc. which reduce the reliability for aircraft critical applications. The new Al–Li alloys along with the efficient structural design provide the options for improved structural performance for next-generation aerospace applications. Improvement of properties based on continuous microstructure modifications has resulted in the third-generation Al–Li alloys, which are lightweight competitors to other series alloys with high static and cyclic performance capabilities and improved fracture resistance. For specific applications they may save between 7% and 14% weight over the 2XXX and 7XXX series aluminum alloys (Dursun and Soutis 2014), are less expensive to manufacture and therefore are attractive alternatives for the conventional aluminum alloys, carbon fiber composites, and metal matrix composites for aerospace applications.

The use of scandium provides alternative opportunities for improvement of strength and is explored by the aircraft industry. In that case, strength is achieved by controlling grain growth, keeping grain size small, and relieving internal stresses in the microstructure. Scandium facilitates the creation of small, equiaxed grain structures in the natural solidification process, without the requirement of specific heat treatments. At the same time scandium is a light element, which may offers weight savings compared to other elements (The Aluminum-Scandium Alloy Advantage

2017). Based on this concept, Al–Mg–Sc alloys (e.g., AA5028 H116) have been found to offer weight savings opportunities in the order of 4–5% compared to conventional 2024 without any design change, excellent weldability (laser welding) without prior treatment and moreover providing limited distortions under creep forming, thus further reducing weight and enabling a cost-effective manufacturing process. Today it may be used in low density fuselage skin sheets with medium strength but excellent damage tolerance and corrosion properties as replacement of the traditional 2024 T3 alloy (Aerospace aluminum AA5028 AlMgSc the strong lightweight 2015). The availability of scandium is limited and it is produced today as a by-product of other metal refining processes. Also, Al–Mg–Sc is more expensive than 2024 aluminum, but it is more cost effective than aluminum–lithium—which requires special casting processes and making sure that the scrap is not mixed with other alloys.

5.1.2 Nanocrystalline Alloys and Hybrid Materials

Nanostructured Aluminum Alloys

Novel fabrication methods of nanocrystalline aluminum such as nanopowder consolidation and severe plastic deformation (Sabirov et al. 2013) associated with microstructure control with grain refinement (grain size <100 nm), have shown that the strength of aluminum alloys of 2XXX and 7XXX series may be significantly enhanced based on the Hall–Petch relationship (Choi et al. 2008). On the contrary, low work hardening and ductility due to the low dislocation accumulation capability of the fine grain structure is problematic. Attempts are focused to encounter the specific drawback in 2XXX and 7XXX series aluminum alloys via precipitation optimization with nanoparticles and texture design (Cheng et al. 2007; Pingwei and Hongyun 2016; Zhao et al. 2006). Also Low Cycle Fatigue performance has shown to be an aspect of concern in ultrafined grain aluminum alloys (Malekjani et al. 2011). Cost is a prohibiting factor for widespread use and search for technological applications of nanocrystalline materials, since availability of large quantities of well-characterized material with reproducible properties is required. Even though nanocrystalline powders are now more expensive than the more commercially available coarse-grained powders, maturing of this technology together with greater usage and production will eventually bring down the cost.

Fiber-Metal Laminates

Fiber-metal laminates (FMLs) consist of permanently combined layers of metal sheets and fiber reinforced polymer matrix composites. FMLs are widely used in aerospace and defense industries, their applications in structural components are consistently growing due to their unique features which combine fatigue and impact resistance (Dhaliwal and Newaz 2017; Yu et al. 2015) with relatively low density, flame (high burn-through) and corrosion resistance. Moreover they possess good damping and insulation properties. Regarding their fatigue performance the bridging effect provided by the fibers prolongs the crack propagation phase in FMLs in comparison to monolithic metals (Schijve 2009; Alderliesten and Homan 2006).

FML such as glass laminate aluminum reinforced epoxy (GLARE) and aramid aluminum laminates (ARALL) were mainly developed to achieve weight reduction in the aircraft structures (Gunnink et al. 2000) and for their application in aircraft components where fatigue resistance is required (Armentani et al. 2011), such as the lower wing and upper fuselage skins of an aircraft. The first commercial aircraft to use GLARE in its structure was Airbus A380, where GLARE panels were used in its upper fuselage skin panel structures (Pora 2001). However, these materials are suitable to be used in other areas such as flap skins, cargo bay liner floors and specialty airline containers (Vlot and Gunnink 2001). FML are composed of metals usually aluminum, with either glass, based on R-glass or S2-glass fibers (Vlot et al. 1999), aramid or carbon (commercially known as CARALL).

Besides the cost compared to aluminum, a basic problem resulting from the use of FMLs is the poor machinability, specifically in the field of drilling, where in general poor hole quality is achieved (Hocheng 2012).

Aluminum Composites

A promising potential for advanced structural material applications comes from hybrid aluminum composites, which at present are only developed at research stage. The idea is to use nanostructured aluminum alloys as described in the previous paragraph reinforced with carbon nanotubes or nanoflakes, in order to achieve superior strength and mechanical behavior (Xin et al. 2010).

Aluminum–matrix composites (AMCs), especially the particulate-reinforced AMCs, are also of particular interest due to their attractive physical and mechanical properties such as low thermal expansion, high specific modulus, and superior specific stiffness (Ibrahim et al. 1991; Bonetti et al. 1997; Flores-Campos et al. 2012). In an effort to improve the poor ductility of nanostructured AMCs, while retaining a moderate strength level, various strategies have been implemented, such as involving the introduction of a bimodal grain size distribution (Ye et al. 2005; Li et al. 2009) and decreasing the size of the reinforcement to nanoscale (Zhang et al. 2011), but a moderate increase in ductility has been obtained.

5.1.3 Hybrid Materials Joining

Apart from the research actions for the development of novel aluminum alloys with superior specific properties, modern and future trends in aircraft industry render the successful joining between polymers and aluminum alloys a necessity for the production of lightweight polymer/metal hybrid structures (Abibe et al. 2016; Seong et al. 2008; Goushegir et al. 2014). Together with traditional methods such as adhesive bonding (Teixeira de Freitas and Sinke 2017; Kang et al. 2007) and mechanical fastening (Lambiase and Ko 2017), new techniques are pursued for efficient joining of aluminum with CFRP, such as "spot" joining ultrasonic staking (Amancio-Filho and Dos Santos 2009) and Injection Clinching Joining (ICJ) (Abibe et al. 2016; Abibe et al. 2013), hybrid ultrasonic welding (Lionetto et al. 2017), Friction Spot Joining (FSpJ) (Goushegir et al. 2014; Goushegir 2016; Buffa et al. 2016). At present certain difficulties associated with the low joining strength obtained with specific methods and limited cost efficiency, make an optimization via research of such methods necessary to fully exploit the potential of future application.

References

Abibe AB, Amancio-Filho ST, Dos Santos JF, Hage E (2013) Mechanical and failure behaviour of hybrid polymer-metal staked joints. Mater Des 46:338–347

Abibe AB, Sônego M, Dos Santos JF, Canto LB, Amancio-Filho ST (2016) On the feasibility of a friction-based staking joining method for polymer-metal hybrid structures. Mater Des 92:632–642

Adachi T, Kimura S, Nagayama T (2004) Age forming technology for aircraft wing skin. Mater Forum 28:320–8564

Aerospace aluminum AA5028 AlMgSc the strong lightweight. (2015). www.aleris.com. Accessed 30 May 2018

Aircraft extrusion—profiles and shapes. (2017). www.aircraftextrusion.com/aircraft_extrusion/. Accessed 30 May 2018

Alderliesten RC, Homan JJ (2006) Fatigue and damage tolerance issues of Glare in aircraft structures. Int J Fatigue 28(10):1116–1123

Alfieri V, Cardaropoli F, Caiazzo F, Sergi V (2011) Porosity evolution n AA 2024 BOP and Butt defocused welding by YB-Yag disc laser. Eng Rev J 31:125

Aluminum forged products for aircraft applications. (2018). www.uacj.co.jp/english/products/cast-forg/aerospace.htm. Accessed 30 May 2018

Aluminum-silicon alloys. (1999–2010). http://www.keytometals.com/Article80.htm. Accessed 30 May 2018

Amancio-Filho ST, Dos Santos JF (2009) Joining of polymers and polymer–metal hybrid structures: recent developments and trends. Polym Eng Sci 49(8):1461–1476

Armentani E, Citarella R, Sepe R (2011) FML full scale aeronautic panel under multiaxial fatigue: experimental test and DBEM simulation. Eng Fract Mech 78:1717–1728

ASM handbook volume 2: properties and selection: nonferrous alloys and special purpose materials. (1990) ASM International, Metals Park

Asmatulu E, Overcash M, Twomey J (2013) Recycling of aircraft: state of the art in 2011. J Ind Eng 2013:1–9

Bariani PF, Bruschi S, Ghiotti A, Michieletto F (2013) Hot stamping of AA5083 aluminium alloy sheets. CIRP Ann 62(1):251–254

Bodily B, Heinimann M, Bray G, Colvin E, Witters J (2012) Advanced aluminum and aluminum–lithium solutions for derivative and next generation aerospace structures. SAE Technical Paper paper no 2012-01-1874

Bonetti E, Pasquini L, Sampaolesi E (1997) The influence of grain size on the mechanical properties of nanocrystalline aluminium. Nanostruct Mater 9(1–8):611–614

Buffa G, Baffari D, Campanella D, Fratini L (2016) An innovative friction stir welding based technique to produce dissimilar light alloys to thermoplastic matrix composite joints. Procedia Manuf 5:319–331

Chakrabarti DJ, Liu J, Sawtell RR, Venema GB (2004) New generation high strength high damage tolerance 7085 thick alloy product with low quench sensitivity. In: Proceedings of the 9th international conference on aluminium alloys. pp 969–974
Charts/data/etc (2009) CES EduPack. Granta Design Limited, Cambridge. www.grantadesign.com
Cheng S, Zhao YH, Zhu YT, Ma E (2007) Optimizing the strength and ductility of fine structured 2024 Al alloy by nano-precipitation. Acta Mater 55(17):5822–5832
Choi HJ, Lee SW, Park JS, Bae DH (2008) Tensile behavior of bulk nanocrystalline aluminum synthesized by hot extrusion of ball-milled powders. Scr Mater 59(10):1123–1126
Davis JR (1999) Corrosion of aluminum and aluminum alloys. ASM International, Metals Park
Davis JR (2001) Aluminum and aluminum alloys, alloying: understanding the basics. ASM International, Metals Park
Dhaliwal GS, Newaz GM (2017) Compression after impact characteristics of carbon fiber reinforced aluminum laminates. Compos Struct 160:1212–1224
Dursun T, Soutis C (2014) Recent developments in advanced aircraft aluminium alloys. Rev Mater Design 56:862–871
Eckelman MJ, Ciacci L, Kavlak G, Nuss P, Reck BK, Graedel TE (2014) Life cycle carbon benefits of aerospace alloy recycling. J Clean Prod 80:38–45
Eswara Prasad N, Rama Rao P (2000) Low cycle fatigue resistance in Al–Li alloys. Mater Sci Technol 16:408–426
Eswara Prasad N, Kamat SV, Malakondaiah G (1993a) Effect of crack deflection and branching on *R*-curve behaviour of an Al–Li alloy 2090 sheet. Int J Fract 61:55–69
Eswara Prasad N, Kamat SV, Prasad KS, Malakondaiah G, Kutumbarao VV (1993b) In-plane anisotropy in fracture toughness of an Al–Li 8090 plate. Eng Fract Mech 46(2):209–223
Eswara Prasad N, Malakondaiah G, Kutumbarao VV, Rama Rao P (1996) In-plane anisotropy in low cycle fatigue properties of and bilinearity in Coffin–Manson plots for quaternary Al–Li–Cu–Mg 8090 alloy plate. Mater Sci Technol 12:563–577
Eswara Prasad N, Malakondaiah G, Kutumbarao VV (1997) On the micromechanisms responsible for bilinearity in fatigue power-law relationships in aluminium–lithium alloys. Scr Mater 37:581–587
Fan G, Gao L, Hussain G, Wu Z (2008) Electric hot incremental forming: a novel technique. Int J Mach Tools Manuf 48(15):1688–1692
Flores-Campos R, Estrada-Guel I, Miki-Yoshida M, Martínez-Sánchez R, Herrera-Ramírez JM (2012) Microstructure and mechanical properties of 7075 aluminum alloy nanostructured composites processed by mechanical milling and indirect hot extrusion. Mater Charact 63:39–46
Flower H, Soutis C (2003) Materials for airframes. Aeronaut J (1968) 107(1072):331–341. https://doi.org/10.1017/S0001924000013658
Foster A, Dean TA, Lin J (2009) European patent specification, process for forming aluminum alloy sheet components, EP 2 324 137 B1
Frazier WE (2014) Metal additive manufacturing: a review. JMEPEG 23:1917–1928
Gardiner FJ (1957) The Springback of metals. Trans ASME 79(1):1–9
Ghainia FM, Sheikhia M, Torkamany M, Sabbaghzadeh J (2009) The relation between liquation and solidification cracks in pulsed laser welding of 2024 aluminium alloy. J Mater Sci Eng A 519:167–171
Gilbert J, Kaufman G (2000) Introduction to aluminum alloys and tempers. ASM International, Metals Park
Goushegir SM (2016) Friction spot joining (FSpJ) of aluminum-CFRP hybrid structures. J Weld World 60:1073–1093
Goushegir SM, Dos Santos JF, Amancio-Filho ST (2014) Friction spot joining of aluminum AA2024/carbon-fiber reinforced poly (phenylene sulfide) composite single lap joints: microstructure and mechanical performance. Mater Des 54:196–206
Gregson PJ, Flower HM (1985) Microstructural control of toughness in aluminium–lithium alloys. Acta Metall 33:527–537

Grujicic M, Arakere G, Yalavarthy HV, He T, Yen CF, Cheeseman BA (2010) Modeling of AA5083 material-macrostructure evolution during butt friction-stir welding. J Mater Eng Perform 19(5):672–684
Gunnink JW, Vlot A, Alderliesten RC, Van der Hoeven W, De Boer A, Hart W, Van Hengel CG, Kuijpers PL, Van Oost RC, Roebroeks GHJJ, Sinke J, Ypma MS, De Vries TJ, Wittenberg Th. (2000) Towards technology readiness of fibre metal laminates. In: Paper presented at international congress of aeronautical sciences, 22nd. Harrogate
Hatch JE (1984) Aluminum: properties and physical metallurgy. Aluminum Association, American Society for Metals, ASM International, Metals Park
Herzog D, Seyda V, Wycisk E, Emmelmann C (2016) Additive manufacturing of metals. Acta Mater 117:371–392
Hocheng H (2012) Machining technology for composite materials: principles and practice. Woodhead Pub, Cambridge
Hu M, Richardson I (2005) Autogenous laser keyhole welding of aluminum alloy 2024. J Laser Appl 17:70
Ibrahim IA, Mohamed FA, Lavernia EJ, Mater J (1991) Particulate reinforced metal matrix composites—a review. J Mater Sci 26(5):1137–1156
Jata KV, Starke EA (1986) Fatigue crack growth and fracture toughness behaviour of an Al–Li–Cu alloy. Metall Trans A 17:1011–1026
Jata KV, Panchandeeswaran S, Vasudevan AK (1998) Evolution of texture, microstructure and mechanical property anisotropy in an Al–Li–Cu alloy. Mater Sci Eng A 257:37–46
Ji YH, Park JJ (2008) Formability of magnesium AZ31 sheet in the incremental forming at warm temperature. J Mater Process Technol 201(1–3):354–358
Jones R, Baker AA, Matthews N, Champagne VK (2017) Aircraft sustainment and repair. Butterworth-Heinemann, Oxford
Kang SG, Kim MG, Kim CG (2007) Evaluation of cryogenic performance of adhesives using composite–aluminum double-lap joints. Compos Struct 78:440–446
Kappes J, Liewald M (2011) Evaluation of pneumatic bulge test experiments and corresponding numerical forming simulations. J Mater Sci Eng B 1:472–478
Kaufman JG (2001) Fracture resistance of aluminum alloys—notch toughness, tear resistance, and fracture toughness. ASM International, Materials Park
Kaufmann N, Imran M, Wischeropp TM, Emmelmann C, Siddique S, Walther F (2016) Influence of process parameters on the quality of aluminum alloy EN AW 7075 using selective laser melting (SLM). Phys Procedia 83:918–926
Kermanidis AT, Pantelakis SG (2011) Prediction of crack growth following a single overload in aluminum alloy with sheet and plate microstructure. Eng Fract Mech 78(11):2325–2337
Kermanidis AT, Tzamtzis A (2017) An experimental approach for estimating the effect of heat affected zone (HAZ) microstructural gradient on fatigue crack growth rate in aluminum alloy FSW. Mater Sci Eng A 691:110–120
Kermanidis AT, Zervaki AD, Haidemenopoulos GN, Pantelakis SG (2010) Effects of temper condition and corrosion on the fatigue performance of a laser-welded Al–Cu–Mg–Ag (2139) alloy. Mater Des 31:42–49
Krajewski PE, Schroth JG (2007) Overview of quick plastic forming technology. Mater Sci Forum 551–552:3–12
Lambiase F, Ko DC (2017) Two-steps clinching of aluminum and carbon fiber reinforced polymer sheets. Compos Struct 164:180–188
Lang L, Danckert J, Nielsen KB (2004a) Investigation into the effect of pre-bulging during hydromechanical deep drawing with uniform pressure onto the blank. Int J Mach Tools Manuf 44(6):649–657
Lang LH, Danckert J, Nielsen KB (2004b) Analysis of key parameters in sheet hydroforming combined with stretching forming and deep drawing. Proc Inst Mech Eng B J Eng Manuf 218(8):845–856

Lavernia EJ, Srivatsan TS, Mohammed FA (1990) Review—strength, deformation, fracture behavior and ductility of aluminium–lithium alloys. J Mater Sci 25:1137–1158

Lenczowski B (2002) New lightweight alloys for welded aircraft structure. In: Paper presented at ICAS congress

Lequeu PH, Smith K, Danielou A (2010) Aluminium–copper–lithium alloy 2050 developed for medium to thick plate. J Mater Eng Perform 19(6):841–847

Lewandowski JJ, Seifi M (2016) Metal additive manufacturing: a review of mechanical properties. Annu Rev Mater Res 46:151–186

Li Y, Zhao YH, Ortalan V, Liu W, Zhang ZH, Vogt RG, Browning ND, Lavernia EJ, Schoenung JM (2009) Investigation of aluminum-based nanocomposites with ultra-high strength. Mater Sci Eng A 527(1–2):305–316

Lin FS, Chakravorty SB, Starke EA (1982) Microstructure-property relationships of two Al-3Li–2Cu–0_2XCd alloys. Metall Trans A 13:401–410

Lionetto F, Balle F, Maffezzoli A (2017) Hybrid ultrasonic spot welding of aluminum to carbon fiber reinforced epoxy composites. J Mater Process Technol 247:289–295

Liu F, Ma Z (2008) Influence of tool dimension and welding parameters on microstructure and mechanical properties of friction-stir-welded 6061-T651 aluminium alloy. Metall Mater Trans A 39(10):2378–2388

Liu H, Zhang H, Yu L (2011) Effect of welding speed on microstructures and mechanical properties of underwater friction stir welded 2219 aluminum alloy. Mater Des 32(3):1548–1553

Ludovico A, Daurelio G, De Filippis L, Scialpi A Squeo F (2005) In: Proceedings of XV international symposium on gas flow, chemical lasers, and high-power lasers, Bellingham. p 887

Lynch SP (1991) Fracture of 8090 Al–Li plate–I. Short transverse fracture toughness. Mater Sci Eng A 136:25–43

Maeno T, Mori K, Yachi R (2017) Hot stamping of high-strength aluminium alloy aircraft parts using quick heating. CIRP Ann Manuf Technol 66:269–272

Malekjani S, Hodgson PD, Cizek P, Sabirov I, Hilditch TB (2011) Cyclic deformation response of UFG 2024 Al alloy. Int J Fatigue 33(5):700–709

Martin JH, Yahata BD, Hundley JM, Mayer JA, Schaedler TA Pollock TM (2017) 3D printing of high strength aluminum alloys. Nature 549:365–369

Mativenga PT, Hon KKB (2005) An experimental study of cutting force in high speed end milling and implications for dynamic force modeling. J Manuf Sci Eng 127(2):251–261

Merica PD, Waltenberg RG, Scott H (1919) Heat treatment of Duralumin. Scientific Papers of the Bureau of Standards, vol. 15. pp 271–316

Moreto JA, Gamboni O, Ruchert COFT, Romagnoli F, Moreira MF, Beneduce F, Bose Filho WW (2011) Corrosion and fatigue behavior of new Al alloys. Procedia Eng 10:1521–1526

Nandan R, DebRoy T, Bhadeshia H (2008) Recent advances in friction-stir welding—Process, weldment structure and properties. Prog Mater Sci 53(6):980–1023

Palumbo G, Piglionico V, Piccininni A, Guglielmi P, Tricarico L (2016) Evaluation of the optimal working conditions for the warm sheet hydroforming taking into account the yielding condition. Mater Des 91(5):411–423

PAMELA—Process for Advanced Management of End of Life of Aircraft LIFE05 ENV/F/000059 (2008)

Peel CJ, McDarmaid D, Evans B (1988) Considerations of critical factors for the design of aerospace structures using current and future aluminium–lithium alloys. In: Kar RJ, Agrawal SP, Quist WE (eds) Aluminium–lithium alloys—design, development and applications update. ASM International, Metals Park, pp 315–337

Pereira DA, Batalha MHF, Carunchio AF, Resende HB (2016) Analysis of superplastic forming process applied to aerospace industry: case study of Al 5083 alloy. In: Full paper, Aerospace technology congress, Solna, Stockholm

Perez I, Madariaga A, Cuesta M, Garay A, Arrazola PJ, Ruiz JJ, Rubio FJ, Sanchez R (2018) Effect of cutting speed on the surface integrity of face milled 7050-T7451 aluminium workpieces. Procedia CIRP 71:460–465

Pingwei X, Hongyun L (2016) Improving the ductility of nanostructured Al alloy using strongly textured nano-laminated structure combined with nano-precipitates. Mater Sci Eng A 675:323–337
Pora J (2001) Composite materials in the airbus A380–from history to future. In: Proceedings of ICCM13. Plenary lecture. CD-ROM
Pourboghrat F, Chandorkar K (1992) Springback calculation for plane strain sheet forming using finite element membrane solution. In: American Society of Mechanical Engineers, Computer Engineering Division, CED, vol. 5. pp 85–93
Prater T (2014) Friction stir welding of metal matrix composites for use in aerospace structures. Acta Astronaut 93:366–373
Preston RV, Shercliff HR, Withers PJ, Smith S (2004) Physically based constitutive modelling of residual stress development in welding of aluminium alloy 2024. Acta Mater 52(17):4973–4983
Rajan R, Kah P, Mvola B, Martikainen J (2016) Trends in aluminum alloy development and their joining methods. Rev Adv Mater Sci 44:383–397
Rambabu P, Eswara Prasad N, Kutumbarao VV, Wanhill RJH (2017) Aluminium alloys for aerospace applications. In: Prasad NE, Wanhill RJH (eds) Aerospace materials and material technologies, Aerospace materials, vol 1. Springer, Singapore
Roberts CE, Bourell D, Watt T, Cohen J (2016) A novel processing approach for additive manufacturing of commercial aluminum alloys. Phys Procedia 83:909–917
Rolling aluminum: from the mine through the mill. (2008). www.aluminum.org. Accessed 30 May 2018
Sabirov I, Murashkin MY, Valiev RZ (2013) Nanostructured aluminium alloys produced by severe plastic deformation: new horizons in development. Mater Sci Eng A 560:1–24
Saha PK (2017) Overview of global aerospace parts manufacturing. In: Boeing Research & Technology the Boeing Company Technologies, symposium on auto parts tech day (auto parts and the future industries: aerospace, electric vehicle, and intelligent system) Bangkok, Thailand
Sanders TH, Starke EA (1982) The effect of slip distribution on the monotonic and cyclic ductility of Al–Li binary alloys. Acta Metall 30:927–939
Schijve J (2009) Fatigue of structures and materials. Springer, Amsterdam
Seong MS, Kim TH, Nguyen KH, Kweon JH, Choi JH (2008) A parametric study on the failure of bonded single-lap joints of carbon composite and aluminum. Compos Struct 86(1–3):135–145
Sercombe TB, Li X (2016) Selective laser melting of aluminium and aluminium metal matrix composites: review. Mater Technol 31(2):77–85
Starke EA Jr, Staley JT (1996) Application of modern aluminum alloys to aircraft. Prog Aerosp Sci 32(2–3):131–172
Starke EA Jr, Sanders TH Jr, Palmer IG (1981) New approaches to alloy development in the Al-Li, system. JOM 33:24–33
Starke EA, Lin FS (1982) The influence of grain structure on the ductility of the Al–Cu–Li–Mn–Cd alloy 2020. Metall Trans A 13(12):2259–2269
Sun J, Liu X, Tong Y, Deng D (2014) A comparative study on welding temperature fields, residual stress distributions and deformations induced by laser beam welding and CO2 gas arc welding. Mater Des 63:519–530
Suresh S, Vasudevan AK, Tosten M, Howell PR (1987) Microscopic and macroscopic aspects of fracture in lithium containing aluminium alloys. Acta Metall 35:25–46
Teixeira de Freitas S, Sinke J (2017) Failure analysis of adhesively-bonded metal-skinto-composite-stiffener: effect of temperature and cyclic loading. Compos Struct 166:27–37
The Aluminum Design Manual (2015). http://www.aluminum.org/aluminum-design-manual-2015. Accessed 30 May 2018
The Aluminum-Scandium Alloy Advantage (2017). http://www.scandiummining.com/s/Home.asp. Accessed 30 May 2018
Thomas W, Nicholas E, Needham J, Church M, Emplesmith P, Dawes C (1991) Friction stir welding, England Patent PCT/GB92102203

Thomas WM, Nicholas ED, Needham JC, Murch MG, Temple-Smith P, Dawes CJ (n.d.) Improvements relating to friction welding. European Patent Specification EP0615480B1
Tzamtzis A, Kermanidis AT (2014) Improvement of fatigue crack growth resistance by controlled overaging in 2024-T3 aluminium alloy. Fatigue Fract Eng Mater Struct 00:1–13
Venkateswara Rao KT, Ritchie RO (1989) Mechanical properties of aluminium–lithium alloys: part– I. Fracture toughness and microstructure. Mater Sci Technol 5:882–895
Venkateswara Rao KT, Ritchie RO (1992) Fatigue in aluminium–lithium alloys. Int Mater Rev 37:153–185
Vlot A, Gunnink JW (2001) Fibre metal laminates: an introduction. Springer, Amsterdam
Vlot A, Vogelesang L, De Vries T (1999) Towards application of fibre metal laminates in large aircraft. Aircr Eng Aerosp Technol 71(6):558–570
Wanhill RJH (1994) Status and prospects for aluminium–lithium alloys in aircraft structures. Int J Fatigue 16:3–20
Webster D (1987) The effect of low melting point impurities on the properties of aluminium–lithium alloys. Metall Mater Trans A 18:2187–2193
Welpmann K, Peters M, Sanders TH (1984) Aluminium–lithium alloys. Aluminium 60:641–646
Wu G, Yang JM (2005) The mechanical behavior of GLARE laminates for aircraft structures. JOM 57(1):72–79
Xin X, Zhiqiang L, Zhang D, Chen Z (2010) In situ synthesis of nanostructured carbon reinforcement in aluminum powders. Mater Lett 64(10):1154–1156
Xu N, Ueji R, Fujii H (2015) Enhanced mechanical properties of 70/30 brass joint by multi-pass friction stir welding with rapid cooling. Sci Technol Weld Join 20(2):91–99
Ye JC, Han BQ, Lee ZH, Anh B, Nutt SR, Schoenung JM (2005) A tri-modal aluminum based composite with super-high strength. Scr Mater 53(5):481–486
Yeomans SR (1990) Successful welding of aluminium and its alloys. Australian Welding J 35:20–24
Yorulmaz B, Demir B, Ulus A (2016) Aluminum sheet production: general principles of cold rolling. In: Paper presented at 18th international metallurgy & materials congress, Instanbul
Yu GC, Wu LZ, Ma L, Xiong J (2015) Low velocity impact of carbon fiber aluminum laminates. Compos Struct 119:757–766
Zampaloni M, Abedrabbo N, Pourboghrat F (2003) Experimental and numerical study of stamp hydroforming of sheet metals. Int J Mech Sci 45(11):1815–1848
Zhang ZH, Topping T, Li Y, Vogt R, Zhou YZ, Haines C, Paras J, Kapoor D, Schoenung JM, Lavernia EJ (2011) Mechanical behavior of ultrafine-grained Al composites reinforced with B_4C nanoparticles. Scripta Mater 65(8):652–655
Zhang H, Zhu H, Qi T, Hu Z, Zeng X (2016) Selective laser melting of high strength Al-Cu-Mg alloys: processing, microstructure and mechanical properties. Mater Sci Eng A 656:47–54
Zhao YH, Liao XZ, Cheng S, Ma E, Zhu YT (2006) Simultaneously increasing the ductility and strength of nanostructured Alloys. Adv Mater 18(17):2280–2283
Zheng K, Politis DJ, Wang L, Lin J (2018) A review on forming techniques for manufacturing lightweight complex-shaped aluminium panel components. Int J Lightweight Mater Manuf 1(2):55–80

Chapter 3
Thermosetting Composite Materials in Aerostructures

Brian G. Falzon and Robert S. Pierce

1 Introduction

Composite materials are generally defined as structural materials that exhibit different characteristics due to the combination of two or more immiscible constituents. These usually consist of a reinforcing material, which provides strength and stiffness to the composite, and a matrix material that supports, protects and facilitates load transfer through the reinforcing material. In aircraft applications, Polymer Matrix Composites (PMCs) with continuous high stiffness and strength fibres are most commonly employed. These are often called 'advanced composites' and are the focus of this chapter.

Thermosetting composites are the most widely used subset of the PMC classification, and employ matrix polymers that undergo an irreversible curing process that enables molecular cross-linking during manufacture. Before processing, the thermosetting resins that make up the composite matrix are commonly composed of independent macromolecules that flow relatively freely from a liquid or soft tacky solid state under heating. These are then cured by polymer cross-linking, often at an elevated temperature and pressure, and sometimes using a chemical catalyst, to solidify the composite material. In order to design an appropriate composite structure for aircraft applications, it is important to understand the properties of both the polymer resins and reinforcing fibres that contribute to the overall composite performance.

B. G. Falzon (✉)
Advanced Composites Research Group, School of Mechanical and Aerospace Engineering, Queen's University Belfast, Belfast, UK
e-mail: b.falzon@qub.ac.uk

R. S. Pierce
Department of Mechanical, Materials and Manufacturing Engineering, University of Nottingham Ningbo China, Ningbo, China
e-mail: Robert.Pierce@nottingham.edu.cn

S. Pantelakis, K. Tserpes (eds.), *Revolutionizing Aircraft Materials and Processes*, https://doi.org/10.1007/978-3-030-35346-9_3

1.1 Resins/Prepolymer

Thermosetting resins for advanced composites tend to be composed of long polymer chains that have a high molecular weight. They are predominantly derived from petroleum sources, although there is an emerging interest in bio-resins that can be farmed from renewable sources. Chemically, the cross-linking reaction that occurs during curing results in the formation of covalent bonds between the polymer chains, often with gaseous by-products. The cross-linking density, and mechanical properties (e.g. strength, stiffness and toughness), are dependent on the prepolymer structure and mechanisms of cross-linking.

Once cured, above their service temperature, these thermoset polymers will usually degrade instead of melt, which can be advantageous for fire scenarios as it prevents dripping, although the resulting toxic fumes can still be hazardous. The strength and stiffness are also better retained by the chemically cross-linked polymers during thermal degradation compared to thermoplastics (which are discussed in the following chapter). Additionally, thermosets tend to show better creep behaviour and fibre wetting, along with improved adhesion, not only with reinforcing fibres but also with paints and non-composite materials. The relatively low viscosity of thermosets makes processing easier and manufacturing can be conducted at lower temperatures than thermoplastics. This allows for a variety of manufacturing approaches, and the capability to form complex shapes.

However, the chemical reaction of cross-linking in thermosetting polymers is relatively slow and often requires heat input, high-pressure or mixing with a catalyst. This means production cycles usually take many hours. The chemical curing reaction can also release various gases (including volatile organic compounds) that can be a safety hazard and affect part quality by introducing voids. Furthermore, because the curing process is irreversible, the polymer matrix and any composite materials cannot be easily recycled (Biron 2014). Gradual curing of the prepolymer resins, even at room temperature, also generally limits the shelf life of these materials, and refrigerated storage is often required.

Among a diverse range of thermosetting resins for polymer matrix composites, polyesters are the most widely used. Their versatility, low cost and good balance of mechanical and chemical properties has made them the primary choice for the marine industry in particular. However, as a lower quality material with limited operating temperatures, polyesters are not so common for aircraft composite applications. Similarly, vinyl esters are often favoured for chemical tanks or marine applications due to greater strength and toughness than polyesters, along with good adhesion, heat resistance and corrosion resistance. They are generally unsuitable for the high-performance requirements of the aerospace industry.

In aircraft applications, several advanced thermosetting resins are commonly used, among which epoxies are the most prevalent and well-studied. Epoxies offer a broad range of attractive properties, including high specific strength and stiffness, and excellent corrosion and fatigue resistance. They can also be fine-tuned to further improve toughness, impact strength, and resistance to moisture absorption by blending with additives, e.g. thermoplastic particles to increase toughness. The right epoxy formulations can be suitable for service temperatures as high as 120 °C, which meets the requirements of subsonic aircraft structures. However, thermal degradation of the epoxy becomes problematic for higher temperature regions such as the engine and nacelle structures, or the aerodynamic surfaces of supersonic aircraft, where specialised resins need to be used. Generally, epoxy resins are more expensive than the more basic polyester and vinyl ester resins, particularly when enhanced with high-performance additives.

For cabin interior applications, where mechanical performance is less important but fire resistance and low smoke emission are essential, phenolic resins have been traditionally employed. These are relatively low cost, compared with comparable fire-resistant systems, and offer good thermal behaviour, high bond strength, dimensional stability and creep resistance at the cost of reduced mechanical performance compared with epoxies.

Cyanate ester resins are another specialised matrix material that are typically employed in composites for aircraft radomes or antennae due to their superior dielectric properties. They offer good strength, toughness, low moisture absorption, high temperature stability and good fire resistance but are relatively expensive, even compared with epoxies. Hence, their use is quite limited.

The drive towards greater resin performance, in terms of damage tolerance and thermal resistance, has led to considerable interest in more exotic polyimide, bismaleimide and benzoxazine resins that can be particularly expensive. Polyimide resins require high temperature and processing pressure, with a long cure profile, and exhibit greater processing challenges in terms of volatile gas and moisture release during cure. However these resins result in a matrix with superior thermal performance compared to epoxies; which combined with good chemical and fire resistance makes them ideal for lightweight tooling, missile parts and jet engine components. The closely related bismalimide (BMI) resins exhibit similar high temperature properties and are used on high-performance aircraft such as the F35 Lightning II Joint Strike Fighter.

Benzoxazines are also being used for structural applications where hot/wet performance or fire resistance is key, although these resins remain relatively immature compared to other matrix systems and were originally intended for aircraft interiors. Benzoxazines also offer great potential for hybridisation with epoxy, phenolic or BMI polymers due to their reactive nature. A high aromatic content and high molecular weight mean that these matrix materials can achieve greater mechanical properties and fire resistance than epoxies and phenolics, respectively, at a lower cost than BMI systems. Accounting for the costs associated with storage, handling, layup, processing, finishing and repairs, it has been suggested that benzoxazines may even be cost-competitive with epoxies (Gardiner 2014a). Primarily this is due to their room temperature stability, which negates the need for frozen storage and thawing. However, at lower processing pressures (below 700 kPa) considerable controls are necessary to vent gaseous products during cure.

1.2 Reinforcements

Reinforcement materials within a composite are typically stronger, stiffer and harder than the matrix material that bonds them together. As such, they act as the main load carrying constituent and can make up 60–70% of the volume of a high-performance composite (fibre volume fraction). Most structural applications rely on continuous, thin diameter, reinforcing fibres, as these provide a relatively large fibre–matrix interfacial area that allows for better bonding and load transfer. A range of different fibrous reinforcement materials can be used to produce thermoset composite parts, depending on cost and performance requirements. Glass fibres are relatively cheap, have a 10–20 μm diameter, and are widely used in the marine and automotive industries. Glass fibre composites are also found on the secondary structures of passenger aircraft, but relatively poor strength and stiffness has prevented their use on load-bearing (primary) large-aircraft structures, beyond being deployed as an electrically insulating layer. One notable exception is the use of GLARE (Glass Aluminium Reinforced Epoxy) on the upper fuselage of the A380. This material is made from the interleaving of alternate thin layers of glass fibre laminates and aluminium. Aramid (Kevlar) fibres, of around 12 μm diameter, provide higher stiffness than glass fibres, and exhibit superior toughness and impact resistance. These have been used for composites on leading and trailing edge panels, along with fairings for some passenger aircraft. Smaller diameter carbon fibres (7–10 μm) exhibit superior strength and stiffness, which prompted their extensive use on 'all-composite' wide-bodied passenger aircraft such as the Boeing 787 and the Airbus A350. Boron fibres, with a diameter of around 100–140 μm, also offer very high strength and stiffness, but their high cost has predominantly restricted their application to use in military aircraft and prefabricated repair patches for both military and commercial aircraft. Figure 3.1 compares the specific strength and stiffness of a selection of reinforcing fibre materials and conventional bulk metals such as steel and aluminium.

The combination of high specific strength and a tailorable specific modulus (intermediate to high) for Carbon Fibre Reinforced Polymers (CFRPs) has resulted in their predominant use on the latest generation of passenger aircraft. Consequently, CFRPs are the focus of this chapter.

1.2.1 Carbon Fibre

Carbon fibres are produced from the carbonisation of spun and highly aligned precursor polymer filaments, made from either polyacrylonitrile (PAN), rayon, or pitch. Synthetic PAN is commonly manufactured from the free radical polymerisation of acrylonitrile (CH_2CHCN) and wound onto bobbins. It is the most widely used precursor material, at around 90% of the precursor market, and it results in the highest quality fibres.

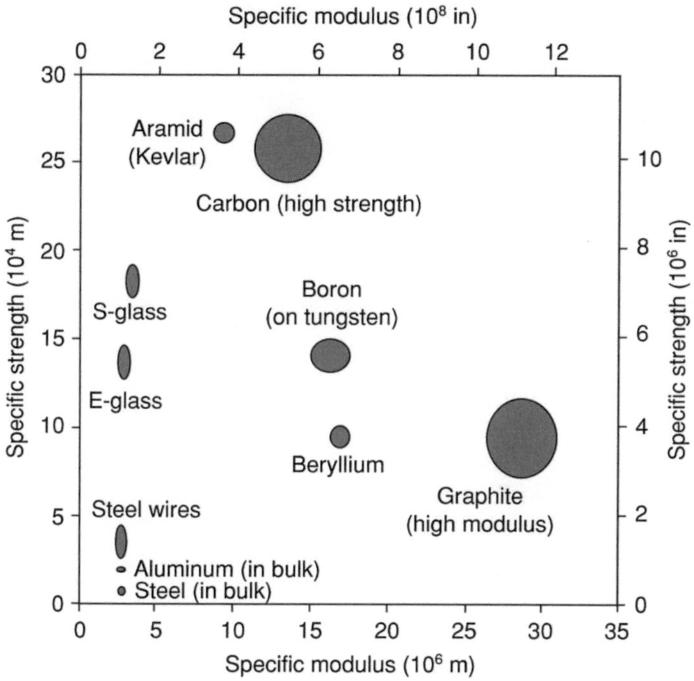

Fig. 3.1 Specific strength and stiffness properties of reinforcing fibres and conventional metals (Daniel and Ishai 2006)

In the production of carbon fibres, the PAN filaments are drawn continuously through a system of tensioning rollers before entering an oxidation oven. Here the cross-linking is initiated between the polymer chains by collecting oxygen molecules from air, at 200–300 °C, for several hours.

Next the stabilised fibres enter a series of inert gas furnaces ranging from 700 to 1500 °C for carbonisation, in an effort to expel all non-carbon atoms from the fibres. After this stage, the fibres are generally comprised of 90–95% carbon atoms and can be further graphitised in furnaces at 1900–2480 °C to attain up to 99% carbon composition. The carbonised or graphitised fibres can then be passed on for final processing where surface etching is performed along with a final polymeric sizing (typically 0.5–5 wt%) that aims to enhance bonding with a matrix. This carbon fibre production process is depicted in Fig. 3.2.

Once fibres have been combined into tows and wound onto rolls, they can be processed using any number of 3D weaving, stitching, knitting or braiding processes to produce composite preforms. Although most often, composite structures tend to consist of a stacked laminate containing a number of 2D preform plies. These include unidirectional (UD) tapes, 2D woven fabrics and non-crimp fabrics (NCF) that contain unidirectional tows finely stitched together. Woven and NCF reinforcements are generally more expensive than UD tapes but can save on labour costs during manufacturing. Woven materials are better for forming complex parts, but also exhibit

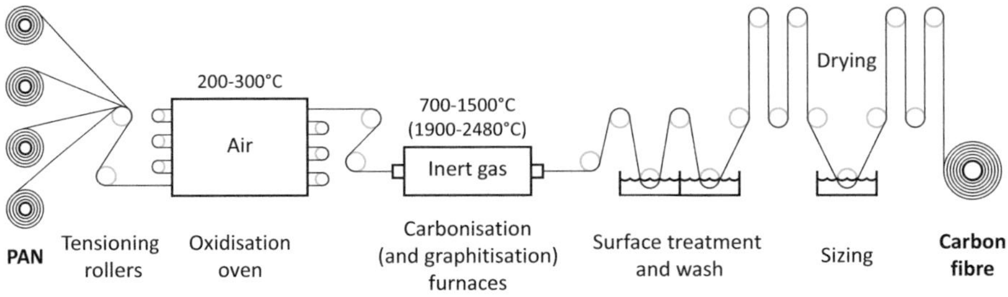

Fig. 3.2 Process to produce carbon fibres from synthetic PAN precursor bobbins

superior toughness and impact resistance, at the cost of some in-plane strength and stiffness compared with UD laminates. All three types of preforms can be pre-impregnated with resin to simplify the manufacturing process; however, the resulting 'prepreg' material will usually need to be kept in frozen storage. Prepregs commonly rely on an elevated temperature and pressure cure cycle, typically using an autoclave, to produce the final composite part. Alternatively, woven and NCF preforms can be used 'dry', such that the matrix resin is added into the system during the manufacture of the final composite part.

1.3 Continuous Fibre Reinforced Composites

The final composite material derives its behaviour from the combination of the cured resin matrix and reinforcing fibres. Its properties are also dependent upon the interfacial bond between its constituents and the content of the reinforcing fibres, along with their alignment, length, distribution and orientation. This form of continuous fibre composite exhibits very high strength and stiffness in the fibre direction but (weaker) resin-dominated properties in the transverse directions, resulting in highly orthotropic behaviour. Hence, composite structures are commonly prepared using a stacked layup of unidirectional plies arranged at different fibre orientations as shown in Fig. 3.3. The stiffness and strength can be tailored using classical laminate theory (Berthelot 1999) to achieve the desired properties. In industry, it is common to create quasi-isotropic laminates, where an equal number of plies are used for each different orientation, with an equal angle between the different orientations. For example, with a 45° angle between ply orientations, a $[45°, 0°, -45°, 90°]_{2s}$ stacking sequence will result in a quasi-isotropic layup. Here the subscript '2' denotes two contiguous stacks of this 4-ply sequence, and the subscript 's' denotes a symmetrical repeat of the layup, for a total of 16 plies in this case.

In high-performance applications like aircraft structures, the critical design constraint is often the stiffness of the composite. Hence, it is common to maximise the longitudinal moduli of the composite while balancing the ultimate strength, impact resistance and fracture toughness properties. However, the laminated nature of these

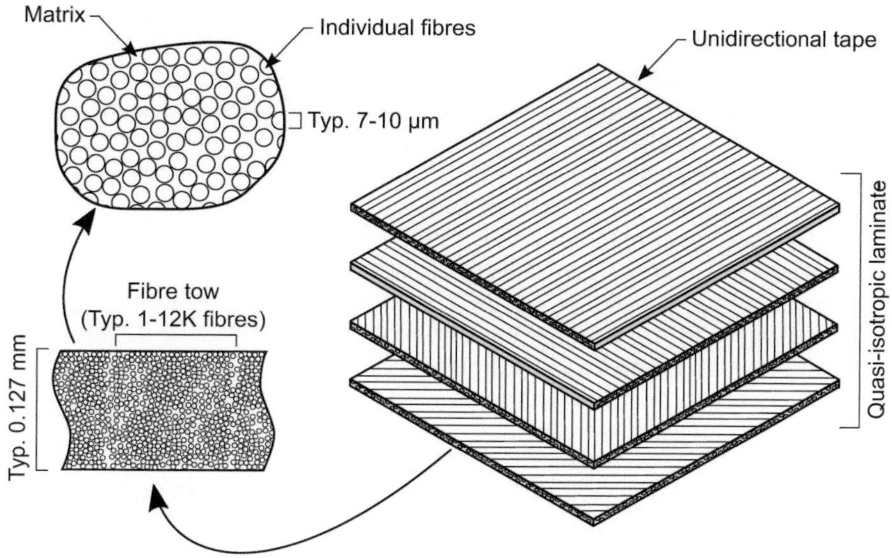

Fig. 3.3 Composite layup made from unidirectional plies

composites leaves them rather susceptible to impact damage and delamination. As a result, the residual compressive strength and stiffness of the composite are also important for high-performance applications. In order to reduce crack propagation and prevent excessive delamination, a high interlaminar fracture toughness is desirable for the matrix materials of aircraft structures.

1.4 Evolution of Composites in Aircraft Structures

The adoption of composite materials for aircraft structures has been a gradual process. Early progress was motivated by military applications where performance outweighed the considerable material costs. However, lightweight business aircraft and, more recently, large commercial aircraft have contributed significantly to the advancement of composite aircraft structures.

One of the earliest examples of the use of fibre reinforced composites in an aircraft was for the Supermarine Spitfire in 1943, where natural hemp fibres and phenolic resin were employed for wing spar and secondary fuselage components (Gay 2014). Around the same time, experimental projects were exploring the use of glass fibre reinforced polymer (GFRP) composites for aircraft fuselage structures at the Wright-Patterson Air Force Base in Ohio (Lubin 1982). GFRP technology then grew rapidly throughout the 1950s, and was introduced in around 2% of the structural weight of the Boeing 707 passenger jet, before also being used in the McDonnel Douglas DC-9.

However in the 1960s, the need for stiffer and stronger composites, driven by military and space applications, led to the development of carbon fibre and boron fibre reinforcement materials. The latter proved more expensive, difficult to process and subsequently, only saw interest for military applications such as the General Dynamics F-111 horizontal stabiliser, which was the first flight-worthy advanced composite component. By the 1970s, Boeing had incorporated composites into the supporting rotor structures of their CH-47 Chinook helicopters, and eventually also replaced the main rotor blades with composites in the 1980s. This increased performance, reduced maintenance costs, and improved ballistic damage tolerance compared with the original metallic rotors (Roeseler et al. 2007). Early in the 1980s, McDonnel Douglas started making CFRP tail assemblies and wings for the AV-8B Harrier II. These were the first combat aircraft to extensively employ carbon fibre composites, comprising around 26% of the airframe, which resulted in a 217 kg (3.4%) reduction in the total aircraft weight compared with that of a conventional metal structure. At around the same time, the National Aeronautics and Space Administration (NASA) also contracted Boeing to explore carbon fibre reinforced polymer (CFRP) composites for large primary wing structures (Mccarty and Roeseler 1984).

In the 1970s the Lear Jet Corporation commenced development of the Lear Fan 2100 business aircraft. This was the first to have a fully pressurised composite fuselage, where composites accounted for 70% of the aircraft's structural weight. Its development was plagued by structural challenges arising from the pressurisation of the fuselage but, ultimately, it was an issue relating to the reliability of the gearbox system which compelled the Federal Aviation Administration (FAA) to refuse the issue of a Certificate of Airworthiness. The programme was terminated soon afterwards. The Beech Aircraft Corporation built on the lessons learnt from the Lear Fan experience, in the development of their Beech Starship business aircraft, which became the first FAA-certified pressurised all-composite business turboprop. Nonetheless, the lack of market acceptance, in a highly competitive market, leads to production coming to an end in 1995, with just over 50 aircraft produced. Eventually the company took the decision to scrap all aircraft that were still under its control, through leasing agreements, as such a small fleet was prohibitively expensive to maintain. At the time of writing only five Starships are still airworthy. The technologies developed under the Starship programme laid the foundations for the more successful Beechcraft Premier I and Hawker Horizon all-composite business jets.

In the mid-1980s both Boeing and Airbus were incorporating CFRP materials into the primary tail structures of their 737 and A300/A310 aircraft, respectively. The advanced composite tail fins in the A300 and A310 commercial aircraft led to a 400 kg weight reduction compared with a conventional aluminium assembly. This composite tail section also achieved a 95% reduction in part count, and was subsequently cheaper to manufacture and assemble (Soutis 2005; Deo et al. 2001). By 1987 McDonnel Douglas were also producing a composite vertical tail for the DC-10 (Roeseler et al. 2007) and Northrop Grumman were manufacturing as much as 80% of the B-2 Spirit bomber from a carbon-graphite composite for high performance and stealth capabilities.

In the mid-1990s, military aircraft such as the Eurofighter Typhoon and RAH-66 Comanche helicopter continued to demonstrate extensive composites use, while Boeing were implementing vertical and horizontal stabilisers made from composites in their commercial 777 aircraft. Further advances in scale were seen in 2007, with the Airbus A380 using an unprecedented 25% composites for large commercial aircraft structures such as the centre wing box, empennage, control surfaces, pressure bulkheads, wing panels and floor beams.

Since the mid-2000s, a new generation of wide-bodied commercial aircraft have been developed and are driving the CFRP composite market with exceptionally large and challenging structures. The Boeing 787 Dreamliner was the first of its kind to reach a 50% composite composition in 2011 with CFRP wing, fuselage and control surface structures, effectively the entire wet surface. This was designed to be 15–20% more fuel efficient than existing metallic aircraft, with a significant reduction in part count and assembly time (3 days, compared to 13–17 days for similar metallic aircraft) (Roeseler et al. 2007). In 2015 Airbus introduced the A350 XWB, with a 53% composite composition.

High-performance requirements and quality control for composite aerostructures have predominantly relied upon autoclave manufacturing techniques using prepreg composite materials. By the mid-to-late 2010s, Bombardier (Canada) and the United Aircraft Company (Russia) had begun producing narrow-body commercial aircraft; the Bombardier CSeries C100/300 (becoming part of the Airbus Group in 2018, with the market designations of A220-100/300) and UAC's Irkut MC-21, making extensive use of composites. Both benefited from the latest resin infusion technologies for reduced production costs, with the MC-21 uniquely pursuing out-of-autoclave manufacturing techniques.

2 State-of-the-Art

2.1 Advantages/Disadvantages of Composites

Since the 1940s, the design of aircraft structures has relied heavily on aluminium alloys. The replacement of these materials with composites has been a gradual process, as new technologies have emerged, costs have reduced, and understanding of the material's behaviour has increased. Initially, CFRP materials were simply manufactured with a 'black aluminium' mentality, to replace existing metallic components with a similar design. However, as the use of composites has matured, aircraft manufacturers have started to employ innovative design and manufacturing methods to better harness the potential of composite materials.

Fundamentally, the transition to composite materials has been motivated by their superior specific strength and stiffness properties compared with their metallic predecessors. CFRP materials have a specific stiffness four times greater than common aerospace metals (2024-T4 and 7075-T6 aluminium alloys, and 6A1-4V titanium), and specific strength eight times greater. Even with, arguably, highly conservative design approaches these properties have enabled aircraft structures to be 20% lighter, resulting in increased efficiency and performance. For example, in a large transport aircraft a 1 kg reduction in structural mass can save around 120 L of fuel each year (Gay 2014).

In addition to weight saving benefits, composite materials can also absorb radar microwaves for improved stealth capabilities, while metallic materials reflect radar signals and are easily detected. These characteristics all contributed to their early adoption in place of metallic parts. However, high acquisition costs compared with aluminium, labour-intensive processing, manufacturing complexity, susceptibility to processing defects, and high capital costs for production equipment are some of the potential drawbacks of composites. Additionally, composite structures require extensive component testing to meet certification standards, particularly as there is not the same existing knowledge base as for aluminium structures.

Another major benefit to composites is their versatility in manufacturing. Tapering and compound contours are relatively easy to achieve with composites, along with greater part integration that potentially reduces the extent of machining and fastening compared with aluminium structures. This can directly contribute to further weight savings, reduced production costs and reduced maintenance. Stacking sequences in composite laminates can also be tailored to achieve the desired directional strength and stiffness properties.

In terms of material understanding, the anisotropic properties and complex internal behaviour of composites has proven to be much more challenging to predict and model than for the isotropic metals they often replace. This is particularly true for composite failure due to the wide range of interacting failure mechanisms that can occur in the fibres, matrix and their interfaces. A limited knowledge and experience base for advanced composite structures is problematic for training, repair and maintenance, where operators may have a limited awareness of the unique challenges associated with composites for aircraft structures.

Laminated composites are also susceptible to impact damage, since their low through-thickness strength leads to delamination. This can be caused by dropped tools, runway debris, hail, bird strike or ballistics and is most critical in compressively loaded structures. Relatively poor thermal and electrical conductivity also limits some of the functionality that can be achieved with aluminium aircraft structures. Despite these limitations, their superior performance overall continues to motivate wider use in aircraft structures, where the emphasis is shifting towards reducing manufacturing costs.

2.2 Manufacturing Processes and Automation

Composites manufacturing processes typically dictate the ultimate performance and quality of a composite component. For aircraft structures that need to meet strict performance and safety requirements, the design of an appropriate manufacturing approach is therefore critical.

Generally, composite manufacturing processes involve three main steps: material placement, forming and cure. Most methods execute the forming stage as the material is placed, to simplify the process. Conventionally, advanced composite structures have been manufactured using stacked laminates of unidirectional prepreg plies cured in an autoclave, as these allow for greater process control, reliability and part quality. For military aircraft, where size, cost and production volume are less of an issue, autoclave-cured prepreg laminates often remain viable. However, for commercial aircraft the manufacturing strategy becomes much more important, since it is necessary to manufacture a considerable volume of very large, high-quality, structures for a competitively priced aircraft. As a result, the commercial aircraft industry is driving most of the innovation in composites manufacturing, much of which focusses on greater process automation. This section discusses the evolution and current practices of composites manufacturing within the aerospace industry.

2.2.1 Material Placement

Hand Layup

Traditionally, hand layup has been the most common method for material placement, where reinforcing plies are cut to size, transported, and positioned by skilled operators. Each layer of composite is placed in-turn, with regular debulking steps, under vacuum, to consolidate the plies and remove entrapped air. This approach is effective for structures of a moderate size, significant complexity, and that contain a variety of reinforcing materials. It requires minimal capital investment, but at larger production scales it can become very expensive due to considerable labour requirements. In many cases, the composite materials can be directly formed by hand over a rigid tool; however, the steps of cutting, transportation, placement and forming are all subject to mishandling and dependent on operator skill. New technologies are helping to eliminate human error by automating the cutting procedure and handling operations. Ultimately, for large commercial aircraft structures, hand layup techniques are now rarely used since the alternative techniques are much more efficient, reliable and cost-effective at the required production volumes.

Filament Winding

An alternative placement method is filament winding where fibre tows, wound on bobbins or spools mounted in creel, are fed through a resin bath (or are pre-impregnated with resin) and automatically wound onto a rotating mandrel (Fig. 3.4). This process is most suited to cylinder-like structures. For example, the European Space Agency's Vega and Ariane 6 launch vehicles both use filament wound rocket motor cases. Filament winding is relatively fast, with laydown rates of up to 23 kg/min for simple parts, or 1.5 kg/min for complex aerospace parts (Strong 2008). Furthermore, by

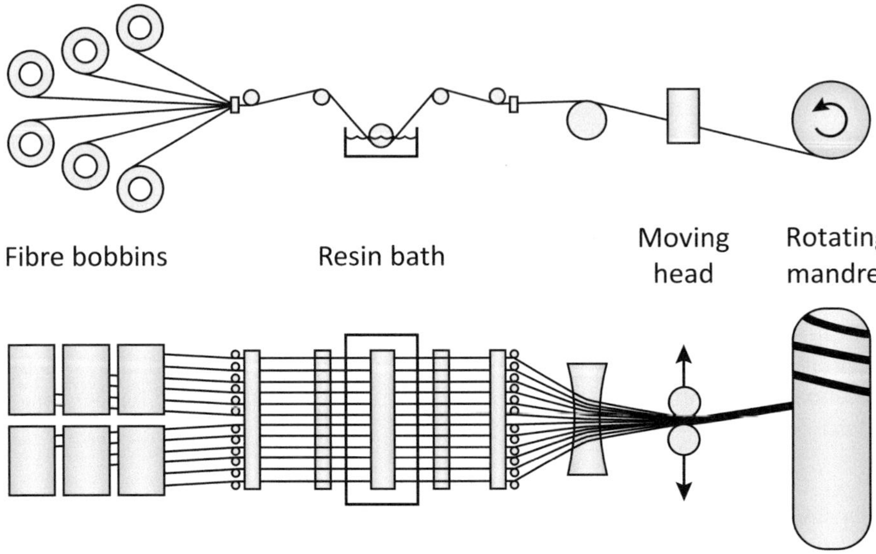

Fig. 3.4 Filament winding

using dry fibre tows, this process can save on cold storage and material costs compared with conventional prepreg-based methods. Since the ultimate form of the structure is generated as the material is placed, there is also no need for secondary forming operations. However, this approach often results in a considerable void content (commonly 3%) (Strong 2008), and due to the rotating mandrel it is not possible to wind 0° fibre orientations (in the direction of the rotating axis). Aside from pressure vessels and rocket motor casings, filament winding does not see a lot of use in aircraft structures, particularly as newer material placement technologies allow for reduced waste and a greater variety of structural geometries.

Tape Laying Machines

Automated Tape Laying (ATL) machines can significantly lower the costs of fabrication for large and relatively simple composite parts. These machines commonly dispense pre-impregnated unidirectional tape from supply rolls, between 75 and 300 mm wide, onto a mould or existing layers of reinforcement tape. As the tape is dispensed, sufficient pressure is applied to tack and debulk the tape in place, which results in a low void content (<1%) in the final parts. The tape laying machine also automatically removes the backing paper and film from the prepreg material. Layers of tape are placed sequentially at different orientations to produce the final preform laminate. Hence, near net-shaped parts are laid up and formed in one automated process. This is a relatively fast production method, since ATL machines are capable of lay-down rates greater than 25 m/min with simultaneous debulking. Automated trimming at the end of each tape pass also means that there is minimal scrap

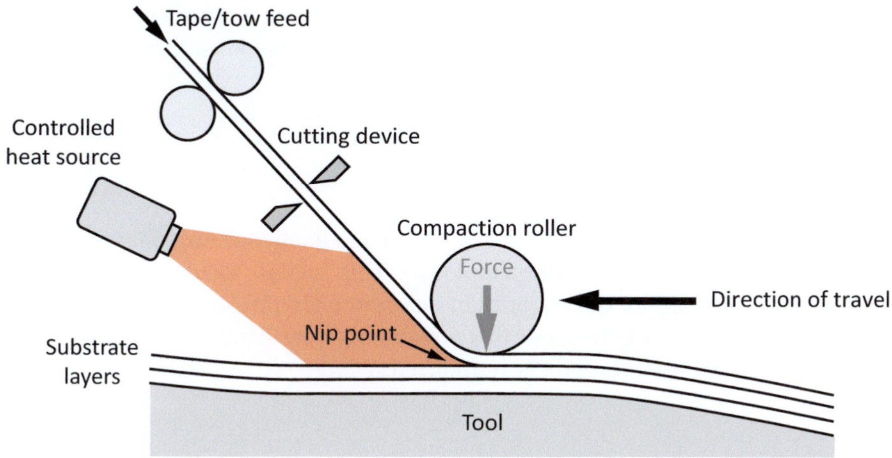

Fig. 3.5 Typical dispensing head configuration for Automated Tape Laying (ATL) and Automated Fibre Placement (AFP) machines

produced. However, the downsides include a high initial capital cost, considerable setup labour and limited contour capabilities.

Figure 3.5 depicts a common dispensing head for ATL, and Automated Fibre Placement (AFP), machines. Here the tape is fed onto the tool, or previous substrate layers, and fixed in place with a combination of heat and compaction at the 'nip point'. As the tape is dispensed, a cutting device can trim each line of tape to the desired length before resetting for the application for the next line of tape. Such dispensing heads are usually attached to robotic arms or large gantries with multiple degrees of freedom in order to lay up complex and large-scale composite aerostructures.

More recently, ATL machines have also been developed to lay dry materials that are not pre-impregnated with resin. This has been made possible by the addition of thermoplastic binder materials, such as thin veils, that provide sufficient tack when heated to hold the dry layers in place. Without this addition, the dry fibres slide, rotate and lift from each other too freely to form a consolidated laminate. By using dry materials, these machines eliminate the need for cold storage and allow for liquid resin manufacturing methods.

Fibre Placement Machines

Automated Fibre Placement (AFP) machines operate on a similar principle to ATL machines; however, instead of laying tape, they place up to 32 individually controlled tows in one path, each with a width of between 3 and 6 mm, at speeds up to 60 m/min. In the commercial aircraft industry, AFP is used in the manufacture of the Boeing 787 Dreamliner fuselage, along with many other large aerostructures. Due to the narrower placement heads, these machines are capable of producing parts with more challenging contours than ATL; however, the overall complexity of the tool must still remain relatively simple. The precision control and integrated debulking of AFP (and ATL) machines result in high-quality composite structures with a low void content and minimal scrap material.

2.2.2 Forming and Cure

Depending on the desired production rates, cost and quality of the composite parts, different forming methods can be employed. This section focusses on autoclave methods that are used for the majority of composite aircraft structures, along with 'out-of-autoclave' (OOA) methods that are gaining considerable interest and traction. Other methods are also briefly introduced.

Autoclave

Autoclave forming is the most common approach for producing composite aerostructures made from prepreg material. Generally, a polymer film covers the composite 'stack' or 'assembly' and vacuum is drawn through the system to consolidate the composite and remove trapped air and volatiles. Autoclave pressure is subsequently applied, and heat is introduced to melt and cure the thermoset matrix. This approach is suitable for both flat and complex structures with variable thickness, even at large scale. The uniform application of pressure and heat within the autoclave results in good dimensional stability of the final composite part. However, a reliance upon consumable bagging materials can contribute to production failures if they are not prepared properly. A schematic of an autoclave, which is basically a pressure vessel, is shown in Fig. 3.6. Cycle times are also long, meaning this approach is only viable for lower production rates and capital investment costs for an autoclave are high, particularly as part size increases. A typical cure cycle is shown in Fig. 3.7 for a standard epoxy-based composite system.

Out-of-Autoclave (OOA)

Rather than using an autoclave, many composites can be formed in ambient conditions or inside an oven. Typically, these 'out-of-autoclave' (OOA) methods involve a single male or female tool and some form of deformable upper tool. Conventionally this would be a thin polymer film such as consumable vacuum bagging material; however, new approaches are making use of semi-rigid upper bags that are more robust. OOA methods are more common for dry preform materials (rather than prepregs), which are formed to the desired part shape and then infused with liquid resin. The resin-infused preforms can then be cured in an oven or in ambient conditions. Typically this is done under vacuum, using ambient pressure to hold the composite in shape against the lower rigid tool. A slightly lower volume fraction is usually achieved in comparison to the use of an autoclave. OOA methods also share some of the same consumable waste and bag failure concerns.

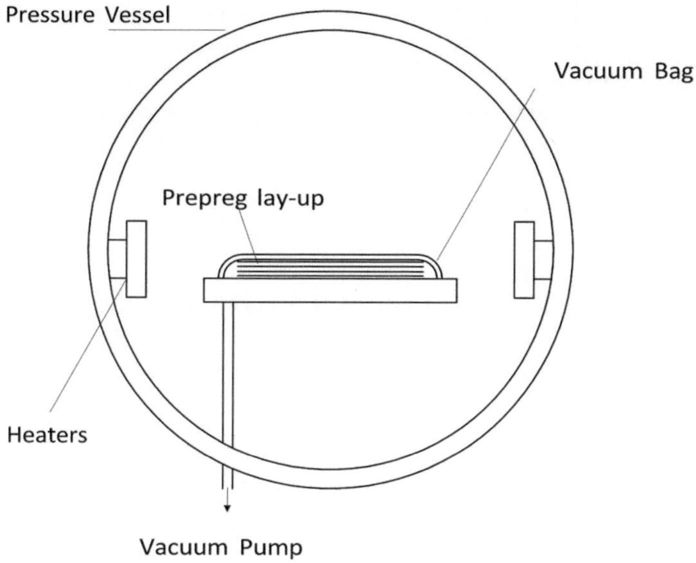

Fig. 3.6 Schematic of an autoclave

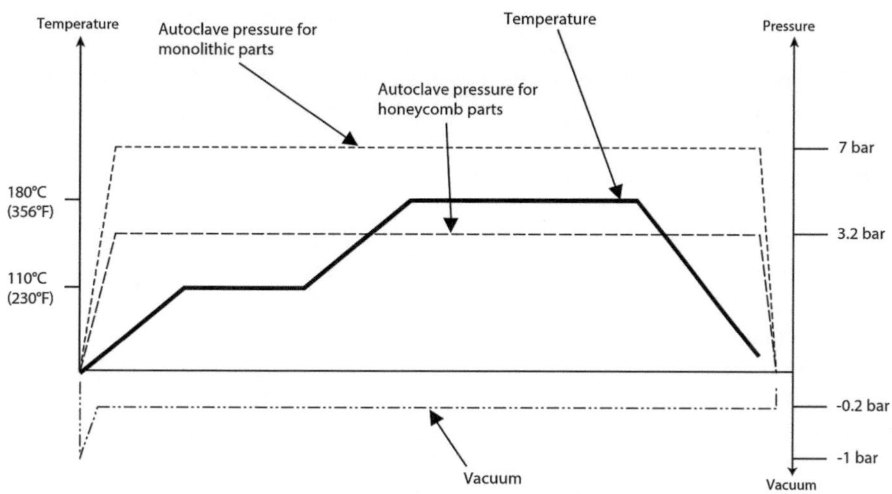

Fig. 3.7 Cure cycle for composite prepreg system (Hexcel Corporation 2016)

Other Methods

Many other composite manufacturing methods can be used to create aircraft components; however, they are more commonly employed in other industries where the scale and quality requirements are not as stringent. For example, Resin Transfer Moulding (RTM) is used for medium production volume, where the fibre preform is placed in a closed mould and injected with resin at around 700 kPa pressure. This process requires rigid tooling on both sides of the part which eliminates the need for consumable materials, but greatly increases tooling cost. RTM is good for complex parts with fine surface detail, and this approach generally eliminates the need for refrigerated storage when using dry preform materials. The net-shaped moulding approach means that minimal post-processing is required after cure; however, it also means that mould tolerances must be tight.

2.2.3 Recent Developments in Composite Aircraft Manufacturing

Over the last 10 years, the advancement of composites manufacturing has been driven by developments for civil aircraft such as the Boeing 787, Airbus A350 XWB, Bombardier CSeries (Airbus A220) and Irkut MC-21. These latest generation airliners demonstrate the recent improvements and trends for the production of advanced composite structures.

Boeing 787 Dreamliner

As the first commercial aircraft to be developed predominantly from composite materials (50% of its structural weight), the 787 Dreamliner represents a major shift in aircraft construction. This has been made possible by new automated tape laying and fibre placement technologies for the production of primary aircraft structures using prepreg materials. These technologies allow for greater speed and economy in the manufacture of large wing skin and fuselage barrel structures than conventional hand layup techniques. However, the fuselage still needs to be manufactured in a number of sections lengthwise, due to autoclave size limitations. Boeing are already using one of the world's largest autoclaves (shown in Fig. 3.8), with a working area 9.26 m in diameter and 23.5 m long, for curing composite fuselage sections. Although, the largest Dreamliner in production (the 787-10) has a fuselage length of 68.3 m that needs to be assembled from a number of barrel sections.

Boeing also employ an out-of-autoclave, Controlled Atmospheric Pressure Resin Infusion (CAPRI) approach to manufacture many of the control surfaces such as the ailerons, flaps and spoilers for the 787 in Melbourne, Australia. This is a lower cost approach that avoids the need for refrigerated storage and autoclave time. A Vacuum-Assisted Process (VAP) is also used to produce the 4.5 m diameter rear pressure bulkhead from composite materials.

Airbus A350 XWB

Airbus are producing their own composite wide-body aircraft, the A350 XWB, which, while not a direct competitor to the Boeing 787 in terms of size, range and seating capacity, it is a flagship aircraft for Airbus in the same way that the B787 is

Fig. 3.8 Large autoclave used for production of Boeing 787's composite fuselage sections

for Boeing. Airbus is taking advantage of advances in automation and process control and these developments are best demonstrated by the manufacture of the A350 forward wing spars. Each C-shaped spar is a 31.2 m long structure, tapering from 1.8 to 0.3 m wide, with sections ranging from a thickness of 25 mm (around 100 plies of CFRP) at the wing root to 5 mm at the wing tip. Due to the size of these structures, they need to be manufactured in three parts (7 m, 12.7 m and 11.5 m long, respectively) in order to fit in their autoclaves. Automated fibre placement (AFP) machines are used to lay down carbon fibre prepreg materials over rotating mandrels at rates up to 60 m/min, to produce chiral spar components for the port and starboard wings. By contrast, the spars in the slightly older A400M transport aircraft relied instead on automated tape laying (ATL) of carbon fibre prepregs on a flat tool, before being hot-draped over a male tool and then transferred to an Invar (36% nickel/64% iron alloy) female tool for the final cure. Hence the one-step forming process for the A350 yields a great improvement in manufacturing efficiency.

The AFP machines for preparing the A350 forward wing spars cost around €5 million each, but can save on considerable waste, cutting operations, and design inefficiencies due to their independent tow cutting and restarting capabilities. Further efficiency in the manufacturing process is also gained by reducing the number of debulking/consolidation cycles necessary to keep the overall void content of the parts below acceptable limits. It is now possible to use only a single vacuum consolidation phase at room temperature at the end of the layup, rather than numerous intermediate vacuum consolidation stages on the mandrel along with infrared heating. Once consolidated, the chiral spar laminates are then moved to a female Invar tool and cured in a large 3.5 m diameter and 16 m long autoclave. Assembly of the six spar sections for an A350 aircraft is still performed manually, but this process is supported by an automated five-axis machine that drills the 16,000 holes necessary for fastening. Other components, such as the rear pressure bulkheads, are manufactured using out-of-autoclave processes similar to those used for the Boeing 787.

Bombardier CSeries (Airbus A220)

In a move to reduce production costs for the latest generation of narrow body aircraft, Bombardier are manufacturing their all-composite wings for the CSeries aircraft using a Resin Transfer Infusion (RTI) method. The composite wing structures for the CSeries are 15.5 m long with integrated stiffeners, and are mechanically fastened to composite spars and aluminium ribs. These structures are made from non-crimp fabrics that are two to four times thicker than equivalent prepreg plies, for a more efficient and reliable layup. Consequently, the material placement phase is faster than for unidirectional prepreg plies and the material shelf life and frozen storage are no longer an issue. This approach also avoids the need for specialised ATL machines, such as those used for the Boeing 787 and Airbus A350 wing skins.

An automated ply cutting machine provides greater precision in the preparation of preform shapes, which are then laid up by hand or by using pick-and-place equipment. The RTI method relies on the forming of reinforcing plies between a flexible upper tool and rigid lower tool under vacuum. The bagged wing structures are then moved into a 5.6 m diameter (21.3 m long) autoclave where epoxy resin is injected into the preform with up to 96.5 kPa of additional pressure. The use of an autoclave provides greater control for the temperature and pressure during infusion but adds considerable cost and limits production rates.

Irkut MC-21

In Russia, further advances are being made by AeroComposit, who are manufacturing composite wing and wingbox structures for the Irkut MC-21 by automated dry fibre placement and an oven cure cycle. This is the first time such OOA techniques have been used for primary wing structures in commercial aircraft. This approach is designed to reduced costs, facilitate greater part integration and reduce preform handling. By using AFP machines with dry carbon fibre tows, rather than performing hand layup with non-crimp fabrics, AeroComposit claim to be able to achieve greater precision and a consistent, predictable processing speed. To avoid the problem of dry fibre movement, a thin thermoplastic veil is heated and laid simultaneously with the reinforcement to act as a binder between the reinforcing plies. This provides sufficient tack to hold each layer in place as it is applied. In addition to its binder function, the grid-like structure of the thermoplastic veil is also claimed to enhance resin flow during infusion and improve interlaminar shear strength.

Using this approach with specialised tape materials made from up to 32 tows, deposition rates up to 60 m/min may be achieved.

Once formed, the dry reinforcement materials are then processed in a 'Thermo Infusion Automated Centre' (TIAC) that is essentially a large oven with in-built resin infusion, process monitoring and process control systems. The TIAC is capable of measuring, mixing, heating and degassing the resin before controlling the injection, infusion and cure of the resin. With greater process control, this approach aims to improve the repeatability and reliability of OOA methods, in addition to the known cost-saving benefits over autoclave methods. Generally, for the manufacture of large composite aerospace parts (8–130 m^2), it has been reported that ovens cost only 10–15% of a similarly sized autoclave (Gardiner 2014b). They also require considerably less power. Furthermore, the cost of dry fibre and liquid resin materials can be as little as 30% of similar prepreg materials.

However, unlike Boeing and other manufacturers, the secondary structures for various control surfaces in the MC-21 are still manufactured from prepreg materials that are batch cured in smaller autoclaves.

2.2.4 Process Modelling

In addition to the technological advancement of composite manufacturing methods, their design, reliability and efficiency can be further improved by predictive modelling. Process modelling aims to simulate the production of composites, including steps such as forming, infusion and cure, based on known processing conditions. This can help to anticipate the success of a new production method or identify potential defects that might result from changes to an existing approach. This is particularly relevant for liquid resin infusion methods and the manufacture of complex or highly integrated composite structures. As the size and complexity of these aircraft structures increases, the prediction of resin flow behaviour and the design of a successful manufacturing strategy becomes more challenging. In the past, much of the development process for new production lines would rely on experience or trial-and-error. This can result in considerable waste, cost and delay. However, as more aircraft manufacturers turn to OOA resin infusion methods for the manufacture of advanced composite structures, process modelling is receiving more attention. A number of commercial software packages already exist for composites process modelling, such as the PAM-COMPOSITES suite by ESI group; however, this also remains an active area for academic research (Pierce and Falzon 2017).

2.3 Design and Assembly

The transition to composite aircraft structures has had a major impact on the design and assembly of structural components, despite minimal changes to the general appearance of commercial airliners over the last 50 years. Conventional aircraft structures were designed from metallic panels, stiffeners, spars and ribs; all assembled using rivets and other mechanical fastening methods. These aircraft were designed to take advantage of well-established metal forming methods such as rolling and extrusion since casting and other methods were not viable for large metal aircraft structures.

Early composite structural design was heavily informed by existing methodologies developed for metallic structures. The term 'black aluminium' was often used to describe the design of composite components to replace their metallic counterparts. However, mechanical fastening, prevalent in metallic aerostructures, can present a number of problems for composites. The effect of drilling in composite laminates can significantly alter their structural performance, since the continuity of fibres is broken and the likelihood of delamination is increased. Water ingress, differences in thermal expansion, additional weight and galvanic corrosion can all cause further issues that should be considered in the design and selection of a fastening system. Hence, traditional design and assembly methods are not ideal as they do not fully exploit the advantages offered by composite materials.

The latest generation of composite aircraft serve as a better demonstration of composite design and assembly, making use of integrated stiffeners and extensive adhesive bonding. This helps to reduce drilling, riveting, part count and assembly times. Recently, GKN Aerospace in the UK have developed an advanced out-of-autoclave wingbox design that incorporates a variety of different stringer geometries and impregnated Pi joints that aim to further reduce the need for riveting.

2.4 Maintenance

In addition to considerable weight savings, composite structures in aircraft are also being adopted due to their superior corrosion and fatigue resistance, which can greatly reduce maintenance requirements compared with aluminium structures. Airbus, for example, claim a 60% reduction in maintenance tasks for the A350 XWB as it requires less time-consuming checks, and requires them less often. Similarly, Boeing claim the lifetime airframe maintenance costs of the 787 are 30% lower than any comparable metallic aircraft (Boeing 2006).

Commercial aircraft maintenance typically consists of a number of regularly scheduled checks, designated A, B, C and D-checks, depending on their frequency and duration (Department for Business Innovation and Skills 2016). 'A' checks are typically performed after around 500 flight hours (or 250 flight cycles), requiring an overnight layover and around 60 h labour. Traditionally 'B' checks were performed less often, around every 6–8 months, over 1–3 days in an airport hangar, although these tasks are now usually integrated into a series of 'A' checks instead. 'C' checks are usually undertaken every 20–24 months, and require the aircraft to be taken out of service for more extensive inspection at a maintenance base, involving up to 10,000 h labour over 1–2 weeks. Lastly, 'D' checks (or Heavy Maintenance Visits) are the most rigorous and costly stage of regular maintenance that occur every 6–10 years, where most of the aircraft is taken apart for inspection and overhaul. This process can take up to 2 months to complete and can require up to 50,000 h labour in a suitable maintenance base.

By taking advantage of composite structures, the inspection intervals for the 787 have been extended beyond those of any previous aircraft, with a first external visual inspection of the structure ('C' check) being performed at 6 years and a first internal inspection ('D' check) at 12 years. This, combined with an average 20% reduction in labour for each check, results in an overall reduction in scheduled maintenance labour of around 60% (Boeing 2006). Specifically, compared with the 767-300ER over a 25-year service life, the 787 is expected to see a 44% reduction in 'A' check labour, a 65% reduction in 'C' check labour and a 63% reduction in 'D' check/HMV labour. This reduction in scheduled maintenance is primarily a result of the improved corrosion and fatigue resistance of composite structures.

However, the use of composite structures also raises challenges for aircraft maintenance. For example, composites can be more difficult to non-destructively inspect for damage than metal alloys. This is particularly true in the case of low-velocity impact damage, which may not be readily visible on an exterior surface despite considerable internal delamination.

In terms of repair, the 787 was designed to accept conventional bolted patching methods, based on Boeing's experience with the repair of 777 composite structures. This allows for a comparable repair time to repairs in metallic airframes and relies on a similar skill set. However, adhesively bonded repairs can offer greater efficiency in terms of load transfer, and a smoother profile for greater aerodynamic performance. Currently, these repairs are only certified for non-critical applications, or where the residual strength of the parent structure continues to exceed the design limit load (Federal Aviation Administration 2014). Bonded repair methods rely on removing damaged material, scarfing the surrounding region, and replacing the removed material with composite plies that are adhesively bonded and cured to the parent laminate. With a manual grinding or sanding scarf method, slow curing adhesive/resin and final non-destructive inspection this whole process can be costly and time consuming. However, recent advances in automation aim to improve the speed and reliability of bonded repairs in composite structures. Several competing robotic systems have been evaluated by Airbus, Boeing, Lufthansa Technik and DMG MORI to perform various combinations of automated repair tasks, including 3D scanning, milling/grinding, plasma surface treatment, ultrasonic testing and laying of the repair tape. For example, the Lufthansa Technik system resulting from the Composite Adaptable Inspection and Repair (CAIRE) project claims to improve scarfing efficiency by 60% compared with existing manual methods, and is capable of more complicated scarfing geometries.

3 Research Challenges

3.1 Repair

Currently, adhesively bonded repairs represent the main research challenge for the repair of composite structures. There remains general concern in this field due to the lack of information, understanding, standardisation, training and awareness related to composite repairs. There is considerable ongoing research to address this lack of knowledge. Improved modelling to predict damaged and repaired composite behaviour is a major area of research that can better inform the design of a repair. The non-destructive inspection (NDI) and characterisation of damage within a composite structure is also an area of considerable interest, since low velocity impact events are capable of producing 'barely visible' surface damage with greater delamination through the thickness of the structure. Additionally, for bonded repairs, similar NDI techniques are necessary to assess the quality of the bond. However, this can be complicated by the presence of kissing bonds that provide intimate contact, hence not presenting as a defect using standard NDI techniques, but exhibiting low strength. Uncertainty of the bonding quality remains the primary impediment to the certification of adhesively bonded repairs for use in critical aircraft structures. Among aircraft manufacturers, this enduring challenge is a significant priority as it could lead to further weight savings and greater efficiency in the design and operation of composite aircraft.

3.2 *Multifunctionality*

In the transition to composite airframes, some of the functional benefits of the original metallic structures have been significantly affected. In particular, highly conductive metallic aerostructures provided sufficient Lightning Strike Protection (LSP) to minimise any damage on the structures and systems of the aircraft. Composite materials exhibit low electrical conductivity and therefore additional measures need to be taken to ensure adequate LSP. The current measures for mitigating lighting strike damage add non-structural weight and complicate aircraft maintenance. To a lesser extent, the low thermal conductivity of composites also reduces the efficiency of conventional anti-icing systems which already incorporate non-structural weight. The desire to eliminate this weight is leading to the development of multifunctional composite structures that incorporate anti-icing/de-icing (AI/DI) and LSP capabilities into the structural composite materials, along with opportunities for new functionality such as in situ Structural Health Monitoring (SHM).

3.2.1 Lightning Strike Protection (LSP)

Statistically, commercial aircraft are expected to be struck by lightning every 3000 flight hours, equivalent to roughly once a year (Larsson 2002). When struck, high-energy electrical current flows through the aircraft structure by the path of least resistance and can cause critical damage to the aerodynamic surfaces or electrical systems of the aircraft. For metallic aircraft, the highly conductive structural materials naturally provide sufficient protection from lightning strike. However, with the transition to composite structures for commercial aircraft, the electrically insulative properties of polymer matrix materials are incapable of replicating this protective functionality. Instead, these new materials are prone to significant thermal decomposition and ablation as a result of lightning strike that can greatly affect the mechanical performance of the composite structure. Subsequently, additional Lightning Strike Protection (LSP) systems are required to prevent structural damage or failure that might result from a lightning strike on composite airframes.

Currently, the most popular approach for LSP is to incorporate a metallic mesh or foil into the outer surface of the aircraft skins to dissipate the lightning strike energy and reducing surface damage. Aluminium and copper are the most common materials for these meshes due to their great specific electrical conductivities and low cost. In the Boeing 787, conductive phosphor bronze wires are interwoven with the outermost reinforcement ply of the fuselage structures. Both the 787 and Airbus A350 also make strategic use of metallic paths to help direct current from lightning strikes. For example, in the A350, components such as the metallic seat rails and cabin floor beams perform both a structural function and also form part of the LSP system of the aircraft in combination with the copper mesh-enhanced composite skins.

The location of aircraft components is also important to the design of their LSP systems. Lightning damage tends to be most severe at entry and exit points of the structure, where the energy density is highest. This is most likely to occur at the most prominent protruding points of the aircraft, such as the wingtips or nose. These areas commonly require the thickest and heaviest mesh materials, while other surfaces are suitable for lighter protective materials. In all areas though, these materials add significant weight and manufacturing complexity to the aircraft structure, purely for the purpose of LSP, which works against some of the weight saving benefits that result from the transition to composites. Additionally, because the existing solutions are only capable of reducing, not eliminating, the damaging effect of lightning strike, there remains a significant desire for more effective and efficient LSP systems (Gagné and Therriault 2014).

A number of alternatives have been investigated for LSP that rely on metallic spraying and coating methods such as electroless plating, Physical Vapour Deposition (PVD), thermal spraying and cold spraying. The use of nanoscale composite particles such as carbon nanotubes and graphene show great potential for enhancing the surface conductivity of aircraft structures. The main research challenge in this field is to develop a system with greater weight-specific electrical conductivity, compared with the existing metallic mesh technologies, without compromising on other design requirements (Gagné and Therriault 2014).

3.2.2 Anti-Icing/De-Icing (AI/DI)

Ice accumulation on the aerodynamic surfaces of an aircraft, either on the ground or in flight, can affect flight performance, lift and control. In the most severe cases this can even cause fatal accidents. It is essential to have preventative (anti-icing) or contingent (de-icing) systems in place to ensure that aircraft surfaces are free from ice build-up. These secondary systems on the aircraft should be as simple, energy efficient and low maintenance as possible. However, with the recent transition to composite structures, it is now also important that these systems are compatible with non-conductive materials.

Conventionally, AI/DI systems rely on the thermal conduction of the aerodynamic surfaces by hot air that is piped from the compressor stage of the jet engines and ultimately released from small exhaust holes. This 'engine bleed air' approach has proven effective for metallic aircraft structures, due to the metal's high thermal conductivity; however, it is still inefficient, heavy and complicates aircraft maintenance (Falzon et al. 2015). Another existing approach for de-icing involves pneumatic bladders that are inflated to remove ice from susceptible leading-edge regions. In addition to the extra weight of the system, the detached ice debris can damage other components of the aircraft once dislodged. For passenger aircraft, de-icing is commonly performed at airports by spraying an anti-freeze fluid over the aircraft.

In newer aircraft, the relatively poor thermal conductivity of composite structures makes an engine-bleed-air approach less effective. A number of alternative anti-icing/de-icing solutions have been investigated for composite airframes. For example, electrothermal systems that behave as resistance heaters within the aerodynamic surfaces can provide greater efficiency, reduce maintenance complexity and eliminate the need for exhaust holes, compared with traditional engine-bleed-air designs. A range of materials have been studied as potential heating elements for electrothermal systems, including constantan wires, sprayed metallic layers, carbon fibres, graphene and carbon nanotubes. In the Boeing 787, a molten metal spray, sandwiched between layers of glass fabric, has been used as a heating element for the leading edge. Alternatively, carbon-based solutions for electrothermal anti-icing/de-icing systems can offer a significant weight reduction and be more easily incorporated into the manufacture of the composite skins. These may even provide a greater contribution to the structural performance than a metal-based approach. Recent research investigating the use of highly aligned carbon nanotube webs as heating elements shows great potential (Yao et al. 2018a, b, 2019). This approach is energy efficient, low weight, and provides uniform heating.

3.3 Recycling

Composite waste materials can come from all stages of an aircraft development and life cycle, although manufacturing contributes as much as 40% of all CFRP waste (Pickering 2006). Out-of-date prepregs, off-cuts, testing materials, production tools, repair materials and end-of-life components are all sources of composite waste. In particular, aircraft with increasing quantities of structural composites will eventually need to be decommissioned. For example, the latest generation passenger aircraft contain more than 50% composites, translating to more than 20 tonnes of thermoset CFRP per aircraft which will need to be disposed of or, preferably, recycled (Pimenta and Pinho 2011).

Unlike aluminium which can be economically melted down and reformed, there is less incentive to recycle polymeric materials, since their repurposed value tends to be low and landfill disposal remains relatively cheap. However, landfill solutions are unsatisfactory for a number of reasons. From an economic perspective, there is a considerable investment in the development and manufacture of composite structures. The raw materials themselves can cost up to £40/kg, and manufacturing can require energy of up to 165 kWh/kg (Pimenta and Pinho 2011). The disposal of CFRP materials at a cost of around £0.20/kg in landfill also represents a considerable loss in potential income, since with the right technology these could be turned into profitable reusable materials. Additionally, tougher legislation on landfill allowances requires alternative solutions to the composite waste problem.

Unfortunately, the recycling of thermoset composites is particularly challenging, due to their inherent composition of fibre, matrix and filler materials. Furthermore, the natural cross-linking of polymer molecules in a cured thermoset material means that they cannot be easily remoulded, remelted or broken down. Aircraft structures often contain a range of additional materials such as lightweight sandwich cores, metallic meshes for LSP, protective coatings/paints, or metal inserts for fastening, that may need to be separated from the composite materials, further complicating the end-of-life (EoL) recycling process. When trying to produce high quality recyclates, the collection, identification and sorting of different EoL composites waste is also an important consideration. Notably, prepreg and manufacturing waste are often easier to recycle than EoL waste, since the source and quality of the materials are easier to trace, and there are not the same issues with the removal of contaminants and inserts.

Currently there are two main categories of thermoset composite recycling methods: mechanical processes and fibre reclamation processes. Both aim to address the needs and challenges associated with existing composite waste. However, researchers are also developing novel thermoset resin systems that can be more easily recycled.

3.3.1 Mechanical Processing

Mechanical recycling methods typically rely on shredding, cutting, milling or crushing processes to reduce composites waste into smaller pieces, and eventually into fine resin-rich powders and fibre-rich clumps (as large as 5–10 mm in size). The resulting recyclate is then graded by size using sorting cyclones or sieves. Mechanical processing is able to recover both the fibre and resin material, without using or producing any hazardous materials. However, this approach results in significant degradation of mechanical properties and produces an unstructured, coarse and inconsistent recyclate that has limited reuse applications (Pimenta and Pinho 2011).

Composites that have been recycled using mechanical processes tend to be used for low-value applications, often as fillers in new composites, artificial woods and asphalts. These recyclates are most often derived from GFRP materials and subsequently regarded as a low value commodity. High volume production applications in the automotive industry, for bulk moulding composites (BMCs) and sheet moulding composites (SMCs) have employed mechanically processed GFRP recyclate materials, but generally these methods are considered undesirable for aerospace CFRP materials as the benefits of the high-cost continuous fibres are lost.

3.3.2 Fibre Reclamation Processing

Fibre reclamation methods attempt to recover the reinforcing fibres from composite waste by breaking down the thermoset polymer matrix with aggressive thermal or chemical processes. The degraded polymer material can either be harvested for energy or for smaller organic molecules, while the reinforcing fibres can be collected for reuse in other materials. In some cases, cleaning or mechanical size reduction is still necessary prior to thermal or chemical processing. These approaches are particularly suitable for carbon fibre composites since there is not a significant degradation of their high-performance mechanical properties, due to superior thermal and chemical stability.

The three most common fibre reclamation methods are: thermal processing within a fluidised bed, pyrolysis and chemical processing.

Fluidised Bed Methods

Fluidised bed methods are designed to separate the high value fibre reinforcement from the thermoset matrix of a scrap composite component via heating and abrasion. Broken down pieces of the composite part, around 25 mm long, are fed into a bed of silica sand that is fluidised by a stream of hot air, at around 550 °C (for epoxy resins). In this hot abrasive environment the polymer is volatilised, releasing oxidised molecules, the fibre material, and any filler material into the air stream for separation and collection. The remaining resin molecules in the air stream can then be fully oxidised in an afterburner and may even be used as a source of energy recovery (Pickering 2006). The short fibre product that results from this recycling method tends to appear fluffy, with a clean surface and no sign of contamination or oxidation, maintaining an equivalent stiffness to virgin fibres and good bonding potential. These recyclates are most suitable for bulk moulding composites (BMCs) or non-woven veils; however, when tested they tend to show a 20% degradation in tensile strength compared to virgin fibres. If used as a substitute for pristine glass fibres in BMCs, the recycled carbon fibres can provide a gain in tensile strength of at least 25% (Pickering 2006). This fluidised bed approach is also highly tolerant of mixed or contaminated scrap materials that might include foam cores, metallic inserts or paint.

Pyrolysis Methods

Pyrolysis relies purely on the decomposition and removal of the polymer matrix from the reinforcing fibres by heating to elevated temperatures, commonly around 500 °C, in an inert environment. This approach is one of the most prevalent recycling methods for carbon fibre composites. The reclaimed carbon fibre materials can exhibit some pitting, charring or residual matrix material, along with some strength degradation as a result, but generally show comparable mechanical properties to virgin fibres (Pimenta and Pinho 2011). However, a high degree of variability in pyrolysis has also been reported, where highly aggressive processing conditions can result in significant surface damage, which reduces fibre diameter by up to 21% and tensile strength by as much as 84% (Pimenta and Pinho 2012). In all cases though, the reinforcing fibre stiffness does not appear to be affected. As with the fluidised bed methods, pyrolysis only requires a small fraction of the energy required to produce virgin fibres and does not require the introduction of any chemical solvents. However, the gaseous by-products can be environmentally hazardous. There is also the potential to recover organic products from the degraded polymer that can be used as chemical feedstock.

Chemical Methods

A range of chemical methods have also been developed for the recovery of thermoset reinforcing fibres based on the solvolysis of the resin material in the presence of catalytic solutions at temperatures below 350 °C. Often these methods will still require initial shredding operations to break down the bulk size of composite waste. In general, chemical methods for fibre reclamation provide greater potential for recovering resin-based recyclates and fibres that retain their mechanical properties. Chemical treatments are not very tolerant of contamination and remove the fibre sizing, which reduces the bonding potential with new polymers (Pimenta and Pinho 2011). These methods can also be limited by scale, and depending on the solvent composition, may be harmful to the environment.

3.3.3 Alternative Resin Chemistries

The development of recycling technologies for existing composite systems is an important challenge to address, while advances in resin chemistry might instead offer a more sustainable alternative for future composite structures. Considerable research also aims to design new resins with similar mechanical performance to existing materials but that can be more easily recycled without the need for destructive methods that degrade or shorten the reinforcing fibres.

Dynamic covalent chemistries are a potential solution to some of the general limitations of thermoset composites, such as recyclability. In particular, vitrimers have shown considerable promise for recycling as they are polymers derived from thermosets, with similar low-temperature performance, that are capable of viscoelastic liquid flow at high temperatures. This allows for novel functionality including complete reuse, recycling, repair or thermoforming, which until recently, has been restricted to softer and non-structural polymer materials. New epoxy-based vitrimers developed from the combination of diglycidyl ether of bisphenol A (DGEBA) and 4-aminophenyl disulfide (AFD) have shown potential as a future matrix material for advanced composite structures, reporting similar mechanical properties to existing thermoset composites (Ruiz De Luzuriaga et al. 2016).

Processing of these vitrimers simply requires the substitution of the standard epoxy hardener for AFD in order to produce the disulfide groups that facilitate the dynamic behaviour, albeit at the cost of reduced thermal and chemical stability.

A wide range of other degradable thermosets have been reviewed recently (Ma and Webster 2018). In particular, a high-performance poly(hexahydrotriazine) (PHT) resin that is prepared with 2,2-bis[4-(4-aminophenoxy)phenyl]propane (BAPP) and formaldehyde shows the most promise (Yuan et al. 2017). This PHT-BAPP resin system has been demonstrated as the matrix for an advanced carbon fibre composite with comparable performance to commercial epoxy and BMI composites, even after multiple cycles of gentle degradation in an acidic solution and reuse. Such technology allows for the repeated non-destructive recovery of the fibre reinforcement materials, offering a near-closed-loop solution to high-performance composites recycling. However, this technology is relatively immature and has not yet been fully tested or considered for larger scale feasibility.

4 Concluding Remarks

Thermosetting composites are a subset of polymer matrix composites that rely on resins with an irreversible curing reaction. For aircraft applications, the most common thermosetting composites are carbon fibre reinforced epoxies, due to their good mechanical performance and versatility. Although other reinforcement materials, such as boron fibre, and matrix materials, such as bismaleimides, are also seen in aerospace applications.

Carbon fibre reinforced epoxies exhibit superior weight-specific strength and stiffness compared with conventional metallic materials for aircraft structures, along with improved fatigue and corrosion resistance. Subsequently, thermosetting composites have been gradually adopted to replace metallic structures, in order to improve aircraft performance, fuel efficiency and to reduce overall maintenance. The transition to composite structures began with military aircraft where performance was the primary design consideration and high production costs could be tolerated. As technology advanced, the transition to composites was also observed for civil aircraft, with a more conservative and iterative adoption of composites into tertiary and secondary structures throughout the 1970s–1990s. However, the latest generation of passenger aircraft have incorporated a greater composition of composite materials in primary structures, making up more than 50% of the aircraft structural weight. The development of larger parts with greater complexity, to address higher production volumes in these latest generation composite aircraft, required significant innovation in the field of composites manufacturing. Primarily, commercial aircraft manufacturers have increasingly focused on automated technologies such as automated fibre placement, although out-of-autoclave manufacturing techniques are receiving considerable attention to produce thermosetting composite structures more economically.

Despite growing adoption and development, there still remains a number of fundamental research challenges for thermosetting composite aircraft structures. The validation and assessment of bonded repairs using non-destructive inspection techniques remains a significant problem for the certification of bonded repairs in flight-critical structures. There is also a desire for better integrated anti-icing/de-icing and lightning strike protection systems to create multifunctional composite structures. Finally, with regards to the inevitable growth in composites waste, thermoset composite recycling is an area in need of further research.

References

Berthelot J-M (1999) Classical laminate theory. Springer, New York, pp 287–311

Biron M (2014) Thermosets and composites: material selection, applications, manufacturing and cost analysis. William Andrew

Boeing (2006) Boeing 787 from the ground up. AERO, pp 17–23

Daniel IM, Ishai O (2006) Engineering mechanics of composite materials, 2nd edn. Oxford University Press, New York

Deo RB, Starnes JH, Holzwart RC (2001) Low-cost composite materials and structures for aircraft applications. In: NATO RTO AVT Panel spring symposium and specialists' meeting. Loen Norway, pp 7–11

Department for Business Innovation and Skills (2016) UK aerospace maintenance, repair, overhaul & logistics industry analysis. London

Falzon BG, Robinson P, Frenz S, Gilbert B (2015) Development and evaluation of a novel integrated anti-icing/de-icing technology for carbon fibre composite aerostructures using an electro-conductive textile. Compos Part A Appl Sci Manuf 68:323–335. https://doi.org/10.1016/j.compositesa.2014.10.023

Federal Aviation Administration (2014) PS-AIR-100-14-130-001: bonded repair size limits

Gagné M, Therriault D (2014) Lightning strike protection of composites. Prog Aerosp Sci 64:1–16. https://doi.org/10.1016/j.paerosci.2013.07.002

Gardiner G (2014a) BMI and benzoxazine battle for future OOA aerocomposites. Compos, World

Gardiner G (2014b) Resin-infused MS-21 wings and wingbox. Compos, World

Gay D (2014) Composite materials: design and applications, 3rd edn. CRC Press, Boca Raton

Hexcel Corporation (2016) HexPly® 8552 Product Data Sheet. https://www.hexcel.com/user_area/content_media/raw/HexPly_8552_eu_DataSheet.pdf. Accessed 10 Dec 2018

Larsson A (2002) The interaction between a lightning flash and an aircraft in flight. C R Phys 3:1423–1444. https://doi.org/10.1016/S1631-0705(02)01410-X

Lubin G (1982) Handbook of composites. Van Nostrand Reinhold, New York

Ma S, Webster DC (2018) Degradable thermosets based on labile bonds or linkages: a review. Prog Polym Sci 76:65–110. https://doi.org/10.1016/j.progpolymsci.2017.07.008

Mccarty JE, Roeseler WG (1984) Durability and damage tolerance of large composite primary aircraft structure (LCPAS). Seattle, Washington

Pickering SJ (2006) Recycling technologies for thermoset composite materials—current status. Compos Part A Appl Sci Manuf 37:1206–1215. https://doi.org/10.1016/B978-1-85573-736-5.50044-3

Pierce RS, Falzon BG (2017) Simulating resin infusion through textile reinforcement materials for the manufacture of complex composite structures. Engineering 3:596–607. https://doi.org/10.1016/J.ENG.2017.04.006

Pimenta S, Pinho ST (2011) Recycling carbon fibre reinforced polymers for structural applications: technology review and market outlook. Waste Manag 31:378–392. https://doi.org/10.1016/j.wasman.2010.09.019

Pimenta S, Pinho ST (2012) The effect of recycling on the mechanical response of carbon fibres and their composites. Compos Struct 94:3669–3684. https://doi.org/10.1016/j.compstruct.2012.05.024

Roeseler WG, Sarh B, Kismarton MU (2007) Composite structures: the first 100 years. In: 16th Int Conf Compos Mater, pp 1–10

Ruiz De Luzuriaga A, Martin R, Markaide N et al (2016) Epoxy resin with exchangeable disulfide crosslinks to obtain reprocessable, repairable and recyclable fiber-reinforced thermoset composites. Mater Horiz 3:241–247. https://doi.org/10.1039/c6mh00029k

Soutis C (2005) Fibre reinforced composites in aircraft construction. Prog Aerosp Sci 41:143–151. https://doi.org/10.1016/j.paerosci.2005.02.004

Strong AB (2008) Fundamentals of composites manufacturing: materials, methods and applications. Society of Manufacturing Engineers, Dearborn

Yao X, Hawkins SC, Falzon BG (2018a) An advanced anti-icing/de-icing system utilizing highly aligned carbon nanotube webs. Carbon N Y 136:130–138. https://doi.org/10.1016/j.carbon.2018.04.039

Yao X, Falzon BG, Hawkins SC, Tsantzalis S (2018b) Aligned carbon nanotube webs embedded in a composite laminate: a route towards a highly tunable electro-thermal system. Carbon N Y 129:486–494. https://doi.org/10.1016/J.CARBON.2017.12.045

Yao X, Falzon BG, Hawkins SC (2019) Orthotropic electro-thermal behaviour of highly-aligned carbon nanotube web based composites. Compos Sci Technol 170:157–164. https://doi.org/10.1016/j.compscitech.2018.11.042

Yuan Y, Sun Y, Yan S et al (2017) Multiply fully recyclable carbon fibre reinforced heat-resistant covalent thermosetting advanced composites. Nat Commun 8:1–11. https://doi.org/10.1038/ncomms14657

Chapter 4
Thermoplastic Composites for Aerospace Applications

Marco Barile, Leonardo Lecce, Michele Iannone, Silvio Pappadà, and Pierluca Roberti

1 Introduction to Thermoplastic Composites

The use of composite materials in transportation sector has progressively increased over the last decades bringing a breakthrough point in the way vehicles are designed and built. With respect to traditional materials, in theory, composites offer unlimited flexibility in designing complex shapes, higher efficiency in manufacturing processes, reduced operational costs and potential notable weight savings. Looking at the aviation industry, which constantly strives to reduce its impact on the environment, the Advisory Council for Aviation Research and innovation in Europe (ACARE) aims for 75% reduction in CO_2 emissions per passenger kilometer compared to 2000 (Flightpath 2050 Europe's Vision for Aviation n.d.). Therefore, the introduction and development of new materials together with related production processes will be one of the key elements to achieve this goal.

Good evidence of this can be found in the current aerospace programs such as the Airbus A350 (Marsh 2007), Boeing B787 Dreamliner or Bombardier C-Series CS-100 (Marsh 2013) which have about 50–53% of their total structural weight (excluding engines) made of composite materials. The first application of

M. Barile (✉) · L. Lecce
NOVOTECH Aerospace Advanced Technology, Naples, Italy
e-mail: marco.barile@novotech.it; leonardo.lecce@novotech.it

M. Iannone
LEONARDO Aircraft Pomigliano, Naples, Italy
e-mail: michele.iannone@leonardocompany.com

S. Pappadà
CETMA—Engineering, Design and Materials Technologies Centre, Brindisi, Italy
e-mail: silvio.pappada@cetma.it

P. Roberti
SOLVAY Aerospace Composite Materials, Milan, Italy
e-mail: pierluca.roberti@solvay.com

S. Pantelakis, K. Tserpes (eds.), *Revolutionizing Aircraft Materials and Processes*, https://doi.org/10.1007/978-3-030-35346-9_4

thermoplastics on aircraft parts dates back to the US military's F-22 jet fighter's landing gear and weapons-bay doors (1980), but today they are not widely used yet, because of some issues related not to the feasibility of the parts, but to the affordability of the manufacturing processes.

Currently flying thermoplastic composite flight-critical parts include the main wing leading edges of the Airbus A380 superjumbo passenger jet and the A340 500/600-series airliners that preceded it, as well as the Gulfstream Aerospace G650's rudder and elevators (Black 2016). In parallel with aerospace field, BMW is leading a step change in automotive market by starting in 2013 the fabrication of the world's first volume-produced vehicle (100 cars/day) with a passenger cell made from CFRP, the i3 model, of which about the 10% (about 13.3 kg with respect to the 140 kg of the car body) is thermoplastic matrix based (Starke 2016).

Composites world is in continue evolution and there has been a progressive change in terms of manufacturing processes, passing from standard wet or prepreg manual layup to automated (preforming) technologies, as Automated Tape Laying (ATL) and Automated Fiber Placement, with the objective to increase production rates and make cheaper manufacturing processes (Fig. 4.1) (Red 2014).

With thermoset composites the use of an autoclave for the curing is seen as a bottleneck to progress and main industrial actors are working to develop or improve out-of-autoclave fabrication processes to produce structural composite parts. Within this context, thermoplastics will give significant cost reductions if the potential advantages in processing (as thermoformability and weldability) will be exploited through suitable and automated processing techniques. A specific opportunity is related to the possibility to produce parts by ATL or AFP without autoclave, processes not yet performed with a suitable quality of the fabricated parts.

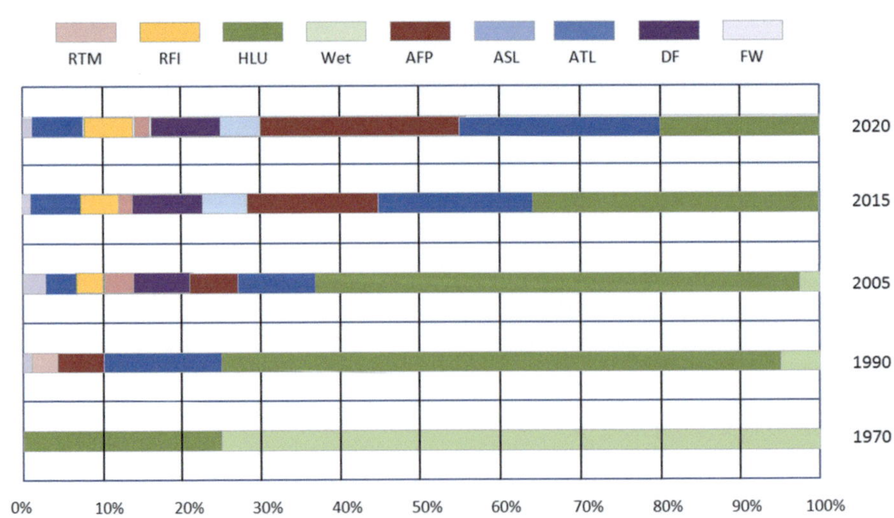

Fig. 4.1 Composites in Aerospace sector, 1970–2020. (Adapted from Red 2014)

This process can give a strong cost reduction which could compensate the high cost of the thermoplastic matrix prepregs. Since thermoplastics are not cross-linked, they are much tougher than thermosetting, hence they are more keen to be used in applications where damage tolerance is the driver. In theory, since thermoplastics are fully reacted, they do not need any reaction during cure so the process is simpler and faster. For this reason thermoplastics can be consolidated in minutes with respect to standard thermoset resin systems which require long curing cycles to build cross-links. Another advantage in terms of economical impact is related to the handling and storage of such materials, since thermoplastic prepregs do not require refrigeration during transportation and storage. Therefore, a wide use of thermoplastics implies a reduction of both investments for storage itself and cost of transportation. Also the material life management is simplified as it is not correlated to the resin expiration having, in theory, an unlimited shelf life. The good thermal stability is important for high-temperature applications (supersonic, high-temperature parts) and the good fire resistance is useful for some specific applications (nacelles, interiors). Figure 4.2 (Wong 2017) reports BMW's manufacturing technology curves highlighting how thermoplastics will impact the production processes and how this transformation will be fast.

The interest for recyclability (discussed in the following) which for thermoplastics is better than for thermosets, is high and it is supposed to increase in the future. On the other side the nature of thermoplastics has some areas of disadvantages respect to the thermosetting. The thermoplastic prepreg is typically supplied in a roll form, stiff and board. As it is fully reacted the material does not have any tack and drapability, resulting in a difficult layup process with respect to the standard

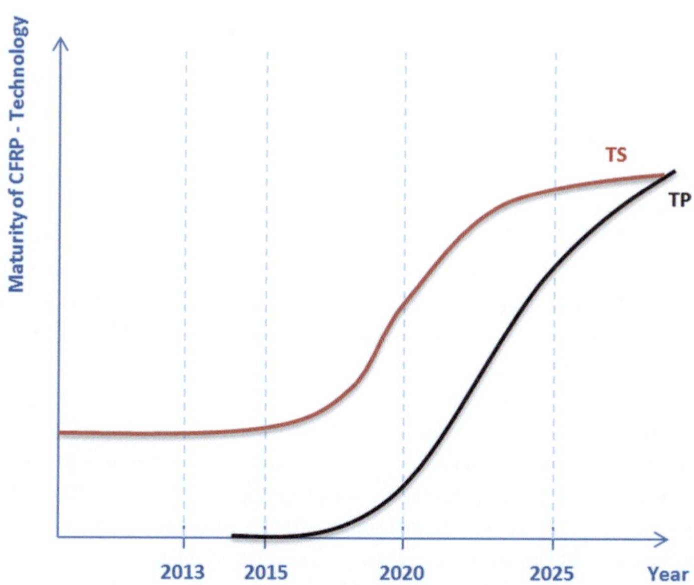

Fig. 4.2 BMW technology curves. (Adapted from Wong 2017)

tacky and drapable thermoset prepreg. In this case, the hand layup process is not a preferred option, especially with complex and contoured shapes. Based on this, the current applications are dedicated to simple geometries even if in the recent years new technologies are focusing on more complex shapes, targeting the standard common applications of epoxy systems. Furthermore, processing temperature of thermoplastics is considerably higher with respect to the standard thermosets. As example a PEEK prepreg composite material should be processed at temperature higher than the melting temperature, around 380 °C, compared to 180 °C of a standard epoxy resin. This means that the equipment to process thermoplastics are more complex, expensive and process toolings with low CTE are needed. Another possible issue is related to resistance to the environment and solvents that is lower than thermosettings (mainly for amorphous thermoplastics).

2 Polymer Chemistry

Molecular structure of polymeric materials is characterized by the presence of long chains. This structure determines all the main chemical, physical and mechanical characteristics of polymers. Depending on the way of creating the chains during polymerization, two major families of polymeric materials can be identified, as thermoplastic and thermosetting polymers. In case of *thermoplastic polymers* the polymerization creates long chains without branching; the obtained polymeric material is characterized by the interaction among such chains which are not interconnected by molecular links. The only interactions between the chains are given by low electrostatic attraction (van der Waals forces) and by temporary links (entanglements). When the temperature decreases, causing the reduction of the molecular motions, van der Waals forces are sufficient to take the macromolecules together and the material keeps its volume and shape, as a solid material. At molecular level the macromolecules cannot be organized in an ordered network and the appearance is typical of the liquid state. This status, characterized by a behavior (e.g., own volume and shape) typical of solids with a disordered molecular state that is typical of fluids is called "glassy" state; some characteristic of glasses (e.g., transparency) depend on such structure. When the temperature is increased, above a certain value the molecular kinetic energy is sufficient to overcome the electrostatic attraction and the movements between macromolecules become possible; in this state the polymer behaves as high viscosity liquid.

The transformation from glassy to liquid state (and vice-versa) is called "glass transition"; the glass transition temperature is generally indicated with T_g. Glass transition is a progressive transition and generally, T_g is an average value of temperature in a transition zone that can be identified by a starting and an ending temperature. In case of *thermosetting polymers* (thermosets) the polymerization is not only linear, but also branching are created. This type of molecular growth gives also the possibility to make molecular stable links between the molecular chains (cross-links), creating a three-dimensional structure. This structure at low temperatures is in glassy state (as the one previously described). Above the T_g molecular motions become possible, but the flow is inhibited by the cross-links that cause a change of properties (e.g., a decrease of the elastic modulus) without flow. This state is called "rubbery state," because is typical of rubbers, that can be considered a thermosetting material with cross-link (namely with a reduced amount of cross-links). Due to the presence of the cross-links, a polymerized thermosetting polymer does not flow also like cracking and/or oxidation, but flow is not possible anymore. On the basis of the previous description some major differences between thermoplastic and thermosetting properties can be identified: Linear growth of molecular chain (Thermoplastics) vs branching and cross-linking (Thermosets); flow possible above T_g (Thermoplastics) vs inhibited flow also above T_g (Thermosets). The behavior of elastic modulus vs temperature for Thermoplastic and Thermosets is shown in Fig. 4.3a, b.

Namely for Thermoplastics the behavior shown is referred to amorphous thermoplastics. In fact in case of molecular flexible chains some portions of macromolecules can reach an arrangement similar to the solid ordered one (crystalline state); in polymers this arrangement can be obtained only for a part of the chains, while the other portions keep a disordered state. A sketch of this structure is shown in Fig. 4.4; the polymeric materials having this structure are called Semicrystalline Thermoplastics and the thermoplastic previously described, without crystalline phase, are called Amorphous Thermoplastics. The crystalline phase is organized like a solid, in crystalline lamellas that frequently are assembled in radial structures (spherulites).

One of the main factors influencing the degree of crystallinity is the rate of cooling during solidification, together with the molecular chemistry of the polymer. It is very difficult that polymers are 100% crystalline, because it is rare to have all chains aligned in all the regions. Heat treating causes crystalline regions to grow and % crystallinity to increase. Degree of crystallinity is usually expressed in % and some physical properties depend on % of crystallinity. The crystalline phase shows also a melting temperature, well above T_g. The presence of a crystalline structure hinders the flow of the amorphous phase above T_g, like in the thermosettings. But when the temperature of crystalline melting T_m is reached, the flow is possible. The behavior

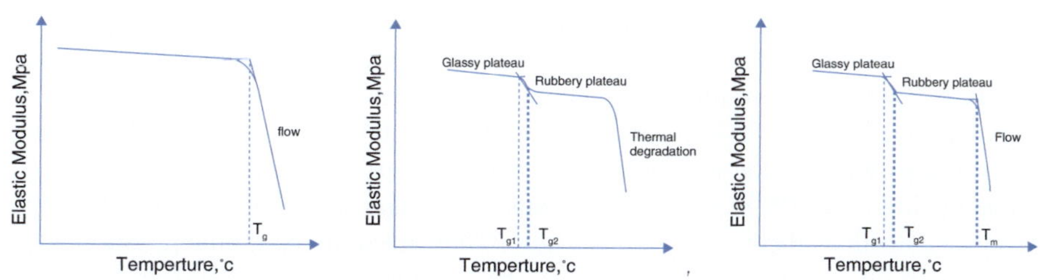

Fig. 4.3 Typical elastic modulus versus temperature curves for amorphous thermoplastics (**a**), thermosets (**b**) and semicrystalline thermoplastics (**c**)

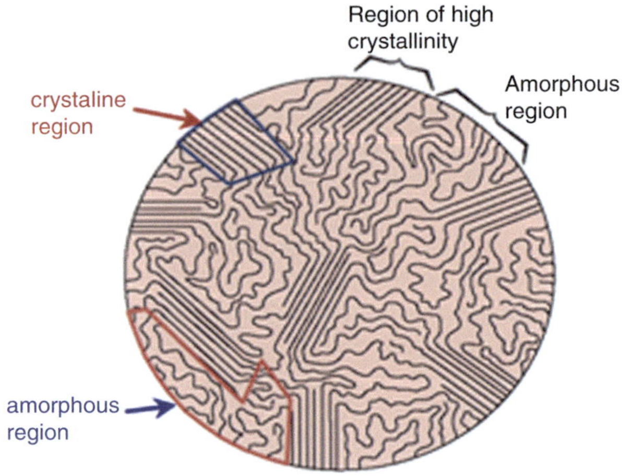

Fig. 4.4 Sketch of a typical semicrystalline thermoplastics structure. (Adapted from Fig. 14.11 H.W. Hayden, W.G. Moffatt, and J. Wulff, The Structure and Properties of Materials, Vol. III, Mechanical Behavior, John Wiley and Sons, Inc., (1965))

of elastic modulus versus temperature for Semicrystalline Thermoplastics is shown in Fig. 4.3c.

With regard to Fig. 4.3c, in correspondence of the T_{g1} the reduction of the modulus is lower than what occurs in case of the amorphous polymers. It is noticed that in case of amorphous polymers, in correspondence of the T_g, there is no change in the arrangement of molecules and only the mobility of the chains changes, which still maintain a disordered state, typical of the amorphous structure. After the T_m it is not possible to measure the modulus since it is in fluid state. By the point of view of structural properties, for the mentioned materials the Glass Transition Temperature T_g is a point where such properties decrease; consequently a material for structural applications must show a T_g suitably above (with a sufficient safety margin) the maximum service temperature, which for commercial aircraft is typically 82 °C. For applications different from the structural ones (e.g., domestic commodities) some semicrystalline thermoplastics (e.g., Poly-ethylene PE or Poly-ethylene terephthalate PET) are used at operating temperature comprised between T_g and T_m (Fig. 4.5). They are flexible, due to the reduced stiffness, but keep the shape and are not brittle.

With regard to the molecular weight, intended as the mass of a mole of chains, tensile strength increases with it because longer chains are entangled better. Furthermore, tensile strength and elastic modulus often increase with the percentage of crystallinity. In terms of high-performance materials, as reported in the thermoplastic polymer pyramid (Fig. 4.6), it is worth mentioning the family of Polyaryletherketones (PAEKs) and their derivatives. PAEKs are obtained by joining ether groups with ketone groups and the way these groups are arranged in the final molecule identifies the various constituents of the family. The percentage of ketone groups determines the melting temperature. They are semicrystalline thermoplastic polymers with a wide temperature range, high strength, stiffness and resistance to hydrolysis, which make them suitable for applications demanding extreme conditions.

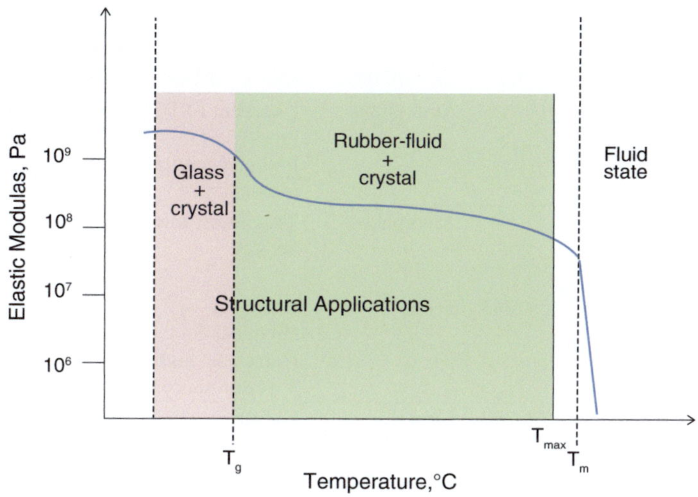

Fig. 4.5 Typical operative temperatures for semicrystalline polymers in applications requiring flexibility

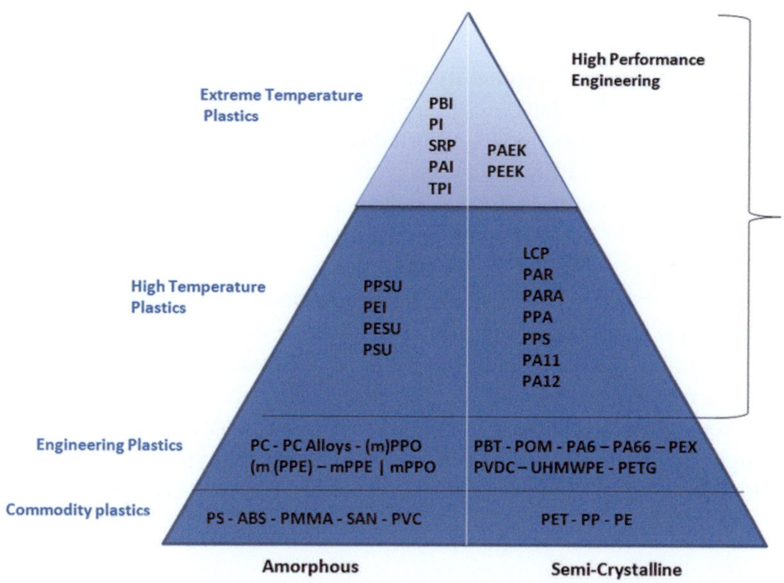

Fig. 4.6 Thermoplastic polymers pyramid

One of the most significant thermoplastic resins used for structural applications in aerospace field is definitely the Polyetheretherketone (PEEK) whose structure is born precisely from the repetition of two ether groups and one ketone. The presence of aromatic rings and ketones confers a certain rigidity to the molecular structure,

Table 4.1 Thermoplastic resins commonly used in aerospace sector

Chemical name	T_g (°C)	T_f (°C)	Crystallinity	Main characteristics	Cost (€/Kg)	Sector
PEI (Poly Ether Imide)	215	–	Amorphous	Excellent FST properties; limited chemical resistance	15–20	Aerospace interiors
PES (Poly Ether Sulfide)	220	–	Amorphous	Toughness; low density	1–3	LFT; flexible pipes
PEEK (Poly Ether Ether Ketone)	140–145	334–343	Semi	Excellent chemical resistance; high T performance	60–80	Aerospace; medical
PPS (Poly Phenil Sulfide)	85–90	275–290	Semi	Excellent chemical resistance	15–25	Aerospace; industrial

dampened however by the introduction of the ether groups. For this reason, PEEK has a lower melting temperature than other compounds of the same family, such as PEK (Polyetherketone) or PEKK (Polyetherketonketone). Table 4.1 lists main properties and cost of the most commonly used thermoplastic resins in aerospace field, also for structural applications.

At last, in addition to polymers reported in Table 4.1, it is worth to mention the semicrystalline polymer PEKK which is recently emerging for a wider spectrum of applications with respect to PEEK, including the potential production of aerospace primary structures by out-of-autoclave manufacturing processes (i.e., Continuous Compression Molding, In Situ Consolidation).

Different properties of PEEK and PEKK, due to different structures (Fig. 4.7), imply that PEKK has up to 80% greater compression strength than PEEK and wider processing window of parameters than PEEK. With regard to the temperatures, PEKK has a melting temperature comprised in the range 280–390 °C and a glass transition temperature of about 150–165 °C. Chemical structure adjustable gives PEKK superior fusion properties with metal and other polymers, induced interesting applications. PEKK can be located at apex of the PAEK class thanks to unique mechanical, physical and chemical properties.

3 Thermoplastic Manufacturing Processes

This paragraph deepens some of the most innovative manufacturing processes already used and under development in aerospace sector to produce structural reinforced thermoplastic-based composites, starting from a brief comparison of the processing aspects with thermosettings.

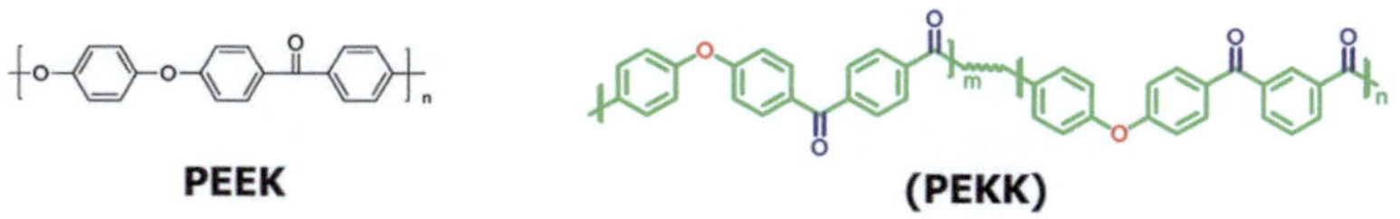

Fig. 4.7 Comparison of PEKK (copolymer) and PEEK (homopolymer) polymer structures

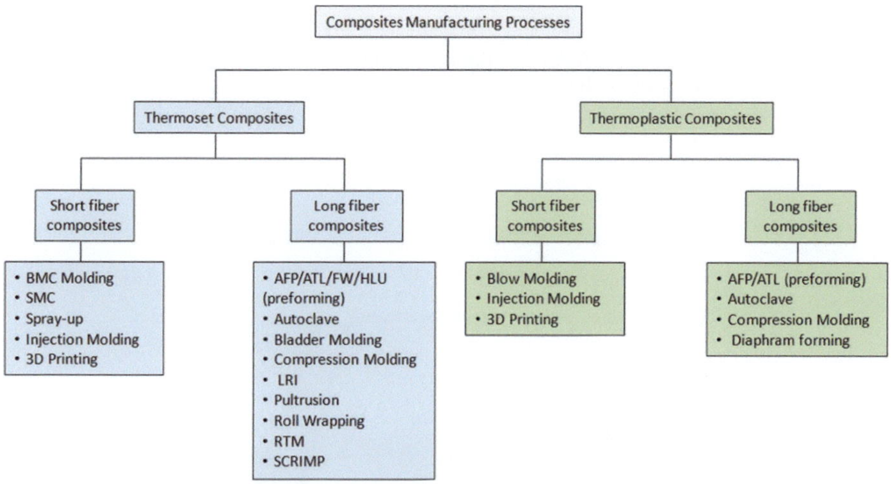

Fig. 4.8 Thermosettings and thermoplastics manufacturing processes

Figure 4.8 summarizes the production processes for thermoset and thermoplastic composites divided by short fiber and continuous long fiber composites.

The most commonly used production process for thermoset composites is based on manual layup of prepregs followed by autoclave curing. This process is based on shaping the part when the prepreg is uncured, consequently tacky and drapable. Afterwards, the material is cured in autoclave with a temperature and pressure cycle (Fig. 4.9a); this cycle requires a controlled heating rate, due to the need to control the curing reaction and the evolution of the resin viscosity with the temperature (rheological curve, Fig. 4.9b). In order to reach a suitable polymerization level, a long period (e.g., from 1 to 2 h) at constant temperature is required before cooling. Generally, the curing cycle is long and requires high pressure, then it is expensive.

With respect to thermosets, thermoplastics are generally purchased in a polymerized form; they are not tacky and their processing requires to reach a temperature sufficient to have the resin in a fluid state (above T_g for amorphous thermoplastics and above T_m for the semicrystalline ones). The holding time at high temperature is only the one needed to shape and consolidate the plies, not for polymerization (the resin is already polymerized), consequently can be short enough. Also cooling can be fast, except for taking in account thermal deformation and (for semicrystalline thermoplastics) the effects on crystallinity of the cooling rate. As a consequence of

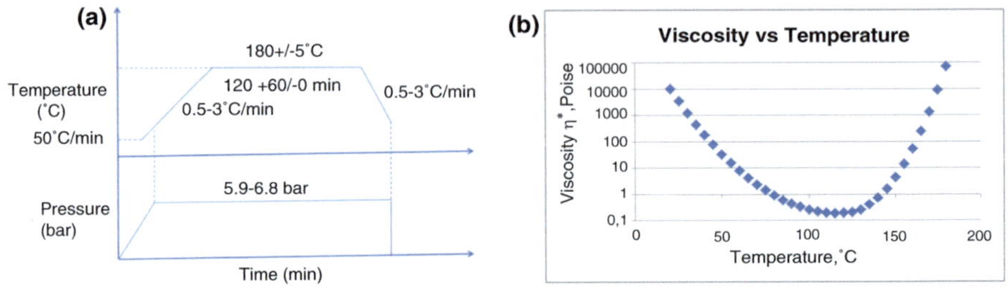

Fig. 4.9 Typical curing cycle of thermoset (epoxy) composites (**a**) and typical viscosity curve of epoxy resin during cure heating (**b**)

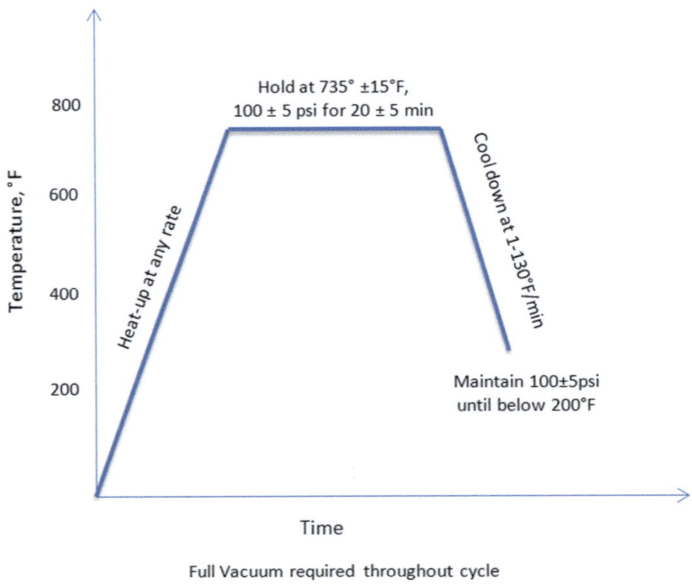

Fig. 4.10 Recommended autoclave cure cycle for APC-2-PEEK Thermoplastic Polymer. (Source: Solvay Cytec website)

the previous considerations, processing of thermoplastics is potentially cheaper and faster than the one of thermosettings, if peculiar fabrication techniques (as thermo-forming and assembly through welding) are developed. As example, recommended autoclave consolidation cycle for APC-2-PEEK Thermoplastic Polymer from Solvay Cytec is reported in Fig. 4.10.

3.1 Thermoforming Process

The fabrication process for thermoplastic composites requires three main steps: heating to melt the polymer matrix; pressurization to consolidate and shape; cooling to bring the polymer to the solid state. When forming and consolidating thermoplastic composite materials, there are three main mechanisms that allow the user to fabricate the parts:

- Polymer percolation (polymer flow through layers)
- Transverse flow (fibers spreading under pressure)
- Intraply shear (ply movement during forming)

Thermoforming process is a rapid manufacturing process used to form a 2D panel into a 3D shape. It is an attractive process thanks to the relative low cycle time. Typically, it requires the use of a semi manufactured item. In order to optimize the process, the thermoforming step is fed using a pre-consolidated sheet.

The pre-consolidated sheet is manufactured by using the thermoplastic composite material, starting from a roll form. The material is cut in a certain number of plies and the stacking sequence is made by holding them together with soldering iron on the edge of the preform. Most recently this activity has been made automatic using numerical control cutting equipment and pick and place stations, where the plies are laid up automatically to get the final panel. The consolidation can be performed with autoclave or press. In any case the temperature to be reached should be higher than the melting point (see Table 4.1). As example, PEKK autoclave consolidation requires a cure cycle with holding phase at 380 °C for 20–30 min and a pressure of 7–10 bar, depending on final part application. The holding time depends on the panel's thickness and it has to be measured on the lagging thermocouple. Experimental studies (Offringa 1996; Scherer and Friedrich 1991) demonstrate that the effect of pre-consolidation has a direct impact on final part quality after the thermoforming process. Free void laminates at this stage can generate good part quality after the final forming. After the pre-consolidation the panel can be stored, shipped, and trimmed. This is the typical part used for the final forming. This stage requires two different equipment. The first one is the *I/R* oven which allows the panel to become "soft" and achieve the melting point. Then the part moves to the press where the tooling is heated. The main phases of the thermoforming process are reported in Fig. 4.11. The transfer time from the *I/R* heater to the press is a key parameter to reach the right crystallization rate. For PEKK the tool should be set at

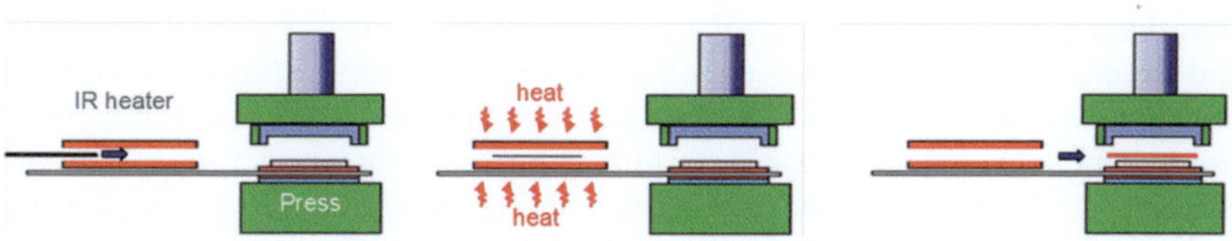

Fig. 4.11 Sketch of thermoforming process

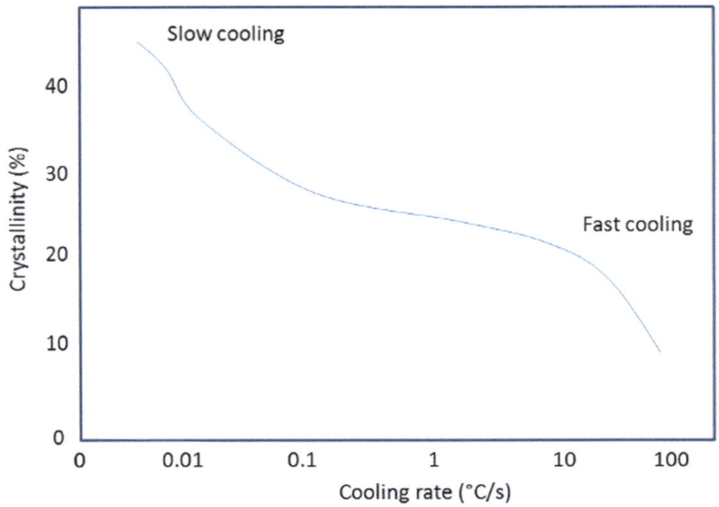

Fig. 4.12 PEEK/CF processing window: relation between degree of crystallinity and cooling rate

about 240° and its function is to anneal the part and obtain the right crystallization rate. The pressure typically reaches 20–30 bar and the time depends on the geometry.

The quality of the final formed parts includes several properties like porosity level, crystallinity and interlaminar bond strength. All the properties depend on local thermal and pressure influence.

While it has been demonstrated by experimental study (McCool et al. 2012; Wakeman and Blanchard 2005) that the porosity level is influenced by the quality of the pre-consolidated sheet, the crystallization rate depends on the cooling rate during the forming process. The polymer morphology depends on the thermal history created during the manufacturing process. To better understand the effect of parameters as temperature and pressure on the mechanical properties of a part produced by thermoforming process, taking as reference a reinforced PEEK-based composite material and related processing window (Fig. 4.12), it was shown (Mallon et al. 1998; O'Bradaigh and Mallon 1989; Davies et al. 1989) that laminates consolidated in the temperature range comprised between 360 and 390 °C have a more uniform impregnation of the matrix and an increased flexural failure.

Furthermore, high cooling rates upon forming decrease the degree of crystallinity of the polymer and increase the void content (Manson et al. 1990; Wakeman and Blanchard 2005). With regard to the fracture toughness, it was demonstrated (Ye and Beehag 1996) that the pressure effect during the cooling returns an increment of the mode II with a higher cooling pressures; with regard to the effect of the temperature, both mode I and mode II fracture toughness of laminates formed between 380 and 400 °C, because of the improved composite interfacial strength, are considerably higher than the ones formed at 360 °C. Despite of these results, it does not mean that increasing the temperature improve properties.

In conclusion, a numerical tool can be used for the design of the mold and in order to predict the effect of processing parameters on the part quality. For instance one of the most important parameters is the formed sheet temperature, but its distribution is also important because it impacts the wall thickness distribution of the part (Pepliński and Mozer 2011).

3.2 Continuous Compression Molding

Continuous Compression Molding (CCM) is an innovative fabrication technique aimed to obtain parts by using a press to overcome some problems of thermal history management that can be experienced in "traditional" thermoforming. Specifically, thermoforming process is made with hot press and thermoplastic matrix composites forming is performed at a temperature above T_g for amorphous thermoplastics, above T_m for the semicrystalline ones. After forming process, the press cannot be opened before cooling up to a temperature where the material is solid. This process then requires long times (each cycle requires heating and cooling) and high energy consumption. Starting from these considerations, a cheaper process can be performed by the heating of pre-consolidated layers, followed by the forming in a press with cold tool. Sometimes, in order to reduce the internal stresses induced by fast cooling and in some cases to prevent amorphization by fast cooling, preheated tools (temperature below the preheating one and above room temperature) are utilized. Both the described processes can produce items with a maximum size limited by the available tool and press. Instead CCM is capable to produce highly shaped profiles or flat panels of effectively unlimited length. Operable by one person, the computer-controlled process yields product at speeds approaching those quoted for pultrusion—as high as 40 m/h (131 ft/h) for shaped profiles and up to 91 m/h (300 ft/h) for flat panels (Gardiner n.d.).

If the need is to produce an angle ply laminate (i.e., quasi-isotropic layup), strips of woven or UD material may be cut and welded together to form a multilayer unconsolidated panel with a specific layup which is used as input material for the continuous press. This sort of preform could also be produced by AFP/ATL machines reaching in this way a very high level of automation of whole manufacturing process. Continuous Compression Molding is based on the utilization of a press having a temperature decreasing from one side to the other; the part to be fabricated can be moved along the platens of the press with an alternate opening/closing cycle, moving the part ahead during the opening. Figure 4.13 shows a sketch of the equipment needed for the CCM process. This process makes possible to obtain a progressive cooling of the part thanks to its movement in the direction of lower temperature zones.

The cooling rate of the material running along the plates of the press depends on the change of temperature along the press (°C/min) and on the medium speed of the part along the press (mt/s); the combination of the two changes gives cooling rate (°C/s) which can be set through the two abovementioned process parameters. As a

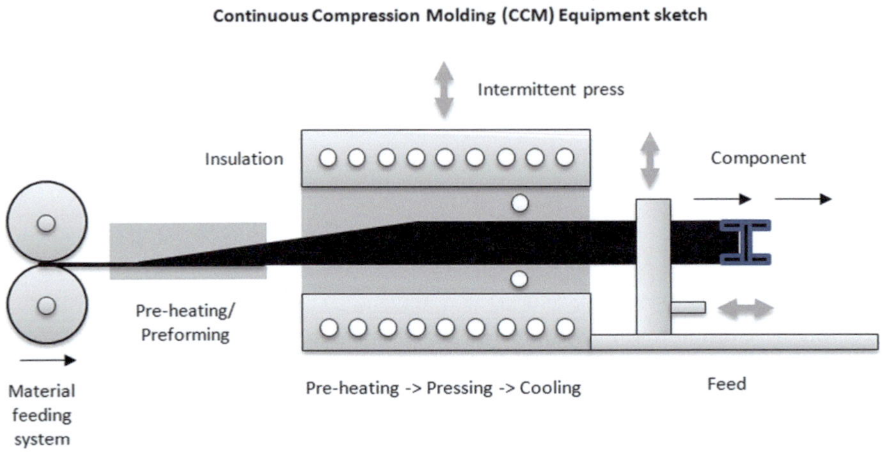

Fig. 4.13 A sketch of the CCM process

consequence of that, the first advantage of CCM is the capability to control cooling rate; a further advantage is given by working with a press at a temperature which is variable through the length but is kept constant in time, with a significant energy saving with respect to the standard thermoforming process. The geometry of the parts that can be fabricated with this technique is characterized by constant thickness and layup, as well as constant section and theoretically unlimited length. The parts are straight, but modifications to obtain constant radius curved parts seem possible. For these parts (straight or constant radius curved, constant section and layup) this technique is very cost effective. Changes in thickness and curvature radius are not presently possible, but they are being presently studied. One of the key features of the CCM process is that a part layup can be modified by adding/reducing layers without stopping the machine, and just adequately replacing the feeding material.

3.3 Pultrusion

Pultrusion is a continuous and automated technology able to produce constant section and rectilinear axis for high rate productions. Thermoset (epoxy, vinyl ester, polyesters) pultrusion is a mature process consisting in four main steps, as: continuous fibers (mat and rovings) are pulled through a guidance system; fibers enter a resin bath carrying out the impregnation step; fibers enter a heated die where exceeding resin/air are removed and starts the curing process (continuous crosslinking); finally, composite profile is pulled and cut at the desired length. A sketch of pultrusion line for thermoset and thermoplastic prepreg composites is reported in Fig. 4.14.

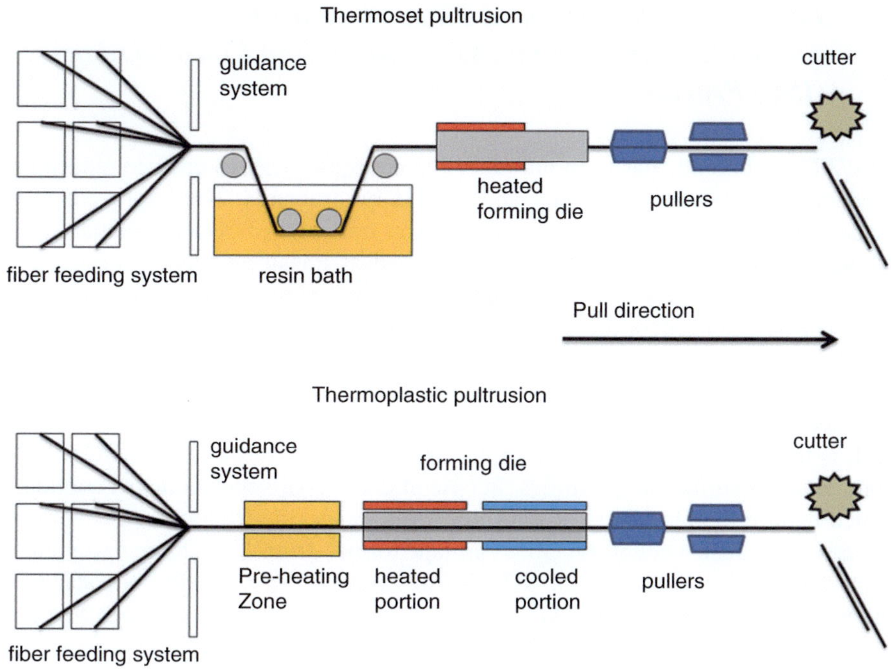

Fig. 4.14 A sketch of pultrusion line for thermoset- and thermoplastic-based composites (Novo et al. 2013)

Thermoplastic pultrusion is a relatively new process and limited research has been done up today. This is because the process has been mainly applied for thermosetting matrices as there are inherent difficulties associated with the thermoplastic matrices, such as high-processing temperatures and high melt viscosities (2 or 3 order of magnitude higher). Despite of this, with recent developments, the use of preforms as pre-consolidated tapes, commingled yarns and towpregs, allowed the thermoplastic pultrusion to gain a great interest (Nguyen-Chung et al. 2007). For the production of thermoplastic based composite profiles, pre-impregnated materials are guided into the preheating furnace to be heated up to the required temperature. Then, they enter in the pultrusion heated die to be heated and consolidated to the required size and, after that cooled down in the cooling die to solidify. Several variables are involved in the impregnation process. The fiber bundles are pulled through the die with an imposed force or velocity. For a given geometrical configuration, if we increase the pulling velocity, the contact time decreases and impregnation can become poor. Conversely, if one decreases the pulling velocity, the process efficiency can become too low and will have no economical interest. As a consequence, a thorough understanding of the pultrusion physics is crucial for a rigorous prediction of impregnation in order to reduce the time and the cost of fabrication (Novo et al. 2013).

3.4 Laser-Assisted Automated Fiber Placement (AFP)/ Automated Tape Laying (ATL) and In Situ Consolidation (ISC) Process

AFP/ATL technology has been introduced in aerospace field in order to reduce the production cost-flows of primary and secondary structures. The improvement is made possible thanks to the higher deposition rate with respect to manual layup (about 4:1); increased buy-to-fly ratios; a seamless transition between design and manufacturing (usage of advanced numerical tools); high repeatability of the process. The base concept of the AFP/ATL technology is quite the same, since the end-effector typically includes a tape feeding system with a cutting mechanism, a compaction roller and a heating source. The usage of AFP or ATL depends on specific needs concerning the geometry and layup of the composite part. More in detail, ATL is faster than AFP if parts have minimal contoured surfaces and exhibit little complexity or surface irregularities. In case of curved parts with complex contours, which require the placement of reinforcements in specific locations, AFP is more effective, because it counts on tapes (slit tape format, from 1/8″), more easily managed than the single tape of an ATL (Barile et al. 2017). Currently, the standard thermoset prepreg-based AFP/ATL process is utilised to produce fuselage sections of Boeing 787 and Airbus A350 XWB. Most recently the AFP/ATL machines are used also to manufacture engineered pre-consolidated sheets in order to reduce the scrap rate with respect to the standard hand layup process.

The scrap rate for a semi-complex part can be somewhere in between 20% and 50% when laid up by traditional Hand Layup. This scrap rate can be brought down to a value below 5% by the usage of ATL/AFP process. The main parameters which affect the quality of the laminate are the heating source power (and so the temperature reaches at the nip point during the deposition), the roller pressure, and the speed of the robot, while the clean room humidity and temperature, as well as the creel setup have no influence, as instead it happens in case of thermoset composites. AFP/ ATL used for processing thermoset materials differ from the ones used for thermoplastics mainly for the heating source used. Due to the need to achieve higher temperatures to melt the polymer and allow the ply–ply adhesion, for thermoplastics usually a diode laser is installed on the end-effector in the place of *I/R* heater, which is commonly used in case of thermoset prepregs.

Furthermore, for the deposition of thermoplastic composites, an infrared camera is used to set up the material heating law and for monitoring the temperature at the nip point, which is at the interface between roller and mold surface (Fig. 4.15a). The data can be used in an open or closed loop system. Figure 4.15a shows the Laser-Assisted AFP head concept and as example, Fig. 4.15b shows its application on a standard anthropomorphic robot (AFP cell) developed by Coriolis Composites (France) and available at NOVOTECH Aerospace Advanced Technology (Italy).

The first step, before fiber placement starts, is to put in relation the tooling with machine head by means of a calibration procedure. After the calibration is per-

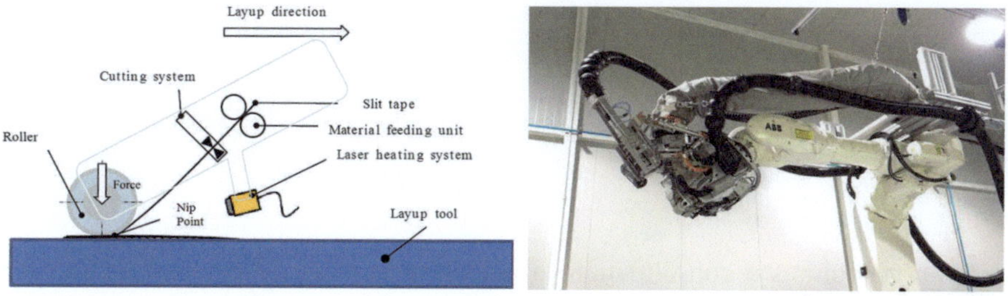

Fig. 4.15 AFP head scheme (**a**); Coriolis Laser assisted AFP robot at NOVOTECH Aerospace Advanced Technology (Italy) (**b**)

formed and the Part Program (layup, contours) is approved by the Design office, before starting fiber placement process, a dry run is performed by the AFPM operators to check the layup strategy (e.g., collisions, etc.). Afterwards, an amorphous layer or a release film is applied on the mold. When the fiber placement process starts, a band of parallel tapes are fed to the AFP head, passed through the cutting system, and placed on the mold according to geodetic curves set via AFP dedicated programming software. When tapes reach the nip point, at the same time they are subjected to the compaction force of the roller together with heating of the laser needed to melt the resin lightly in order to produce the right tackiness for the band deposition. Being able to adapt the width of the deposited band, AFPM is able to eliminate the excessive overlaps between adjacent courses. At the end of each trajectory, remaining tapes are cut so as to respect the contour of the part. Each tape is cut at 90° angle. The head is then positioned at the beginning of the next band. During the deposition of a band each tape is placed with its own speed so as to ensure that each other is in conformity with the others on the surface of the part. Therefore, the fibers can satisfy specific design requests. The action of the compaction roller enables to remove the air pockets. The positioning of the bands is previously set by a dedicated programming software associated with the fiber placement machine. The software enables to set the ply boundaries and orientations, the material parameters, gap/overlap, and other tolerances to be considered during the deposition. Basically, main technology parameters are: layup speed, roller's compaction force and heating source power (temperature at nip point). The software can also analyse fiber direction and perform specific simulations useful to optimize the deposition strategy. Specifically, it can include several possible analyses useful to assess the quality of the part that will be produced (i.e., steering analysis, angular deviation, etc.) and avoid collisions of the robot with mandrel. Figure 4.16 shows more in detail the Laser-Assisted Automated Fiber Placement process flowchart, in case of thermoplastic prepreg materials. At the current state of art, the main issue to fabricate thermoplastic-based composite aerostructures is not in the feasibility of the parts, but in the affordability of the production process. In particular, in case of primary structures, for sizes for which hot press is not suitable, the AFP/ATL

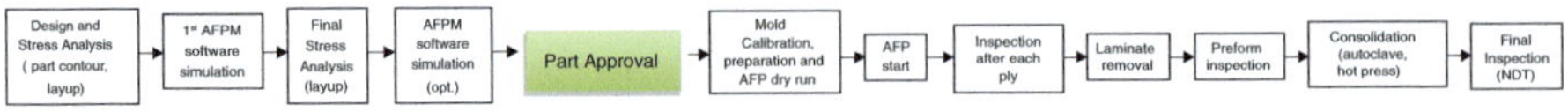

Fig. 4.16 State of art for Thermoplastic composite structures produced by automated preforming and next consolidation step

deposition requires a consolidation in autoclave at high temperatures with high energy costs and the usage of very expensive "not flying materials." In such context there is a technology which targets to have a consolidation during the layup phase known as In Situ Consolidation (ISC) by AFP process, mainly driven by a lower cost of part fabrication thanks to the reduced process flow. The concept of the ISC is depicted in Fig. 4.17. The consolidation temperature must be higher than the T_g for amorphous polymers and higher than T_m for semicrystalline polymers.

Generally, because of the higher viscosity, the temperature shall be at least 150–200 °C higher than the T_g in amorphous polymers and up to 100 °C higher than T_m in semicrystalline polymers. Complete ISC foresees the thermoplastic tape melted and consolidation during the machine lay up in order to not require any further process like consolidation in autoclave. The material quality plays an important role in this process as the resin-rich side improves the bonding and consolidation. Furthermore due to the nature of instantaneous consolidation and low amount of time to remove the porosity, the ply should be intraply void free, resulting fully impregnated.

Basically, sources of voids have to be searched in an insufficient level of intimate contact between the layers of a laminated. In particular, voids can migrate from the prepreg tape or can be created by the roughness of the tape and presence of gaps/overlaps during the placement. In order to improve the material adhesion this can be preheated but then the melting point can be reached at the interface between the new material layer and the previous one. The intimate contact occurs in fraction of seconds and the consolidation occurs only using the roller force application. Cooling rate is driven by the mandrel temperature and ambient temperature. Current limitations to commercial adoption of ISC processes in aircraft structures include: lack of proven data on materials and part performance with these new processing techniques; the need to demonstrate robustness process parameters for integral stiffened panels. If the applied pressure is less than a critical value, de-consolidation will occur and if it is higher than critical value, re-consolidation will happen. Also decompaction behavior of the fiber and applied external pressure determine final deconsolidation state while resin melt viscosity which depends on processing temperature affects the time for de-consolidation to reach a stable state. Therefore, the general perception to avoid a two-phase process (AFP preforming process followed by autoclave or press consolidation) and obtain parts with mechanical properties comparable to the ones of parts produced with autoclave/press consolidation, is to optimize the machine parameters and material behavior when subjected to heating.

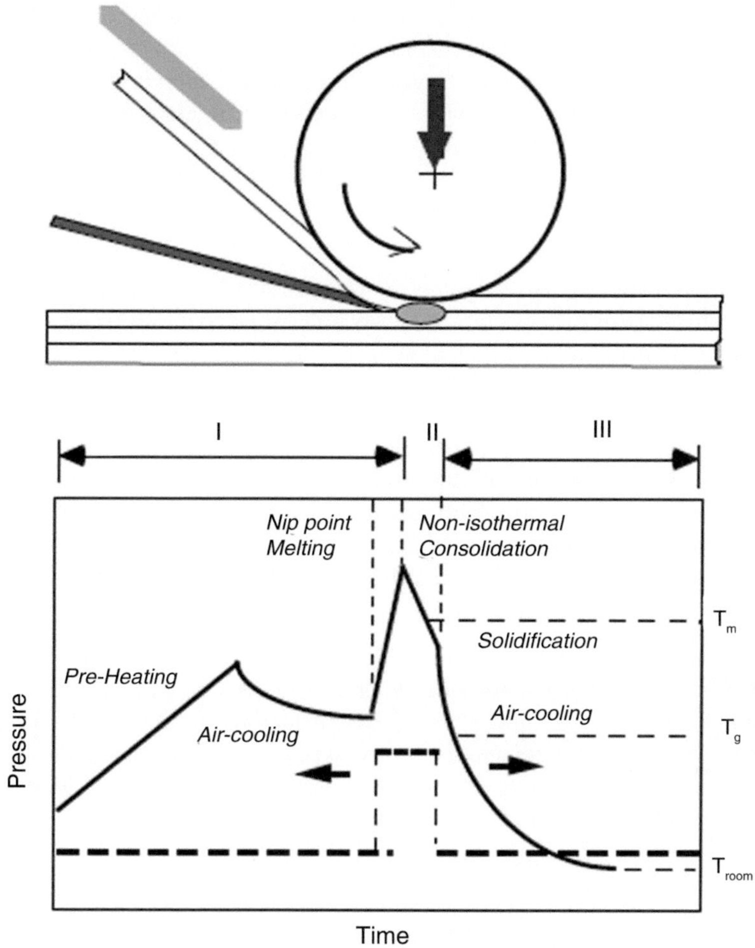

Fig. 4.17 Concept of the in situ consolidation by AFP process

4 Joining Processes

The choice of a material for an aerospace application depends most on the time of assembling procedures of the final structure (Vodicka 1996; MIL-HDBK-17-3F n.d.). This consideration is even more relevant for thermoplastic composites since their most important advantages are high damage tolerance and fast process rate (Biron 2007): for these materials the development of fast and reliable assembling procedures is a need coming from their requested production rate. In addition, it is important to consider that for thermoplastic composites higher limits in terms of complexity of the geometries that can be manufactured are experienced and that often a solution of the designer is the development of structures made of several subcomponents with easier geometries. Thermoplastic composites, differently than thermosets, can be melted and reformed. In this way delaminations and other damages may be re-welded and repaired heating the matrix and applying a consolidation pressure. At the same time a complex component can be assembled through the fusion of some portion of the subcomponents to be jointed.

Several joining techniques are available for composites and some of these have been specifically developed for thermoplastics. All these techniques can be divided into three classes:

- **Mechanical fastening**: Bolted joints are characterized by easier manufacturing, lower cost, and ease of inspection. Anyway holes weaken the structure since they act as stress concentrators (Thoppul et al. 2009).
- **Adhesive bonding**: Adhesive bonding requires controlled and delicate procedures for surface treatment and adhesive application method. Thermoplastic matrices are characterized by lower surface energies when compared to thermosets which makes difficult for adhesives to wet the thermoplastic adherend surfaces and creates a bond with high mechanical strength (Hart-Smith 1973).
- **Fusion bonding or welding**: In this case bonding is obtained by heating the region above melting temperature of thermoplastic matrix under applied pressure (Ageorges et al. 2001).

Joining by welding can provide a significant step in the further exploitation of thermoplastic composites in aerospace sector. In fact the welded joining of thermoplastic composites are often characterized by outstanding mechanical performance (high single lap shear strength and resistance to peel stress). Moreover the welding techniques can be easily automated, thus allowing the development of joining procedures characterized by fast production rate.

4.1 Fusion Bonding or Welding

In Fusion bonding or welding the property of thermoplastics to be melted and reformed is used to create a joining. There are different welding techniques, depending on the way to apply heat for matrix melting:

- ***Frictional heating*** *(spin welding, vibration welding, ultrasonic welding)*: In this case matrix heating is obtained using the friction between the two adherends. In frictional heating the motion of the substrates may cause deterioration of the microstructure, such as fiber breaking. Moreover, there is a limit with this heating principle regarding the area that can be welded. Ultrasonic welding seems to be less affected by these disadvantages, making this technique of interest for aerospace sector.
- ***Hot plate welding***: In this technique the heating device can be a hot plate that has to be removed before welding of adherends. This technique has limitations on size of the component that can be welded since the whole joining surface must be heated in a single step.

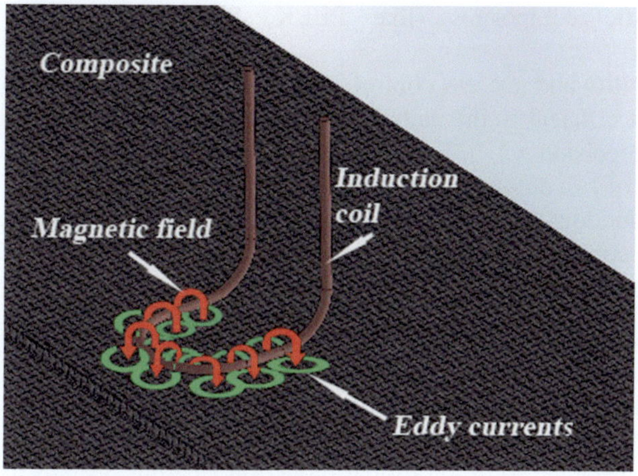

Fig. 4.18 Working principle of induction welding

- ***Resistance welding:*** In this technique the matrix is heated thanks to the Joule heat produced by a resistance placed between the adherend surfaces. The limits of resistance welding are that the resistance remains embedded into the joining (causing a decrease of mechanical performance) and the possible current leakages for carbon composites.
- ***Induction welding:*** In this process the material is heated by means of a high frequency alternate magnetic field (from 100 kHz to 1 MHz for plastic materials) generated by an induction coil. The magnetic field produces eddy currents within the conductive composite material, which heat the material due to Joule effect (Fig. 4.18). Joining consolidation is usually obtained thanks to the pressure applied with a consolidation cylinder. Among the materials than can be welded with this technique there are carbon-reinforced composites, but also glass-reinforced composites in which conductive layers were embedded near the joining surface (like, for example, metallic meshes or carbon fabric layers).

In this chapter special focus will be given to Induction Welding Technology, since this technique has a high potential of exploitation for different aerospace applications.

4.2 Induction Welding (IW)

Induction welding is one of the most promising techniques to join thermoplastic composites (Rudolf et al. 2000; Ahmed et al. 2006), for different reasons:

- High mechanical properties of the joining, for both static and dynamic loads: Single lap shear strengths between 30 and 50 MPa can be achieved with advanced

thermoplastic composites (like PEEK, PEKK, PPS and PEI carbon composites).
- High reliability and process control: Induction heat transfer phenomenon is one of the most efficient in the nature, so it is possible to heat the material fast and with high precision.
- Possibility of full automation of the welding process: There are different examples of Induction Welding equipment able to carry out automated continuous welding of complex geometries thanks to the use of robotic arms (Pappadà et al. 2015).

The main disadvantage of the induction welding of thermoplastic composite is related to the difficulty to optimize the temperature distribution through the thickness of the joining, avoiding, for example, too high temperatures on the surface directly exposed to the induction coil (the electromagnetic field decreases with the square of the distance, so in traditional Induction Heating systems the temperature under the induction coil is higher than the temperature in the joining interface) and near the edges of the joining (for the currents concentration due to the "point effect"). Anyway different approaches were developed to avoid this effect, like the use of metallic mesh in the joining interface for heat concentration, the use of low-melting film in the surface to be welded, or the introduction of cooling systems to decrease the temperature under the induction coil. In Fig. 4.19 the working principle of the induction welding machine developed by CETMA is described (Pappadà et al. 2015; Patent EP3017931A1 2014). In addition to the induction coil for material heating one cooling nozzle is introduced, to direct an air flux to cool the composite on its surface or near the edges, in order to optimize the thermal gradient during the welding. The cooling is adjusted by software, and the temperature on the upper surface of the joining is controlled by means of a thermo-camera. This cooling system makes possible to melt the composite matrix just near the joining surface, avoiding the melting of the same matrix in the upper surface (Fig. 4.19b) and its excessive heating near the borders.

The key parameters to understand the potential of Induction Welding techniques are reported below:

- Time to heat a carbon composite panel from room temperature up to 400 °C: 3–5 s
- Maximum welding velocities for common aerospace structures (skin and stringer for example): 2–5 mm/s
- Maximum working power: 3.3 kWh
- Maximum single-lap shear strength: 30–50 MPa depending on thermoplastic matrix
- Cost of a fully automated induction welding equipment (induction head equipped on a robotic arm): 200–300 k€

It is important to consider that since in the Induction Welding joining can be obtained thanks to a consolidation cylinder, the cost of the tools required for welding are low, since they are required just to keep subcomponents in the right position (tools to apply pressure during welding are not necessary).

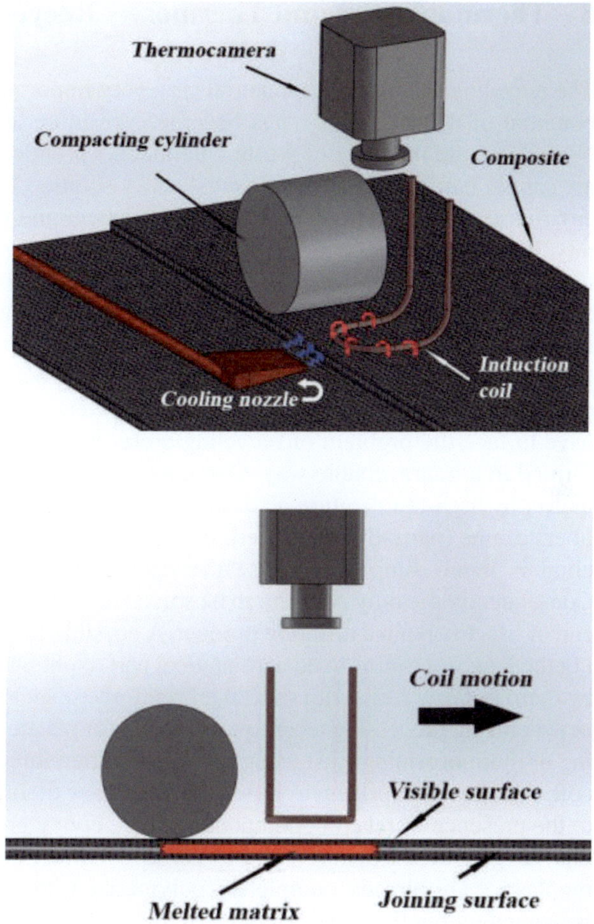

Fig. 4.19 Working scheme of the induction welding process in the induction welding technology developed by CETMA

The most important aerospace companies are currently carrying out research and development activities on different applications regarding Induction Welding, from TRL3 up to TRL9. Different approaches are investigated to optimize the thermal gradient within joining thickness during welding, in order to maximize performance and minimize defects (like voids and delaminations). Regardless of the followed approach, the fundamental aim is always the same: to develop reliable and efficient Induction Welding procedures, able to ensure high performance for welding of most used aerospace structures. According to the opinion of the author, this fundamental step for thermoplastic composites could be achieved through a realistic finite element modeling of the continuous welding process, and thanks to the definition of "Design for Assembling" procedures, able to optimize the design of the components to be welded, thus facilitating the development of the joining process.

5 Thermoplastics and Thermosets Recycling

The recycling of composite materials is a very important topic, which thanks to the potential of thermoplastics, has become even more attractive in the last decades. Nowadays, the majority of waste components produced with thermosetting resins are sent to landfill. This is an unsatisfactory solution from the point of view of the environmental, legislative, resource management and potential economic opportunity, linked to the recycling of waste. Taking into account that companies involved in production of composites for aerospace sector must be compliant with Aerospace Standard 9100, which includes also specific requirements of quality and safety aspects. It is evident that for an actual recycling there is a need to carry out suitable testing campaign useful to rewrite standards for products for their acceptance with recycled content. From the technical point of view, the thermoset-based composites have to face the problem of recycling cross-linked resins, which have been transformed in a nonreversible way. Consequently recycling requires dissolution of the matrix by high-temperature acids or thermal degradation of the resin, and only the fibers can be (partially) recovered. In case of thermoplastic-based composites recycling is much simpler, because the resin can be remelted by simple heating. Consequently the only problem to be solved is to transform the produced part, at the end of life, to obtain a different product. A possible recycling is performed by grinding the long fiber-reinforced aeronautical part to obtain pellets, which are composed by thermoplastic resin with short fiber reinforcement. Pellets can be used to produce new items by processing techniques suitable for plastic pellets (e.g., injection molding or thermoforming) just using the right temperature, but basically without major concerns. By the work environment point of view thermoset give some problem due to the presence of reacting components (e.g., epoxy groups, amine, and catalyzers) that can give health damages if they come in contact with the workers. Instead thermoplastics are already completely polymerized, thus no unreacted component is contained within them and handling of them is much safer. Furthermore, thermosetting processing requires a consolidated and cross-linking reaction, which is strongly exothermic and related to the chemo-rheology of the resin. Therefore, a very long process is needed, requiring slow heating (about 2 °C/min) and a long isothermal curing (generally about 2 h); this process is very demanding for energy consumption and, consequently, for CO_2 generation. Instead for thermoplastic heating is only required to make the part workable (no limits for heating rate) and the holding period is only needed for consolidation, and a short time is sufficient.

6 Conclusions and Future Perspectives

This chapter is aimed to give an overview on the recent advancements and applications in thermoplastic composites, focusing on the reasons why for aerospace sector, they are increasingly representing a more viable option for structural components.

Starting from thermoplastics polymer structure and difference with respect to thermoset-based composites, the most promising automated and continuous out-of-autoclave manufacturing concepts and processes, including assembling methods, have been deepened and some test cases have been illustrated. At last, an overview on different recycling concepts related to thermoplastics and thermoset composites has been provided. Current barriers to a wide application of thermoplastic composites can be overcome by the responding to the following needs:

- Development of affordable automated and fast processes to manufacture thermoplastic complex structures (aerospace sector)
- Development of efficient joining thermoplastic composite structures

At European level, Airbus in parallel with a variety of national aerospace consortia/programs are supporting the thermoplastic development roadmap. These activities are possible also thanks to the public–private partnership programs part of the Clean Sky, the largest European research program funded by EU's Horizon 2020 program (2014–2021) aimed to develop innovative, cutting-edge technology aimed at reducing CO_2, gas emissions and noise levels for the next generation of aircrafts. As previously stated one of the problems affecting a larger diffusion of thermoplastic-based composites for aeronautical applications is the need to process the material at very high temperature (above T_m for semicrystalline thermoplastics); these needs cause the increase of processing costs, both in the phase of prepreg and part fabrication. Furthermore, also cooling rate must be controlled to be in a window (e.g., for PEEK a cooling rate between 10 and 300 °C/min) (Kenny et al. 1989). A possible evolution aimed to solve this problem is the utilization of thermoplastics with lower melting temperature and more forgiving cooling rate requirements, as PEKK (Poly-Ether-Ketone-Ketone), material with a molecular structure very close to the PEEK one (except than the unit sequence is modified). PEKK shows some modification like the ones previously envisaged. Tm of PEKK is 337 °C, the one of PEEK is 343 °C; Tg of PEKK 160 °C is higher than the one of PEEK 145 °C. In general these slight changes can help, but the changes are minor. Major advancements are expected by more radical changes, like the utilization of uncured polymers which develop chains without cross-links ("curing" thermoplastic).

Another possible solution is the use of hybrid semicrystalline/amorphous polymers. Based on this, a new concept was developed (Preimpregnated materials with semi-crystalline matrix and amorphous surface layers 2011) to keep the advantages of working on the amorphous state without negative effects on crystallinity: the amorphous-semicrystalline thermoplastic matrix prepreg. This is obtained by making a sandwich with a PEEK matrix carbon prepreg (Solvay APC-2) between two amorphous films, and consolidating it in a process which gives full consolidation and controlled cooling rate, in order to obtain the right PEEK crystallinity (both amount and morphology). The good integration of the two abovementioned materials is obtained through a transition zone with different percentages of the two materials, progressively ranging from 100–0% to 0–100%, which is possible due to the capability of the two materials to make blends (Maffezzoli et al. 1989). A multinational research program was activated with the aim to develop fuselage thermoplastic composite demonstrators by means of hybrid prepreg manufacturing process; skin panel fabrication by Automated Fiber Placement In Situ Consolidation; stringers production by Continuous Compression Moulding; final aeronautical structure assembly by Induction Welding. This research project named NHYTE (New Hybrid Thermoplastic Composite Aerostructures manufactured by Out of Autoclave Continuous Automated Technologies) is in progress, supported by the European Commission (Barile et al. 2018; NHYTE Project n.d.).

With regard to future perspectives, a fabrication manufacturing process known as "Additive Manufacturing" or "3D Printing" deserves mention, by which digital 3D design data is used to build up a component in layers by depositing material. There are different additive manufacturing processes, but they have as common characteristics the capability to produce complex shapes by addition of melt layers; consequently this technique is used for metals and for thermoplastic polymers. The points that need to be improved are productivity, in terms of mass per time, and maximum size of the producible parts.

The thermoplastic parts can be produced both by amorphous and semicrystalline resins; in case of semicrystalline the cooling rate issue and its effect on final crystallinity must be considered. Continuous fiber composites cannot be processed by additive manufacturing; discontinuous fiber composite is under investigation, as well as the utilization of different resins, including the recycled ones. Injection molding is a relatively old technique; additive manufacturing is a new technique being developed, but both produce small and complex shapes. The main difference between these two techniques is in the tooling: injection molding requires expensive female close mold tools, with injection path and demolding pins, and additive manufacturing requires very simple tools or no tool at all. Production rate is presently much higher for injection molding, but additive manufacturing productivity could be improved in the future. Both for injection molding and additive manufacturing cannot be used continuous fiber composites. For injection molding a technique has been invented based on insertion of a continuous fiber composite part inside the injection molding shape, and then injecting the short fiber reinforced resin inside the mold. This technique is called "overmolding," and allows to combine the fabrication of complex shapes with the partial utilization of continuous fibers. A similar technique (additive molding on continuous formed consolidate parts) can be considered also for additive manufacturing. In conclusion, the continue research on thermoplastics will be the key point for the development of the next-generation transportation systems only if automated, fast, and reliable manufacturing methods will be improved according to the chemical of the materials and the specific application field.

References

Ageorges C, Ye L, Hou M (2001) Advances in fusion bonding techniques for joining thermoplastic matrix composites: a review. Compos Part A 32:839–857

Ahmed TJ, Stavrov D, Bersee HEN, Beukers A (2006) Induction welding of thermoplastic composites—an overview. Compos Part A 37:1638–1651

Barile M, Lecce L, Sportelli A, Iagulli G, Raffone M (2017) Systematic down selection approach to automated composites lay-down processes. In: The third international symposium on automated composites manufacturing, Montreal (Canada), April 2017

Barile M, Iannone V, Lecce L (2018) Automated fabrication of hybrid thermoplastic prepreg material to be processed by In-Situ Consolidation Automated Fiber Placement process. In: 5th international conference of engineering against failure, Chios, 20–22 June 2018

Biron M (2007) Thermoplastics and thermoplastic composites. Elsevier, Amsterdam

Black S (2016) Thermoplastic composites in flight...for decades. Compositesworld.com—https://www.compositesworld.com/articles/thermoplastic-composites-in-flight-for-decades

Davies P, Cantwell W et al (1989) Cooling rate effects in carbon fibre/PEEK composites. Presented at 3rd ASTM symp. on fatigue & fracture, Orlando, Nov.

Flightpath 2050 Europe's Vision for Aviation—Report of the High Level Group on Aviation Research (n.d.)

Gardiner G (n.d.) Aerospace-grade compression molding. Compositesworld.com—https://www.compositesworld.com/articles/aerospace-grade-compression-molding

Hart-Smith (1973) Adhesive bonded single lap joints. NASA CR-112236, Jan 1973

Kenny J, D'Amore A, Nicolais L, Iannone M, Scatteia B (1989) Processing of amorphous PEEK and amorphous PEEK based composites. SAMPE J 25(4):21 https://www.osti.gov/biblio/5422250

Maffezzoli A, Kenny JM, Nicolais L (1989) SAMPE J 25(4):35

Mallon PJ, O'Bradaigh CM et al (1998) Polymeric diaphragm forming of complex curvatures thermoplastic composite parts. Composites 20(1):48–56

Manson JE, Schneider T, Seferis JC (1990) Press-forming of continuous-fiber-reinforced thermoplastic composites. Polym Compos 11(2):114–120

Marsh G (2007) Airbus takes on Boeing with reinforced plastic. Reinf Plast 51(11):26–27, 29. https://doi.org/10.1016/S0034-3617(07)70383-1

Marsh G (2013) Bombardier throws down the gauntlet with CSeries airliner. Reinf Plast 55(6):22–26. https://doi.org/10.1016/S0034-3617(11)70181-3

McCool R, Murphy A et al (2012) Thermoforming carbon fiber-reinforced thermoplastic composites. Proc Inst Mech Eng Part L J Mater Des Appl 226(2):91–102. https://doi.org/10.1177/14644207124373

MIL-HDBK-17-3F: Composite materials handbook, vol 3. Polymer matrix composites materials usage, design. Cap. 6. Department of Defense

Nguyen-Chung T, Friedrich K, Mennig G (2007) Processability of pultrusion using natural fiber and thermoplastic matrix. Res Lett Mater Sci 2007:1–6. https://doi.org/10.1155/2007/37123

NHYTE Project (n.d.). https://ec.europa.eu/inea/en/horizon-2020/projects/h2020-transport/aviation/nhyte

Novo PJ, Nunes JP, Silva JF, Tinoco V, Marques AT (2013) Production of thermoplastic matrix pre-impregnated materials to manufacture composite pultruded profiles. Ciênc Tecnol Mater 25:84–90

O'Bradaigh CM, Mallon PJ (1989) Compos Sci Technol 35:235

Offringa AR (1996) Thermoplastic applications composites-rapid processing applications. Compos Part A 27(A):329–336

Pappadà S, Salomi A, Montanaro J et al (2015) Fabrication of a thermoplastic matrix stiffened panel by induction welding. Aerosp Sci Technol 43:314–320

Patent EP3017931A1 (2014) Induction machine for bonding polymeric matrix conductive composite material and bonding method for said machine

Pepliński K, Mozer A (2011) Ansys Polyflow software use to optimize the sheet thickness distribution in thermoforming process. J Pol CIMAC 6:215–220

Preimpregnated materials with semi-crystalline matrix and amorphous surface layers (2011) EP 2 109 532 B1 of 02/03/2011, LEONARDO S.p.A.

Red C (2014) The outlook for thermoplastics in aerospace composites, 2014-2023. In: High-performance composites. Gardner Business Media, Inc., Cincinatti, pp 54–63

Rudolf R, Mitschang P, Neitzel M (2000) Induction heating of continuous carbon-fibre reinforced thermoplastics. Compos Part A 31:1191–1202

Scherer R, Friedrich K (1991) Inter- and intraply-slip flow processes during thermoforming of cf/pp-laminates. Compos Manuf 2(2):92–96

Starke J (2016) Carbon composites in automotive structural applications, Eucia: composites and sustainability 19.03.16. http://www.eucia.eu/userfiles/files/Starke-Eucia%20 2016-V4-Druck%20b.pdf

Thoppul SD, Finegan J, Gibson RF (2009) Mechanics of mechanically fastened joints in polymer–matrix composite structures—a review. Compos Sci Technol 69(3–4):301–329

Vodicka R (1996) Thermoplastics for airframe application, a review of the properties and repair methods for thermoplastic composites. Department of Defence, DSTO Aeronautical and Maritime Research Laboratory, Melbourne Victoria

Wakeman MD, Blanchard P (2005) Void evolution during stamp-forming of thermoplastic composites. In: 15th international conference on composite materials (ICCM-15)

Wong J (2017) Processing of high performance thermoplastic composites, 131-5048-00L manufacturing of polymer composite, CMAS Lab ETH Zurich, 01.03.2017

Ye L, Beehag A (1996) Role of cooling pressure on interlaminar fracture properties of commingled CF/PEEK composites. Compos A: Appl Sci Manuf 27(3):175–182

Chapter 5
Additive Manufacturing: Design (Topology Optimization), Materials, and Processes

George Lampeas

1 Introduction

Additive Manufacturing is defined as the process of joining materials to make objects from 3D model data, usually layer upon layer, as opposed to subtractive manufacturing methods (definition from ASTM International Committee F42 on Additive Manufacturing Technologies). Synonyms include additive fabrication, additive processes, additive techniques, additive layer manufacturing, freeform fabrication, and 3D printing. AM methods exhibit considerable advantages over conventional manufacturing techniques, especially in cases of complex geometry components. The adoption of AM techniques can offer the ability to design and produce complex and demanding components that have optimal material topology and therefore optimal behavior in terms of functionality, load transfer, strength, and mechanical behavior.

In parallel to the progress of AM technology, appropriate topology optimization design methodologies have also been evolved, usually leading to complex components with optimal design and optimal material exploitation, which however usually have so high complexity that their production is only possible by exploiting the potential of AM methodologies.

Despite the great potential of AM in high added-value manufacturing sectors, like the aerospace, automotive, and biomedical, AM is still not well established to those or other industrial sectors. The most important technological barriers to the implementation of AM, not only as a rapid prototyping method, but also as a mass production method, among others are:

G. Lampeas (✉)
Laboratory of Technology and Strength of Materials, Department of Mechanical Engineering and Aeronautics, University of Patras, Patras, Greece
e-mail: labeas@mech.upatras.gr

S. Pantelakis, K. Tserpes (eds.), *Revolutionizing Aircraft Materials and Processes*, https://doi.org/10.1007/978-3-030-35346-9_5

- The difficulty of (almost) first-time-right definition of process parameters that would result in the best quality of the produced component, without the need to follow a long trial-and-error process. As an example, in the case of powder bed fusion AM process, these important parameters comprise the powder material type, the powder particle size and distribution, the layer thickness, the scanning speed and pattern, as well as the laser-beam power.
- The difficulty to take into account the particular features of AM during part design and analysis, such as in-homogeneity and anisotropy of the material caused by the layer-wise nature of the part creation.
- The development of residual stresses and distortions in the final component, making a postproduction heat treatment in many cases inevitable.
- The lack of satisfactory precision in the dimensions of the produced components, as well as the poor quality on some or all the part surfaces, requiring rework; the requirement of high precision machining in order to achieve the desired surface quality and to improve material properties (especially fatigue) is not the exemption for many AM aeronautical parts.
- The requirement for many AM methods to design a support structure for each particular component, which is used during the printing phase and must be removed afterward.
- The lack of properly adapted destructive and nondestructive methodologies for the quality control and certification of complex components manufactured by AM.

This chapter is structured as following: initially an introduction to structural optimization is provided; different types of structural optimization techniques are presented, focusing on topology optimization that is better suited to AM parts. The basic concepts and techniques used for the topological optimization of a part are presented and applied in two characteristic aeronautical structural parts of different scale, namely a bracket connection and a commercial aircraft fuselage airframe section. Consequently, the various types of AM manufacturing process are described, with special emphasis in the powder bed fusion techniques, which are more commonly applied for the production of aeronautical parts and components. This chapter is completed with an overview of the main active research subjects in the scientific field of additive manufacturing, especially relative to the aerospace sector.

2 Topology Optimization

Structural optimization is the most essential part of the design phase of any structure, especially in the aeronautical sector, in which high specific stiffness and strength are required. The objective of this process is to define the optimal material distribution in the structure, leading to minimum values of specific component's magnitudes, usually its mass and compliance, while fulfilling specific constraints like material stress, strain, and/or displacement limits.

Structural optimization techniques can be classified into three characteristic categories, namely sizing optimization, shape optimization, and topology optimization. In sizing optimization, the configuration of the structure is known and the objective is to optimize the size variables. For example in a truss structure, the positions of spars are predefined, while their cross-sectional dimensions have to be appropriately selected, in a way that the objective function is minimized (see Fig. 5.1a). This is the simplest and oldest category of structural optimization.

Shape optimization is applicable on continuum structures that do not consist of discrete parts (e.g., trusses). Here, the structural features (e.g., boundaries and openings) are predefined and the objective is to optimize variables like thickness distribution, dimensions of openings, and other geometrical properties. In the cantilever web of Fig. 5.1b, the size and direction of the elliptical openings are the subject of optimization.

Topology optimization is the most generic form of structural optimization. The objective is to define the optimum distribution of material in any available design space. It differs from the other types of structural optimization, as the optimized design can obtain any shape inside the design domain, i.e., the final shape or topology is not known a priori. In Fig. 5.2 the application of topology optimization for the design of a cantilever beam loaded with a concentrated force is presented. It is interesting to observe that the final shape of the topologically optimized structure in Fig. 5.2 does not look very different to that obtained by sizing optimization, shown in Fig. 5.1a; this can be usually observed in structures subjected to single or simple loading and boundary conditions, for which various design variations have been parametrically studied and analyzed, such that the "optimal" configuration is already known from engineering experience and historical data.

The obtained optimized design mainly depends on the design space, material properties, loading and boundary conditions. Compared to sizing and shape optimization techniques, topology optimization is by far more challenging technically and at the same time more rewarding, especially in cases of complex loading and shape constraints. Rather than limiting the changes to the sizes and features of structural components, topology optimization provides much more freedom and allows the

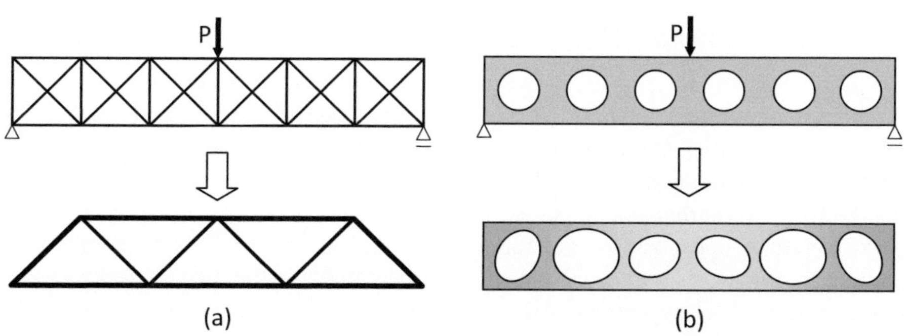

Fig. 5.1 Schematic of structural optimization categories, (**a**) sizing optimization and (**b**) shape optimization

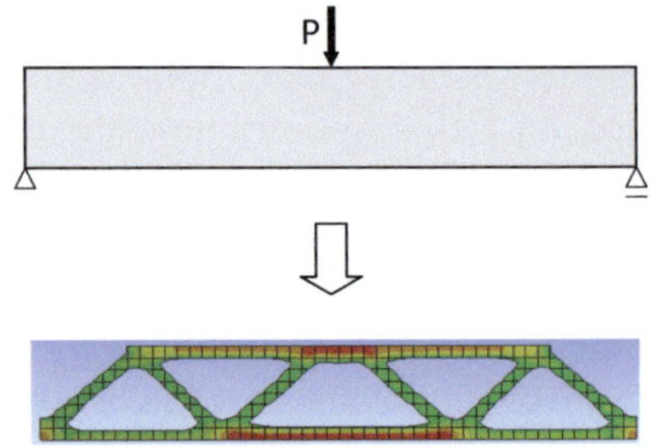

Fig. 5.2 Schematic of topology optimization

engineer to create totally novel and highly efficient conceptual designs for continuum structures. The need for novel manufacturing techniques, such as AM, required in the realization of these topologically optimized designs should be kept in mind.

The Finite Element Method (FEM) is generally used in combination with topology optimization methods in order to assess the design performance of the component or structure. In most of the topology optimization variations, an FE model of the design space is developed and the objective of topology optimization algorithm is to define which of the model elements should be solid and which should be void, i.e., to define the areas of the design space where material should or should not exist in order to optimally distribute material into the structural space. The most mature numerical methods for performing topology optimization are the Solid Isotropic Material with Penalization (SIMP) method, the Evolutionary Structural Optimization (ESO), and the Bidirectional Evolutionary Structural Optimization (BESO) methods. These methods are presented hereunder in brief. Detail information on these methods can be found in specialized books for topology optimization, e.g., Bendsøe (1995), Xie and Steven (1997), Hassani and Hinton (1999), Bendsøe and Sigmund (2003), and Huang and Xie (2010), while applications of these methods in aeronautical structures may be found in Remouchamps et al. (2011), Krog et al. (2002), and Tomlin and Meyer (2011).

2.1 Solid Isotropic Material with Penalization (SIMP)

As already mentioned, the basic idea of topology optimization is to find the optimal layout of material in a component by defining material points and voids (blanks) in a reference domain, which depend on the available space, the boundary and loading conditions, and (perhaps) some prescribed openings.

In topology optimization, the topology of the structure is not represented by continuum functions, but the design space is discretized in the form of an FE model mesh. The material properties are individually defined for each element of the FE model, thus forming a set of discretized property values in the design space, from which the stiffness tensor of the FE model is determined and the topology optimization problem is formulated.

The defined optimization problem is solved using numerical techniques aiming to find which elements of the FE model should be filled with material and which should be transformed to voids; such a formulation comprises a "zero – one" problem, in which elements either exist or not, i.e., an integer problem with two different states for each element is defined, resulting to an Isotropic Solid or Empty (ISE) topology. The number of potential solutions is 2^N, where N is the element number of the FE model. As a usual FE model can consist of many thousands of elements and in some cases millions, this discrete optimization problem can hardly be solved in many practical optimization cases of aeronautical structures.

An appropriate strategy for solving an ISE topology problem is the Density Method, in which the material density is allowed to take any value between zero and one, i.e., 0–100% density. By such relaxation it is possible to use gradient-based optimization methods and find the minimum of the objective function. The design variable of the optimization problem is therefore the density function over the design domain; when the density is approximated as constant over each element, the resulting problem has one design variable per element, i.e., its density value. In practical terms, such a solution approach looks similar to the sizing optimization, for which the sizing variables are the densities of the elements.

In the ISE approach, elements stiffness is considered to be a function of the respective material density. In the simplest approach the stiffness–material density relationship is considered linear, i.e., $E = \rho E_0$ where E is the elasticity tensor, ρ is the density ($0 \leq \rho \leq 1$), and E_0 is the elasticity tensor of the fully dense material.

An advancement of the Density Method is the "Solid Isotropic Material with Penalization" (SIMP) method. According to SIMP method intermediate densities are penalized, i.e., the cost of intermediate densities is higher compared to the relative stiffness, which makes intermediate densities unfavorable. This is obtained by expressing material stiffness as a power function of the density, i.e., $E = \rho^p E_0$, where the exponent p is a penalization factor. In the SIMP method, p is chosen to be larger than unity; it has been empirically derived that values of the factor p higher than three lead to very good designs, consisting mainly of areas with or without material, while areas with intermediate material densities tend to be eliminated. Practically, in cases that the densities are assumed constant over each element, the density–stiffness relation can be implemented by simply scaling the individual elements stiffness matrices before assembling them into the global stiffness matrix. In order to solve the optimization problem described above, several methods are available, including the Optimality Criteria method (Bendsøe and Sigmund 2003), the Method of Moving Asymptotes (Svanberg 1987), and the Sequential Convex Programming method (Zillober 2002).

An application of the SIMP method for the case of a long cantilever beam, fully constrained at its left edge and loaded by a concentrated force applied at the middle of the right edge, is presented in Fig. 5.3. In a preprocessing phase, the rectangular design space is determined and loads/boundary conditions are

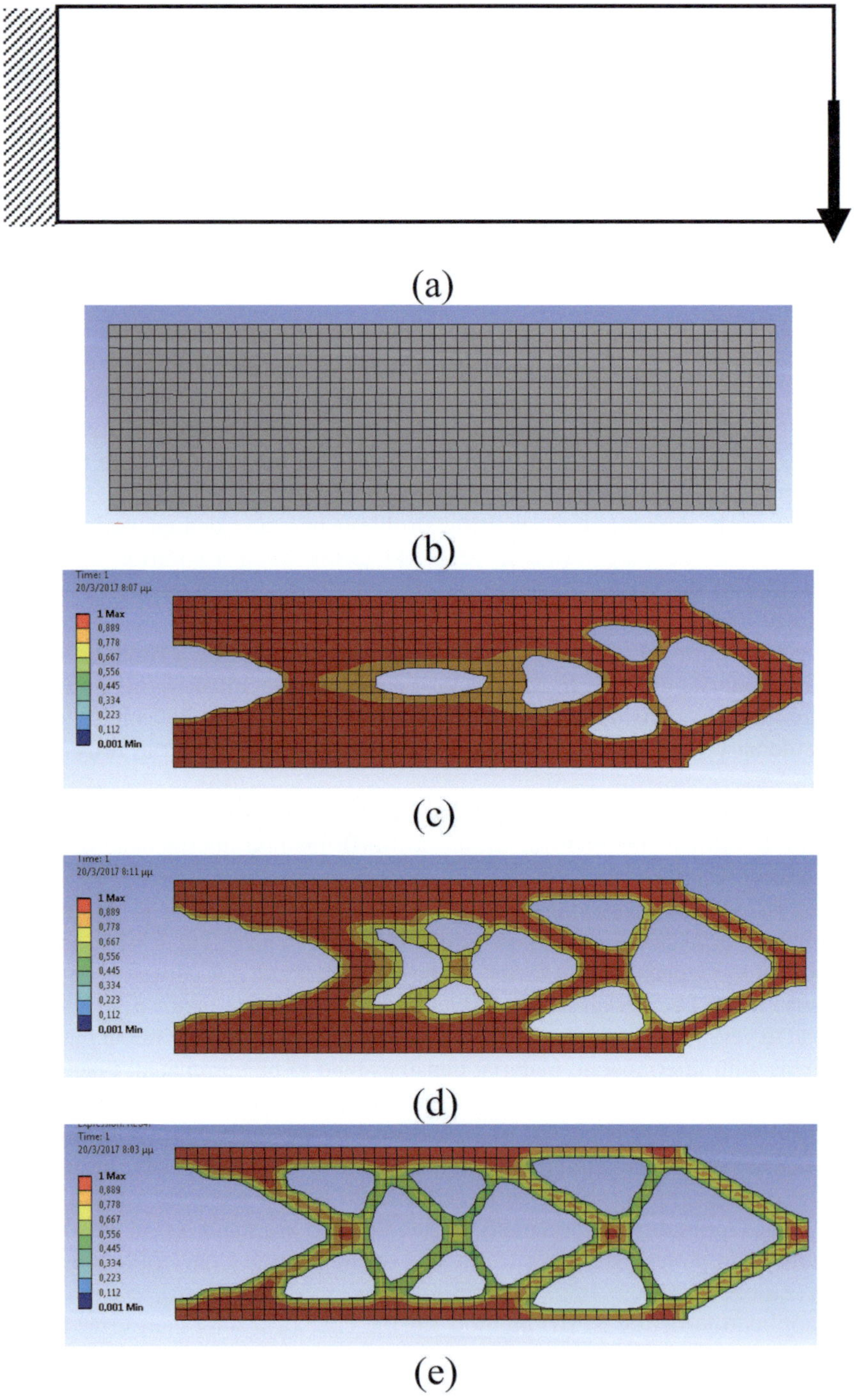

Fig. 5.3 (**a**) Cantilever beam under concentrated edge load, (**b**) FE discretization of design space; results of topology optimization using SIMP method for different volume constraints: (**c**) 80%, (**d**) 60% and (**e**) 40%

defined. The design space is then meshed by finite elements, as shown in Fig. 5.3b for which a uniform distribution of material is initially considered. In the optimization phase, depending on the density distribution, displacements, and deformations are iteratively calculated from the FE model solution and the compliance (the inverse of stiffness) is calculated. The elements updated densities are determined based on the conditions of optimality. The optimization phase steps are repeated until the calculated compliance is not improved significantly any more. In Fig. 5.3c–e the results of topology optimization are presented for different volume constraints.

2.2 Evolutionary Structural Optimization Method

The Evolutionary Structural Optimization (ESO) method was first introduced by Xie and Steven (1997) and is based on the simple idea of gradually removing inefficient material from the design space, until a structure of optimal shape and topology is obtained.

In the ESO method, stresses in each point of the component are calculated by finite element analysis, as in the case of the other topology optimization variations. The efficiency of the material utilization in every location of the design space is indicated by its stress level, with high stress levels obviously indicating justified material use. Ideally, the stresses at every point of the structure must be at the same level and should tend to reach the allowable material strength. By taking advantage of this concept, the low-stressed material is considered underutilized and is gradually and iteratively removed by deleting the corresponding elements from the FE model. The process of elements rejection is iteratively performed until no additional elements are rejected. At this point the rejection ratio is increased by a specific value and the iterative process is repeated until the model comes at a steady state again. The optimization process is finalized when the stress level of all elements lie within a predefined range around the maximum stress of the entire structure.

In Fig. 5.4, the ESO method is presented for the case of a short beam under bending. In Fig. 5.4a, the problem design space is shown, comprising a short beam fully constrained at its left edge, with a concentrated force applied in the middle of the right edge. The FE model of the structure is developed and solved, in order to compute elements stresses. An initial rejection ratio of 5% is considered and the elements having a ratio of stress over the maximum stress (developing at any point in the structure) lower than this rejection ratio have been removed from the FE model. The FE model is updated, solved again, and the element rejection process is repeated until no additional elements are removed. The rejection ratio is gradually increased by steps of, e.g., 5% and the process is repeated until the final optimized design is obtained. In Fig. 5.4b–d the obtained optimized topology is presented for increasing rejection ratios.

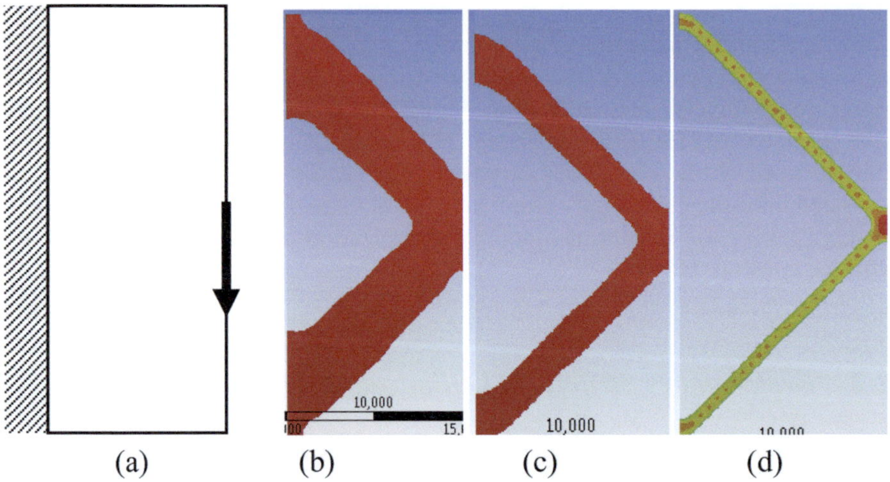

Fig. 5.4 Short beam under bending problem (**a**) and optimized topologies for different rejection ratios (**b**) 10%, (**c**) 15% and (**d**) 30%

2.3 Bidirectional Evolutionary Structural Optimization (BESO)

A more sophisticated ESO algorithm has been developed by Huang and Xie, known as Bidirectional Evolutionary Structural Optimization, or BESO method (Huang and Xie 2010). According to this topology optimization method, it is possible to remove or introduce new material in the structure, i.e., remove and/or add elements in the FE model in every optimization iteration. Two independent ratios, the rejection ratio and the insertion ratio, are used in each optimization loop to select the elements to be removed or added into the FE model. The BESO method can also be implemented using as criterion the ratio of element stress over maximum stress of the design space, as in the case of ESO. Practically, when BESO is applied, any elements with low stress levels are removed, while the elementary void areas in the vicinity of elements with high stresses are transformed into new FE elements.

The result of BESO method is very sensitive to the selection of the rejection and the insertion ratios, which if are not appropriately chosen, the optimal solution cannot be obtained. The computational performance of BESO method is lower as compared to ESO, due to the large number of iterations required to obtain a steady state solution.

A problem relevant to BESO technique is its mesh dependency, i.e., the numerical instability of the solution leading to a higher number of voids when the FE mesh utilized in the topology optimization becomes denser. In order to overcome the mesh dependency problem various approaches have been proposed like sensitivity filters from Sigmund (1997) and perimeter control technique by Harber et al. (1996). Yang et al. (2003) used the BESO method with perimeter control technique to obtain mesh-independent solutions using the perimeter length as an extra constraint in the topology optimization problem.

2.4 Demonstration of Topology Optimization Processes in Aeronautical Parts and Structures

The application of topology optimization is demonstrated in two characteristic aeronautical examples, i.e., a bracket connection of an Airbus A350 flight crew rest compartment, as well as a fuselage airframe section of a civil aircraft. Both cases have been extensively studied in the open literature to demonstrate the benefits of topology optimization in combination to AM production of aeronautical structural components and parts of different scale and geometry.

The conventional design of the bracket connection consists of two plates perpendicular to each other, a base plate with ten mounting holes and a second plate with a large hole, where load is applied (see Fig. 5.5).

In order to demonstrate the significant effect of the applied loading type to the derived optimal topology, different load cases typical for the bracket have been examined separately, comprising various combinations of forces in *x*-, *y*-, and *z*-axis that are applied in the bracket main bore, where *x* is the bracket longitudinal axis, *y* is the horizontal transversal axis, and *z* is the vertical axis. The topology optimizations have been performed by SIMP method and analyzed using the optimality criteria with the compliance minimization as objective, while retaining 30% of the initial volume was set as constraint. The penalization factor has been set to 3 and the convergence tolerance to 1%. Some of the characteristic calculated optimized topologies for the different load cases analyzed are presented in Fig. 5.6.

The design which satisfies the combination of all examined load cases can result from the combination of the individual topologies calculated for each separate load scenario. In any case, structural FE analysis taking into account the final topology has to be performed under all different load cases, in order to verify the component integrity. It should be mentioned that the final optimized component design is also affected by the choice of topology optimization parameters made by the structural engineer performing the optimization process. In Fig. 5.7 an optimized bracket design, for Airbus A350 aircraft, is presented. It is obvious that it has similar geo-

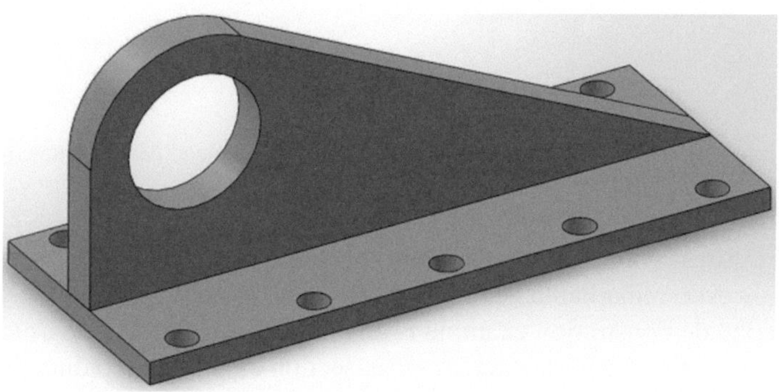

Fig. 5.5 Bracket connection conventional geometry

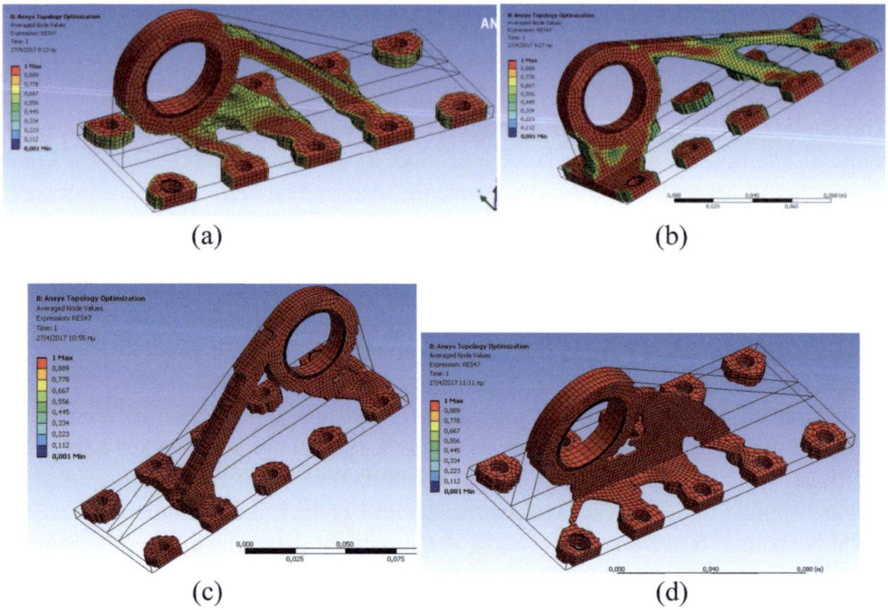

Fig. 5.6 Bracket characteristic optimized topologies for different load cases: (**a**) Fx = 10 kN and Fz = 6 kN, (**b**) Fx = −10 kN and Fz = 6 kN, (**c**) Fx = 11.66 kN, and (**d**) Fx = 10 kN Fy = 1 kN and Fz = 6 N

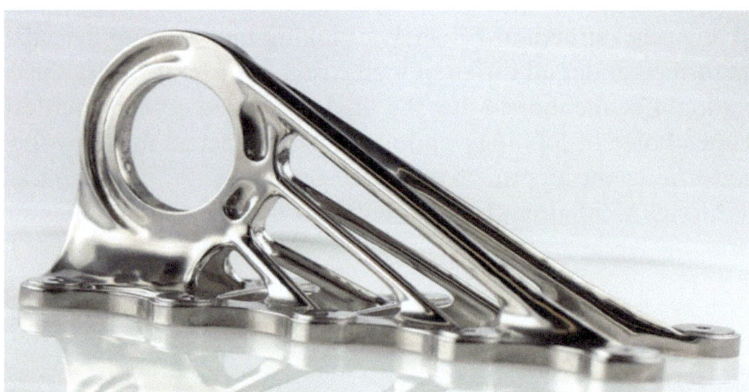

Fig. 5.7 Airbus A350 FCRC bracket connection topology optimized design (https://idw-online.de/en/image?id=266073 2018b)

metrical features with the combination of calculated topologies presented in Fig. 5.6; however, remarkable geometrical differences exist as well.

The second demonstration example is an aircraft fuselage section as taken from Niemann et al. (2013); it has a circular cross-section of 2000 mm radius, 13,652 mm length. At both ends there are load and support introduction bays of 399.8 mm length, while the rest of the fuselage comprises 23 bays of 558.8 mm pitch. The

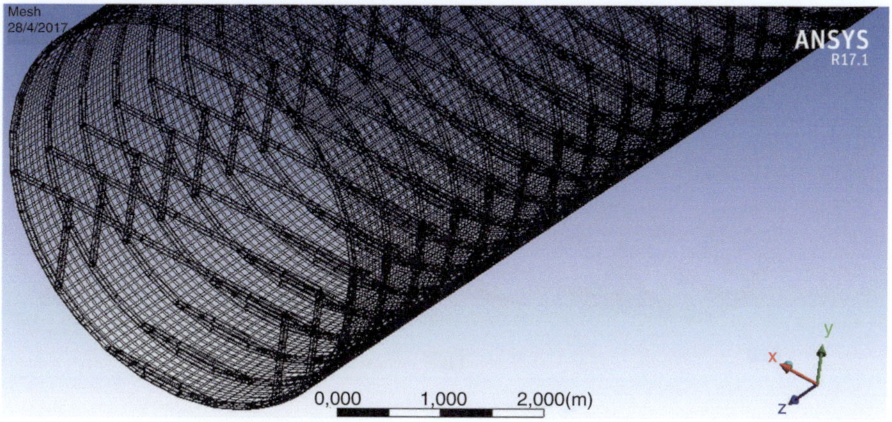

Fig. 5.8 Fuselage barrel section FE model

developed FE model of the fuselage barrel section is presented in Fig. 5.8. Two typical load cases have been studied: load case "A" consists of shear forces Qx = 80 kN and Qy = 212 kN, bending moments Mx = 447 kNm and My = 250 kNm and torsional moment Mz = 280 kNm, while load case "B" comprises shear force Qy = 212 kN, bending moment Mx = 447 kNm and torsional moment Mz = 280 kNm.

The SIMP method in combination to optimality criteria has been utilized in order to perform the topology optimization of the fuselage structure. A penalization factor equal to 3.3 and a convergence tolerance equal to 0.5% have been used for the analysis. In Fig. 5.9 the optimized topology is presented, as obtained by the topology optimization process of the fuselage barrel section for the two different load cases analyzed.

The optimization process of Fig. 5.9 reveals the basic topological characteristics of an optimized fuselage structure. More specifically, longitudinal backbones of significant size are required in the longitudinal direction for carrying the bending loads of the structure, complemented by secondary stiffening members originating from the backbones at an angle of about 30°, which transfer the shear and torsion loads at each section.

3 Additive Manufacturing (AM) Process Variations

Parts optimized by Topology Optimization methods have usually extremely complex shapes and may also include internal voids, which means that an AM production process is the most appropriate (if not the only possible) manufacturing route. Although all AM processes join materials to make objects from 3D model data, there are several discretely different techniques to obtain the final object shape. ASTM Committee F42 on Additive Manufacturing Technologies has published the specification F2792-12a in which AM technologies have been classified into seven

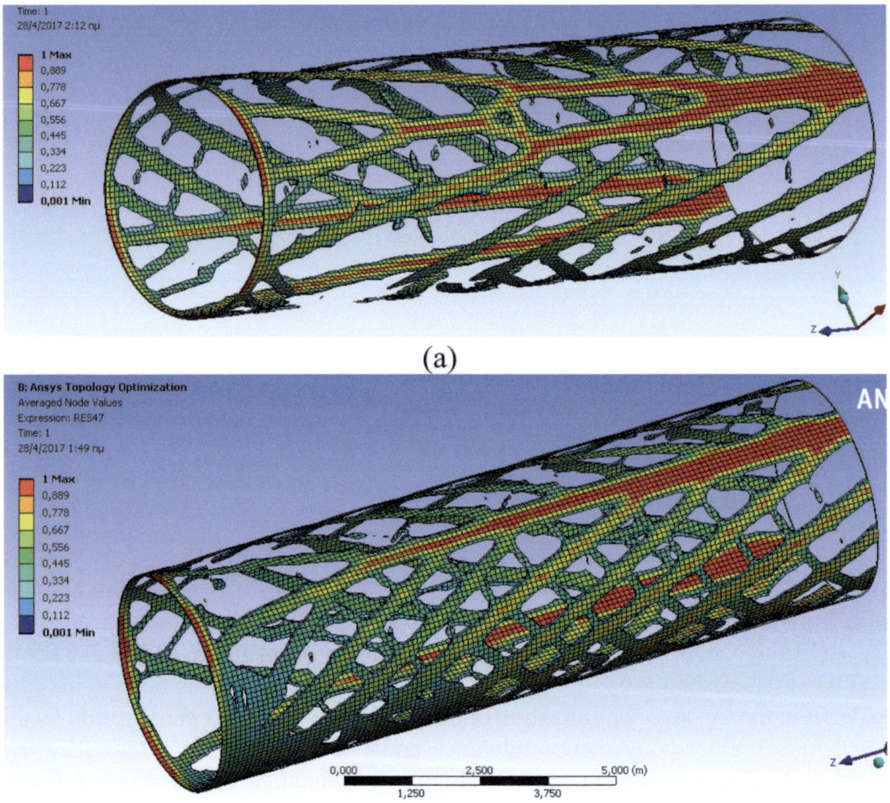

(a)

(b)

Fig. 5.9 Fuselage barrel section optimized topology for two different load cases: (**a**) load case A and (**b**) load case B

generic categories. Classified in order of interest for aeronautical applications, these categories are as described below.

Powder bed fusion An additive manufacturing process in which thermal energy selectively fuses regions of a powder bed.

Directed energy deposition An additive manufacturing process in which focused thermal energy is used to fuse materials by melting as they are being deposited.

Vat photopolymerization An additive manufacturing process in which liquid photopolymer in a vat is selectively cured by light-activated polymerization.

Material extrusion An additive manufacturing process in which material is selectively dispensed through a nozzle or orifice.

Binder jetting An additive manufacturing process in which a liquid bonding agent is selectively deposited to join powder materials.

Sheet lamination An additive manufacturing process in which sheets of material are bonded to form an object.

Material jetting An additive manufacturing process in which droplets of build material are selectively deposited.

Below, the most important AM processes for the aerospace sector are briefly presented, along with characteristic cases of aeronautical parts production.

3.1 Powder Bed Fusion

Powder bed fusion is a process in which thermal energy fuses selective regions of a powder bed. Variations of this category include Selective Laser Melting (SLM), Selective Laser Sintering (SLS), Direct Selective Laser Melting (DSLM), Direct Metal Laser Sintering (DMLS), and Electron Beam Melting (EBM) depending on the thermal power source and the developed material temperature level. Powder bed fusion processes are based on powder deposition on previous deposited layers (or on the substrate for the first layer) using a rolling mechanism to create the so-called powder bed. A laser- or electron-beam selectively scans the powder bed according to the designed part geometry until the powder reaches melting temperature and consequently achieves consolidation. When a new layer is to be scanned, a piston positioned under the part moves downward in a predefined distance equal to the layer thickness and a new layer of powder is deposited over the existing consolidated structure. The above process is repeated until building of the entire part is completed. A schematic representation of the process is presented in Fig. 5.10. The entire system is included in an isolated chamber operating at elevated temperature in order that residual stresses and distortions developing in the final part due to large thermal gradients developed during the process are reduced. In cases of metallic materials reaching their melting temperature, the process takes place within a controlled atmosphere of inert gas in a low oxygen environment in order to prevent oxidation phenomena.

Polymer and metal powder materials are offered to be used with powder bed fusion processes. Available metal powders include all the material families interesting for aeronautical applications including different alloys of Aluminum, Titanium, as well as, stainless steels. Metal powders used in AM of aerospace parts are made exclusively through an atomization process. The particles size varies from 30 to 100 μm depending on the material and the powder bed fusion technique used.

A support structure is usually necessary to hold the parts to the substrate plate and to support the down-facing surfaces. This is required especially for cases of metal powders, as their higher melting point require the use of increased thermal energy leading to very high thermal gradients in the build chamber, which induce thermal stresses and warping if supporting columns are not used.

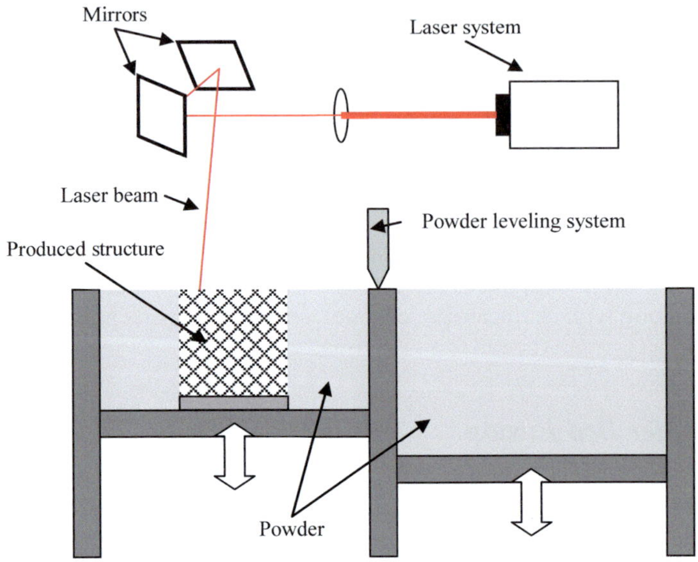

Fig. 5.10 Schematic representation of SLM for manufacturing of cellular materials

For polymeric part manufacturing, usually no supporting structure is needed, as the powder surrounding a fused part serves as a supporting system. However, each time the polymer powder is exposed to elevated temperatures it degrades. Therefore, un-sintered powder can only be reused for a finite number of times and requires also mixing with unused powder material.

Electron-beam powder bed fusion systems are faster than laser-beam systems and generally lead to the development of less residual stresses and distortions of the produced parts. On the other hand, when a better surface finish and finer feature detail are required, laser-beam powder bed fusion systems are preferred, as they generally result to better quality.

AM machines utilizing powder bed fusion processes are more expensive and complex compared to systems implementing other AM processes. The price is even higher for systems that use metal powders. Moreover, the respective production costs are comparatively high mainly due to materials cost, machines maintenance, and potential requirements for inert gas. On the other hand, powder bed fusion can be used for production of extremely complex parts with very fine accuracy, as demonstrated by numerous applications from the aerospace sector. Characteristic examples of aeronautical components manufactured by powder bed fusion processes are the bracket presented in Fig. 5.7 and the LEAP jet engine fuel nozzle (https://www.ge.com/reports/mind-meld-ge-3d-printing-visionary-joined-forces/), shown in Fig. 5.11; this nozzle combines the 20 parts of the conventional design into a single part that weighs 25% less than its predecessors and is more than five times as durable.

Fig. 5.11 LEAP engine fuel nozzle manufactured by AM (GE 2017)

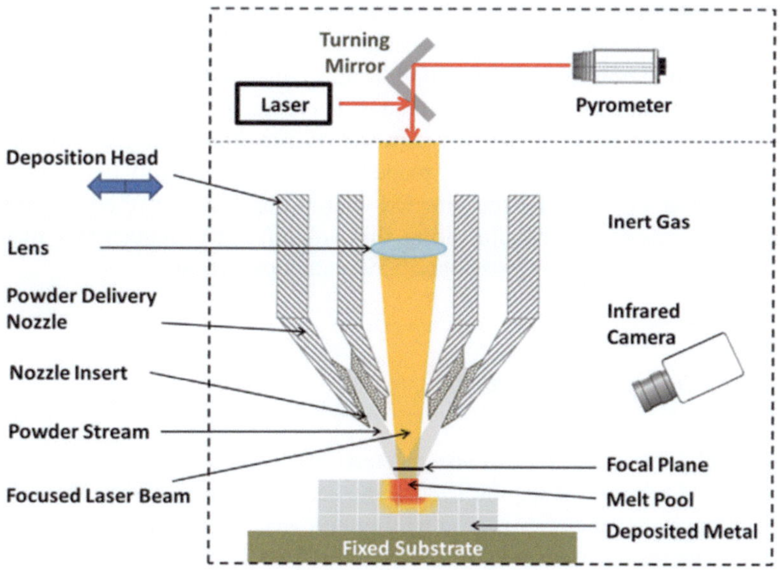

Fig. 5.12 Schematic representation of the directed energy deposition process principle (Thompson et al. 2015)

3.2 Directed Energy Deposition

Directed energy deposition is an additive manufacturing process in which focused thermal energy is used to fuse materials by melting as they are being deposited. This process is also referred as blown powder AM and laser cladding. A laser- or electron-beam is commonly used as thermal energy source and metal powders or wires as building materials. The principle of the specific AM process is schematically presented in Fig. 5.12.

Fig. 5.13 3D printed titanium propulsion tank (https://3dprint.com/142052/lockheed-martin-sciaky-ebam/ 2018a)

Directed energy deposition systems have had relatively limited success in the AM market, although the process offers unique capabilities. For example, more than one material can be deposited simultaneously, making functionally graded parts possible, which is an interesting feature providing the potential for aeronautical applications. Also, most directed energy deposition systems use a four- or five-axis motion system or a robotic arm to position the deposition head, so the build process is not limited to successive horizontal layers on parallel planes. This capability makes the process suitable for adding material to an existing part, such as repairing damaged parts or tools. In Fig. 5.13, a satellite propellant tank made from a light and highly durable titanium material (Ti-6Al-4V) built by the directed energy deposition method is presented. Compared to conventional manufacturing, this tank offers 80% reduction of manufacturing time, 75% reduction in waste, and 55% reduction in cost.

3.3 *Vat Photopolymerization*

Vat photopolymerization, known also as stereolithography, has been the first commercialized AM process since the 1980s. The basic concept of this process is that a liquid photopolymer material is selectively polymerized using a form of light (see Fig. 5.14). The resin is contained in a vat and the part is usually built upside-down, starting from the top of the vat. The light source to activate polymerization can be either a laser directed by galvanometer mirrors or Light Emitting Diodes (LEDs) controlled by Digital Light Processing (DLP) technology. This AM process is utilized in the aerospace sector mainly to build wind tunnel models and demonstration parts. Some applications of stereolithography in combination with investment casting have also been reported, e.g., Zhou et al. (2015), as shown in Fig. 5.15.

3.4 *Material Extrusion*

Material extrusion is an AM process in which material is drawn through a nozzle, in which it is heated and then is deposited layer by layer. It is also known as Fused Deposition Modeling (FDM). The nozzle and the build platform can move rela-

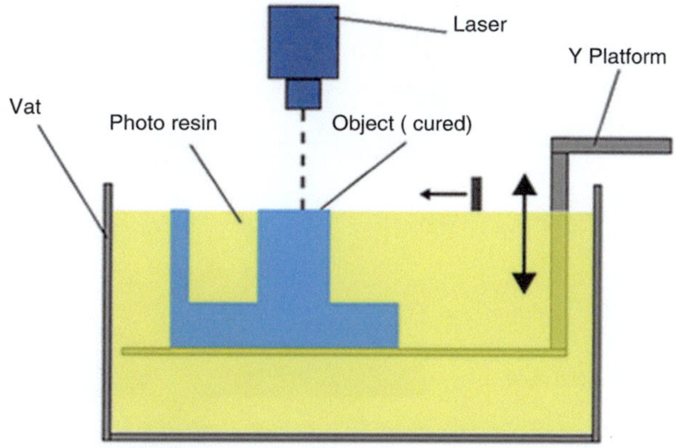

Fig. 5.14 Schematic of vat photopolymerization printing process (Loughborough University 2019)

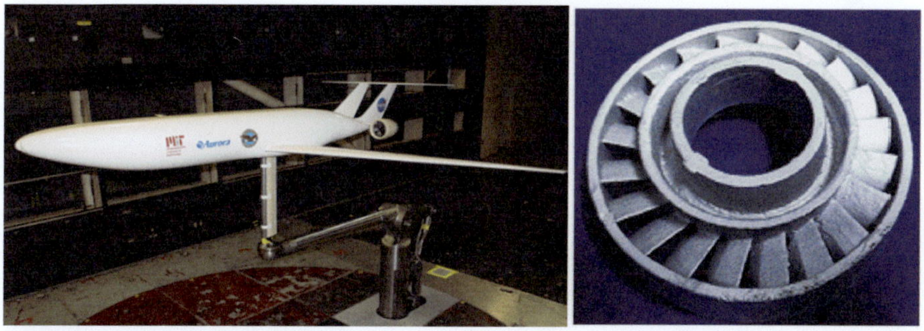

Fig. 5.15 (**a**) NASA's N + 3 wind tunnel model (https://www.stratasysdirect.com/resources/case-studies/fdm-nasa-n3-model-aurora-flight-sciences 2018), (**b**) metal turbine stator

tively to each other in all three directions. Thus, after a layer is completed in the plane, the build platform moves down, or the extrusion head moves up and the next layer is deposited to the previous one. Material bonding is achieved as the material is melted in the extrusion head and when it comes into contact with the previously deposited material it adheres to it.

Typical materials used in FDM manufacturing are thermoplastic materials that are provided in the form of filament in the extrusion head. The filament form provides a convenient way of controlling the quantity of material extruded. Other materials used with material extrusion AM processes of interest in aeronautics are composites and ceramics (Fig. 5.16).

Material extrusion AM machines are widely available in the market, due to the fact that they are quite cheaper than any other machine type, while their operation requires minimum specialization. Most material extrusion machines employ a single extrusion head, but there are also machines with two or three extruders. The additional heads are used to mix different materials or colors in the component, or to build the support structure that is required, when the design to be produced includes bridging features or bottom surfaces with an easy to remove material.

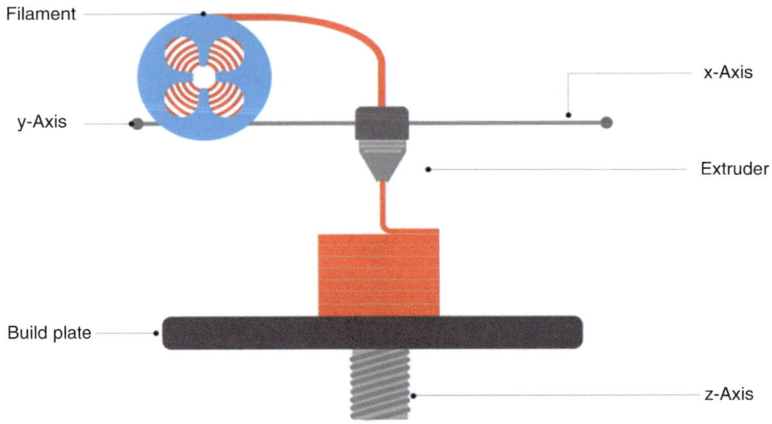

Fig. 5.16 Schematic representation of FDM process (https://druckwege.de/en/home-en/technology/fused-deposition-modelling-fdm 2019)

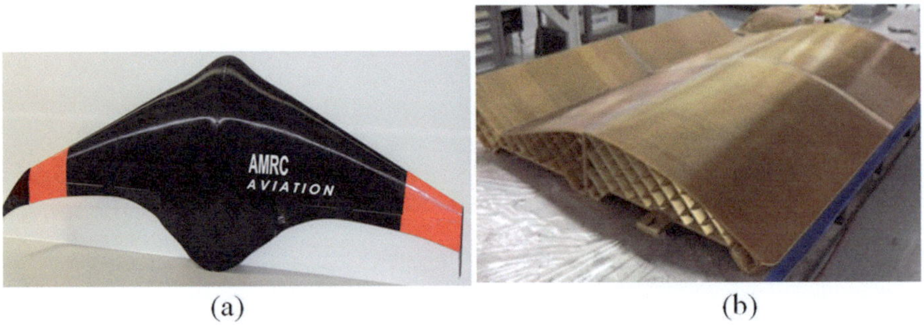

Fig. 5.17 (**a**) UAV of AMRC aviation (AMRC 2018) and (**b**) Falcon jet cabinet headliner lay-up mold (http://www.stratasys.com/resources/search/white-papers/dassault-falcon-jet 2018)

Due to the low cost of the FDM machines and the ease of method application, many examples of aeronautical parts made by this technique are available; drone structures and manufacturing molds to be used for composite production are two of the most typical. In Fig. 5.17a an Unmanned Aerial Vehicle (UAV) airframe constructed entirely of ABS plastic using the FDM technology is presented (wingspan of 1.5 m and weight under 2 kg). In Fig. 5.17b the Falcon jet cabinet headliner lay-up mold is shown.

3.5 Material Jetting

In material jetting processes objects are created similarly to a two dimensional ink jet printer. One or more inkjet printing heads are used to selectively deposit droplets of build material and create layers on the build area. The material layers are then cured

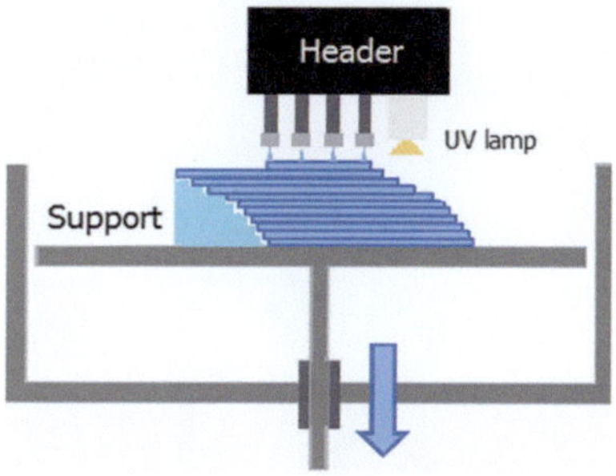

Fig. 5.18 Schematic representation of the material jetting

or hardened using ultraviolet (UV) light. A schematic of a material jetting 3D printing is shown in Fig. 5.18. As material must be in liquid state, the number of materials available to use with material jetting AM systems is limited. Photopolymers and waxes are suitable and commonly used materials, due to their viscous nature and ability to form drops. Due to the nature of materials used with material jetting, that are susceptible to UV light, the process has limited applicability in the aeronautical sector.

With material jetting process, multiple materials can be used or even combined in the same part when multi-nozzle print heads are utilized. Thus, customized material properties or colors can be obtained. Additionally, one of the materials can be used to create support structures, while the others are used to actually build the component. There are applications of material jetting where the print materials can be metal in the form of aerosol gas. However, it still remains an emerging technology that may have high potential in the efficient production of aeronautical parts.

3.6 *Sheet Lamination*

Sheet lamination is the AM process in which the final part is constructed by sheets of material that are bonded together. The material can be either paper or metal. In the case of paper, the sheets are cut in the required form and adhesive is used to bond the successive layers. Metal sheets in the form of foils or tapes are welded together, utilizing laser-beam or ultrasonic welding. In Fig. 5.19 sheet lamination method principle is schematically shown.

3.7 *Binder Jetting*

In binder jetting a liquid adhesive material is used to bond powder particles in a powder bed (Fig. 5.20). The adhesive material is selectively deposited utilizing ink-jet print head nozzles to hold the powder in the desired shape. Polymer, ceramic, and metal powders are available to be used with binder jetting machines. No residual stresses are developed in the created parts, as binder jetting does not employ heat

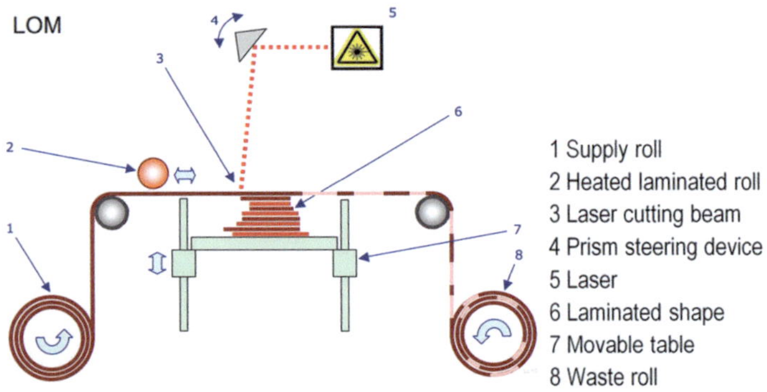

Fig. 5.19 Principle of the sheet lamination manufacturing technique (wikimedia 2018)

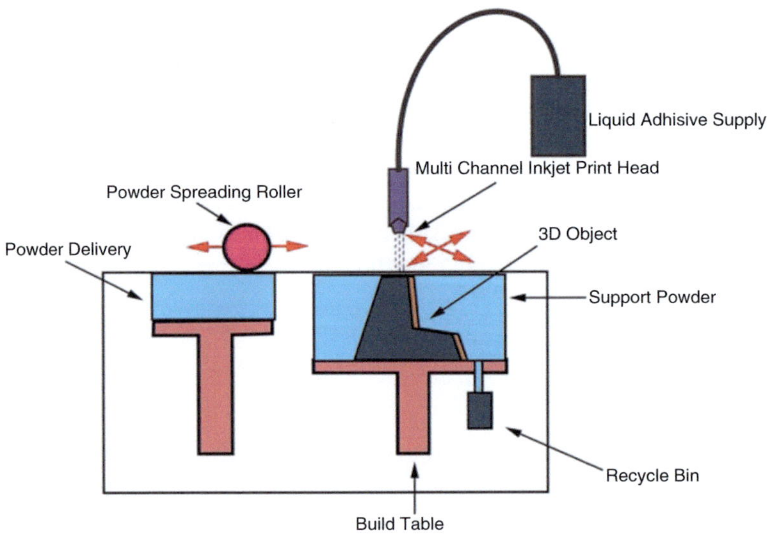

Fig. 5.20 Illustration of the binder jetting process (wikimedia 2018)

during the build process. Additionally, there is no need for creating any supporting structure, as the parts are supported by the loose powder in the job box. However, metallic components created by binder jetting after the AM build procedure require the application of a sintering process and sometimes to be infiltrated with another material, depending on the application. Hot Isostatic Pressing (HIP) may be employed to achieve high densities in solid metals. This AM process is mainly used to create printed sand cores and molds, while, even if printing of metal parts is possible, limited aeronautical applications exist.

4 Outlook

Despite the huge steps performed until now in the development of AM processes and their combination to Topology Optimization principles toward the efficient production of aeronautical components and parts, many relevant areas still need more attention. In this frame, recent research efforts in the field of AM aim to address the both fundamental as well as technological AM issues, in order to succeed a wider and efficient utilization of AM for the production of complex aeronautical components reliably and efficiently; this is expected to be achieved through the improvement and combination of advances in AM methodologies and the progress in topology optimization techniques, assisted by AM process simulation and experimental Nondestructive Evaluation. In this frame, many research efforts are currently focused on:

- Advancement of the different AM production processes, focusing in increase of production rates, part quality and process control, as requirements for the achievement of high-tech and added-value parts or components, mainly for the Aerospace, Automotive, and Biomedical Technology sectors.
- Material related issues, focusing on the understanding and control of locally varying material properties, part inhomogeneity, anisotropy, residual stresses and distortions.
- Optimal part designs derived by topology optimization methodologies, which take into account the local material property distributions.
- First time right selection of optimal AM process parameters, by establishing the correlation between process parameters and produced component quality, through multi-scale numerical simulation of the different AM techniques.
- Adaptation of the available destructive and nondestructive testing methods for the determination of the quality of final AM products and the validation of the multiscale simulation methodologies.
- The exploitation of the AM potential in the design and production of complex specialized components in large industrial scale, mainly in the aeronautical sector.

References

FDM Printed Fixed Wing UAV—AMRC. http://www.amrc.co.uk. Accessed 17 July 2018

Bendsøe MP (1995) Optimization of structural topology, shape and material. Springer, Berlin

Bendsøe MP, Sigmund O (2003) Topology optimization: theory, method and application. Springer, Berlin

Demonstration of Additive Manufacturing (FDM) for Production Composite Tooling at Dassault Falcon Jet (2018). http://www.stratasys.com/resources/search/white-papers/dassault-falcon-jet

Harber RB, Jog CS, Bendsøe MP (1996) A new approach to variable-topology shape design using a constraint on the perimeter. Struct Optim 11:1–11

Hassani B, Hinton E (1999) Homogenization and structural topology optimization. Springer, Berlin

https://3dprint.com/142052/lockheed-martin-sciaky-ebam/. Accessed 17 July 2018a
https://druckwege.de/en/home-en/technology/fused-deposition-modelling-fdm. Accessed 15 July 2019
https://idw-online.de/en/image?id=266073. Accessed 17 July 2018b
https://www.ge.com/reports/mind-meld-ge-3d-printing-visionary-joined-forces/. Accessed 15 May 2019
https://www.stratasysdirect.com/resources/case-studies/fdm-nasa-n3-model-aurora-flight-sciences. Accessed 17 July 2018c
Huang X, Xie YM (2010) Evolutionary topology optimization of continuum structures: methods and applications. Wiley, Chichester, West Sussex
Krog L, Tucker A, Rollema G (2002) Application of topology, sizing and shape optimization methods to optimal design of aircraft components. In: Proceedings of 3rd Altair UK Hyper-Works users conference
Loughborough University, VAT Photopolymerisation. https://www.lboro.ac.uk/research/amrg/about/the7categoriesofadditivemanufacturing/vatphotopolymerisation/. Accessed 16 June 2019
Niemann S, Kolesnikov B, Lohse-Busch H, Hühne C, Querin OM, Toropov VV, Liu D (2013) The use of topology optimisation in the conceptual design of next generation lattice composite aircraft fuselage structures. Aeronaut J 117:1139–1154
Remouchamps A, Bruyneel M, Fleury C, Grihon S (2011) Application of a bi-level scheme including topology optimization to the design of an aircraft pylon. Struct Multidiscip Optim 44:739–750
Sigmund O (1997) On the design of compliant mechanisms using topology optimization. Mech Struct Mach 25:495–526
Svanberg K (1987) The method of moving asymptotes. A new method for structural optimization. Int J Numer Methods Eng 24:359–373
Thompson SM, Bian L, Shamsaei N, Yadollahi A (2015) An overview of direct laser deposition for additive manufacturing; part I: transport phenomena, modeling and diagnostics. Addit Manuf 8:36–62
Tomlin M, Meyer J (2011) Topology optimization of an additive layer manufactured (ALM) aerospace part. In: The 7th Altair CAE technology conference
wikimedia. https://commons.wikimedia.org/wiki/File:Laminated_object_manufacturing.png. Accessed 17 July 2018
Xie YM, Steven GP (1997) Evolutionary structural optimization. Springer, London
Yang XY, Xie YM, Liu GT, Clarkson PJ (2003) Perimeter control of the bidirectional evolutionary optimization method. Struct Multidiscip Optim 24:430–440
Zhou J, Lu Z, Miao K, Ji Z, Dong Y, Li D (2015) Quick fabrication of aeronautical complicated structural parts based on stereolithography. Propuls Power Res 4:63–71
Zillober C (2002) SCPIP an efficient software tool for the solution of structural optimization problems. Struct Multidiscip Optim 24(5):362–371

Chapter 6
Cellular and Sandwich Materials

George Lampeas

1 Introduction to Cellular and Sandwich Materials for Aeronautical Structures

There is a continuous trend for introduction of novel material systems in aeronautical structures, aiming to lighter solutions, resulting to less fuel burn, and consequently to more cost efficient and environmentally friendly structures. A well-known outcome on this trend is the use of composite materials in aircraft structural components. The initial applications of composite materials were in structural parts, such as spoilers and rudders in the early 1980s. Since then, the percentage of composite materials has steadily increased in both primary and secondary aircraft structures. In Airbus A380 some major parts of the fuselage, such as the un-pressurized tail, the pressure bulkhead, the center wing box, and the keel beam are made out of composites. In latest aircraft models like Boeing 787 and Airbus A350, the percentage of composite materials has reached more than 50% of the total aircraft structure.

Novel aeronautical metallic or composite designs comprise small thickness structural elements, offering improved specific stiffness and strength compared to conventional ones. A major challenge related to the use of novel metallic and composite materials systems is to find ways to effectively increase in a cost-effective way their out-of-plane stiffness and their impact energy absorption capability. Both polymeric and metallic sandwich skinned constructions have the specific structural features to succeed this goal. Taking as an example the aircraft fuselage design, the development of a complete pressurized fuselage in composites represents a big challenge, particularly because of increased sensitivity of composite materials in impact damage, such as soft and hard body impacts, e.g., tire, hail, runway debris,

G. Lampeas (✉)
Laboratory of Technology and Strength of Materials, Department of Mechanical Engineering and Aeronautics, University of Patras, Patras, Greece
e-mail: labeas@mech.upatras.gr

S. Pantelakis, K. Tserpes (eds.), *Revolutionizing Aircraft Materials and Processes*, https://doi.org/10.1007/978-3-030-35346-9_6

resulting to impact delamination damage; sandwich constructions comprising cellular cores offer the opportunity for large improvement of the impact performance and the residual mechanical properties after impact.

Sandwich construction is particularly efficient when the main loading is bending, since the stronger skins placed at the external panel surfaces, where flexure stresses are high, can resist increased applied bending moments by skin loading in tension or compression. The core acts as separator/spacer for the skins increasing the panel bending stiffness due to the increase of second moment of area inertia; as the core is a region or relatively lower stresses it may have low mass density and lower elastic properties, as far as it is sufficient in carrying the out-of-plane transverse normal and shear loads. In addition, when the panel is subjected to impact loading, core crushing provides the geometry, space, and time for potentially high energy absorption. In particular, the loads developing during an impact event around the impact area are distributed in the vicinity of the impact location of the outer skin and through this into the core volume up to the internal skin. The sandwich panel impact resistance depends on both the mechanical properties of the skins and the core of the sandwich material, which should be properly matched. In addition to bending and impact, the core may also provide other functional properties such as thermal insulation, vibration, or acoustic damping. By proper choice of skins and core materials, sandwich constructions of high stiffness-to-weight and strength-to-weight ratios can be designed. Both metals and composites may be used as skin materials, depending on the structural requirements, with aluminum been more traditional, while carbon, glass, and aramid composite have been recently used in sandwich aerospace constructions. The skins are attached to the core by adhesive bonding, brazing, or other techniques, depending on the skin/core combination.

However, sandwich constructions have also disadvantages, mainly with respect to their joining techniques, the difficulty to form hard points of sufficient strength and in case of closed cell types moisture and water ingression. For these reasons, the initial composite fuselage structures for commercial aircraft used composites in a conventional structural layout, comprising monolithic skins stiffened by frames and stringers. Traditional aircraft sandwich materials that have been used in secondary wing structures like radomes and fairings are aramid fiber paper (Nomex) honeycomb with composite facings and aluminum honeycomb with aluminum or composite facings. They both have drawbacks for primary fuselage structures, since they have closed cells, are not ventilated, and have low damage tolerance. Folded composite cores and Z-pinned foam materials have been investigated as candidate core materials with improved damage tolerance seem to be more promising core materials for aircraft fuselage panel structures.

In future structural concepts, a high potential of sandwich cellular or lattice structures exists, due to their much higher bending stiffness and their better impact resistance, as well as their ability to combine structural stiffness with thermal insulation, as well as acoustic and vibration damping; such features provide opportunities for functional integration of systems, interior facilities, and fuselage cabin equipment, which lead to significant weight reduction and also increase of the inside space of the aircraft, allowing for a reduction of the fuselage diameter, which would result to less drag. A cellular material consists of an interconnected network of plates or beams suitably connected to one another. The joining of these beams or plates leads to the creation of an elementary volume called unit cell that creates the cellular material, when properly copied in the three directions of the space. In Fig. 6.1, the three most typical types of cellular materials are presented: honeycomb, open foam, and lattice.

Cellular materials can be classified according to the size of their unit cells (small, medium, and large cell sizes), the unit cell homogeneity throughout the cellular material (stochastic or periodic), and the type of unit cell (open and closed type). For example, foams can be characterized to be stochastic, open or closed, with small, medium or large unit cell sizes. Open type foams consist of random geometry shells joined in a way creating trusses as shown in Fig. 6.2a, while closed-type foams comprise interconnected polyhedrical enclosures, as shown in Fig. 6.2b. Foams are a special category of cellular materials, in which a particular for which an elementary cell that if repeated in the space will result to the foam core geometry does not exist, as each polyhedron has different dimensions and orientation; therefore, it is of stochastic nature. The investigation of such stochastic cellular materials can be performed by considering a characteristic small material volume, which includes enough enclosures, such that macro-mechanical properties may be defined; this characteristic volume is referred to as Representative Volume Element (RVE).

The most widely used form of cellular material in aeronautical applications is the cellular honeycomb core, which belongs in the category of periodic cellular materials and more specifically in the subclass of prismatic cellular materials. A feature of this category is that the elementary cell, which has the form of a prism with a hexagonal or generally polygonal base is inspired by the beehives, from which also took its name.

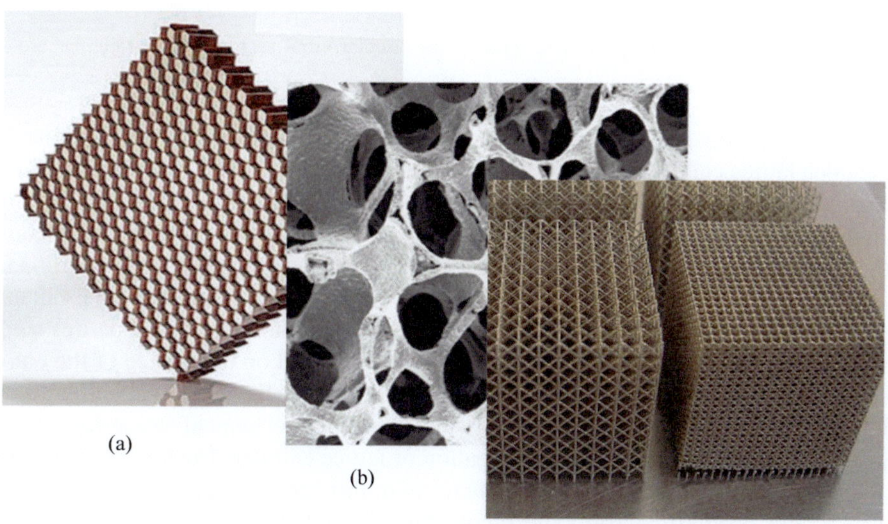

(a) (b) (c)

Fig. 6.1 Typical formats of cellular materials: (**a**) honeycomb; (**b**) open foam; (**c**) lattice

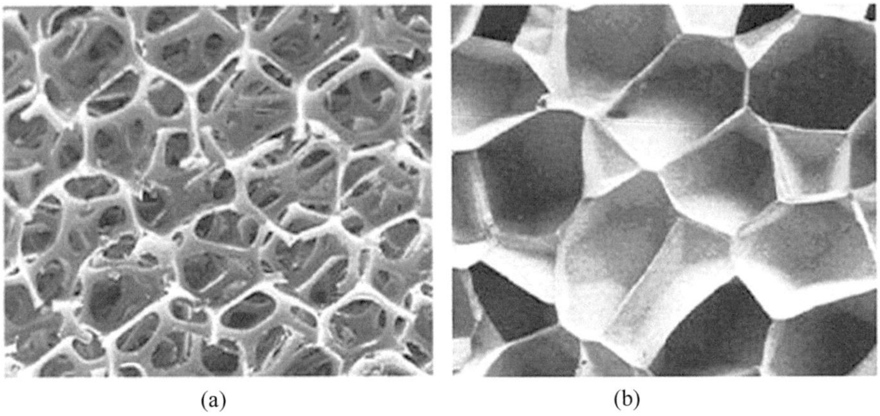

Fig. 6.2 Examples of cellular foam materials: (a) open type, (b) closed type (Goga 2010)

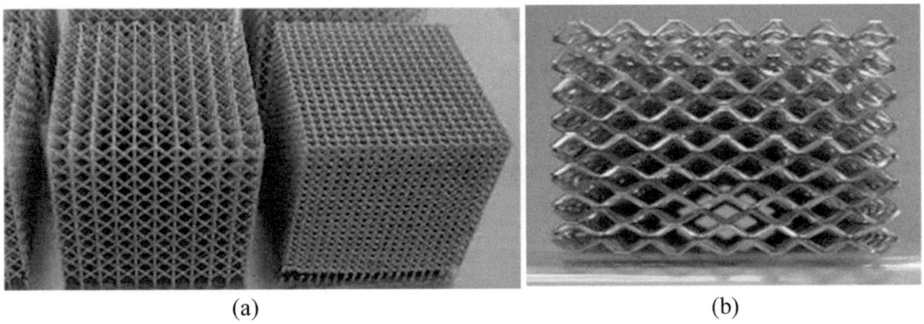

Fig. 6.3 Open beam type beams with beams: (a) circular cross section, (b) rectangular cross section

A more promising format of cellular materials for aeronautical applications is the open lattice cellular category, which mainly includes periodic cellular materials. Examples of open lattice structural cores are shown in Fig. 6.3 comprising an interconnected network of beams or tubes suitably connected to one another. These configurations are characterized by the precision of their geometry and the additional (in relation to other types) ability of their ventilation, which eliminates the problem of moisture absorption and consequent degradation of material properties and weight penalty. The properties of open cell types depend on the choice of the construction material, its density, the shape and geometry of the elemental cell, allowing the creation of materials with a wide range of mechanical properties. This justifies the growing interest in the creation of lattice cellular structures with desirable mechanical properties. The main features of this category are the precise geometry of the beams that form the elemental cell and their ability to tailor their mechanical properties to the required in the different directions in space; the beams of the elementary cell may have either a circular or rectangular cross section and the unit cell may have many different topologies, as will be presented later.

The dimensions of the elementary cell geometry, the dimensions and the geometry of the plates or beams forming the elementary cell and their material types comprise the most important factors affecting the mechanical properties of the cellular structure. For aeronautical applications, of major importance is the relative density of the elementary cell, defined as the ratio of the density of the cellular material to the density of the cell, as if it was made of solid material of the same type and external dimensions. The relative density values that the elementary cells may have range from 0.02 up to close to unity. The structural cores for aeronautical applications, e.g., open lattice cellular cores, which seem to be the most promising for future aeronautical structures, may be designed to have relative density values ranging between 0.2 and 0.4; in contrary the well-known polymeric foams used as protective or insulating materials, e.g., in packaging, have relative density between 0.03 and 0.15.

2 Types of Regular Lattice Cellular Cores

Cellular materials have been extensively researched for their application as cores in twin-skinned sandwich structures, where specific bending stiffness, strength, and energy absorption capability are important structural properties. In "conventional" cellular materials, the micro-architecture variations are unlimited. For example, foam cores are available in different densities and parent materials, while honeycombs are available in different cell sizes, wall thicknesses, and parent material (Gibson and Ashby 1997). Several works have been presented for the study of honeycomb (Zhu et al. 2010; Caserta et al. 2010; Lorato et al. 2009) and foam (Caserta et al. 2010; Flores-Johnson and Li 2010; Jang and Kyriakides 2009) cellular materials. The main subject of these works is the mechanical behavior of the cellular structures under either quasi-static (Caserta et al. 2010) or dynamic loading (Zhu et al. 2010; Caserta et al. 2010; Jang and Kyriakides 2009). Different approaches have been adopted including experimental (Zhu et al. 2010; Caserta et al. 2010), analytical (Zhu et al. 2010; Lorato et al. 2009; Flores-Johnson and Li 2010; Jang and Kyriakides 2009), and numerical simulation (Caserta et al. 2010; Jang and Kyriakides 2009).

However, as mentioned above the most interesting category of cellular cores for the aerospace sector is the open lattice or open micro-truss cores, which offer specific advantages compared to the conventional honeycomb or foam cores, with most important the capability of core ventilation, which eliminates the problem of moisture absorption and the resulting material properties degradation. The fineness of the structure relies on the realization of the exact cell geometry and depends on the limitations of the production methods described in Sect. 3.

A number of novel core topologies for open lattice structure have emerged, offering structural advantages over traditional constructions. Some of the unit cells characterized by cubic material symmetry are the Simple Cubic (SC), the Gibson Ashby (GA), the Body-Centered Cubic (BCC), and the Reinforced Body-Centered Cubic (RBCC) and are made of struts with circular cross sections. In Fig. 6.4 the

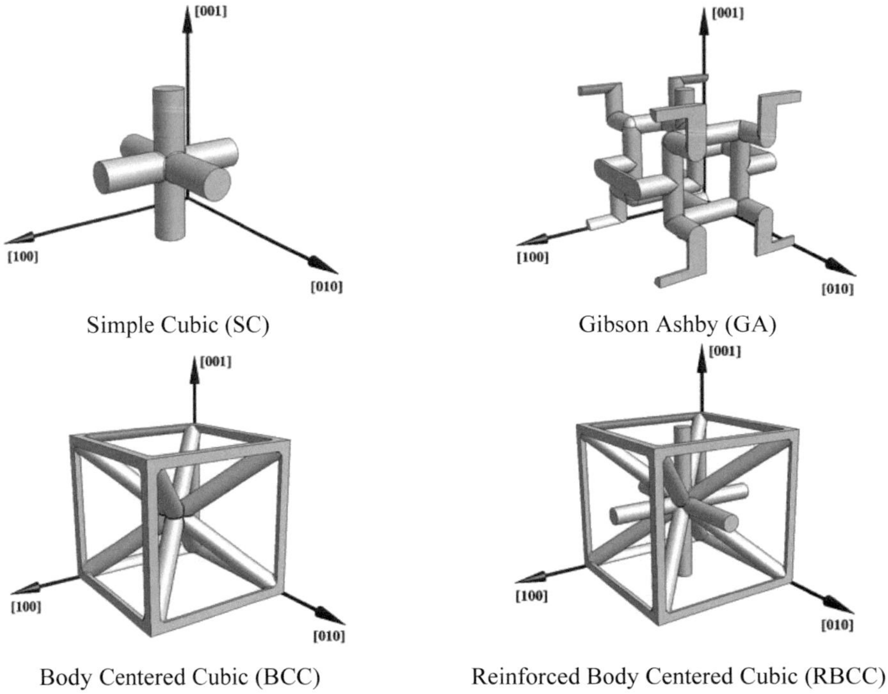

Fig. 6.4 Unit cells of (**a**) simple cubic (SB), (**b**) Gibson Ashby (GA), (**c**) body-centered cubic (BCC), and (**d**) reinforced body-centered cubic (RBCC) after Pettermann and Hüsing (2012)

unit cells of these topologies, the strut diameter of which is constant, are presented. By repeating unit cells in the three principal directions, periodic open lattice cellular structures are obtained.

Other types of fully open lattice core structures can also be created from slender beams (trusses) of any cross-sectional shape: circular (Deshpande and Fleck 2001; Wang et al. 2003), square (Kooistra et al. 2004; Rathbun et al. 2004), rectangular, I-beam, or hollow (Queheillalt and Wadley 2005a, b). The trusses can be arranged in many different configurations depending upon the intended application (Evans et al. 2001). In Fig. 6.5, six examples of micro-truss cellular topologies used as the cores of sandwich panels are presented.

The tetrahedral structure, Fig. 6.5a, has three trusses meeting at a face sheet node (Deshpande and Fleck 2001; Kooistra et al. 2004; Wadley et al. 2003; Rathbun et al. 2005; Chiras et al. 2002; Bouwhuis et al. 2008; Liu et al. 2006). In Kooistra et al. (2008) flat sandwich panels with tetrahedral metallic core of relative densities between 2 and 10% have been fabricated and tested in the minimum shear strength orientation for the tetrahedral lattice; analytical stiffness and strength predictions agree well with measured values for all relative densities and parent alloy heat treatments investigated. In Sugimura (2004) the mechanical performance of single-layer tetrahedral truss core under shear loading is investigated. In Kooistra et al. (2004)

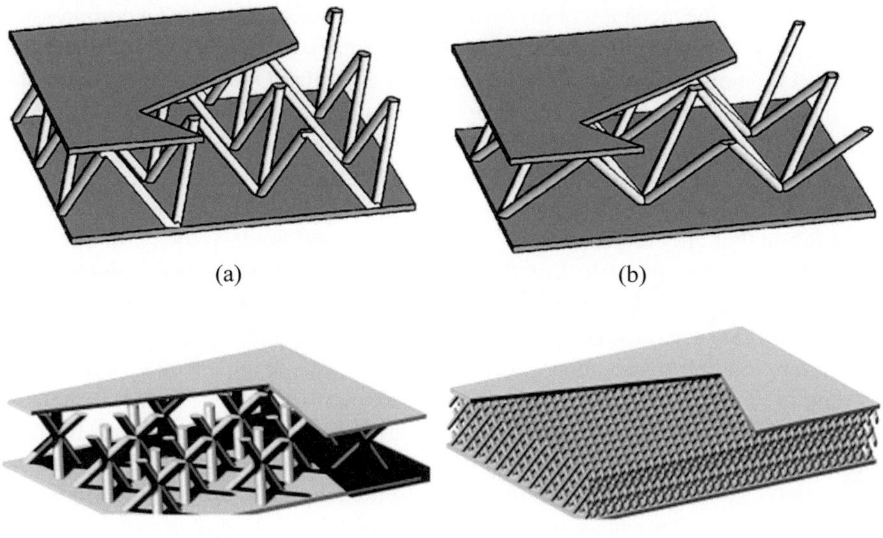

Fig. 6.5 Typical micro-truss cellular topologies for sandwich cores, (**a**) tetrahedral, (**b**) pyramidal (Wadley 2006), (**c**) three-dimensional Kagome, (**d**) diamond textile (after Zhang et al. 2017)

the compressive behavior of age hardenable tetrahedral lattice truss structures made from aluminum material is investigated. In Xue and Hutchinson (2003) preliminary assessment of sandwich circular panels clamped at their edges subjected to blast loads is performed.

The pyramidal structure has four trusses meeting at a face sheet node, Fig. 6.5b. In Xiong et al. (2010) the crushing response of the truss core was investigated and the corresponding failure modes were studied and compared with the analytic model of the core crushing response. In Queheillalt et al. (2008) the mechanical properties of an extruded pyramidal lattice truss sandwich structure was investigated. In Cote et al. (2007) an experimental and analytical investigation was carried out to examine the in-plane compressive response of pyramidal truss core sandwich columns; the identified failure mechanisms include Euler buckling, shear buckling, and face wrinkling. The subject of Biagi and Bart-Smith (2007) is the examination of the mechanical response of pyramid structures under compression and shear tests.

A slightly different topology has been proposed by Salvatore Torquato at Princeton and is referred to as a three-dimensional Kagome topology (Wang et al. 2003; Queheillalt and Wadley 2005a; Liu et al. 2007), Fig. 6.5c. Kagome is a Japanese term for the basket weave pattern created by in-plane weaving in three directions. In Torquato's structure, the nodes that are formed at the face sheets have the two-dimensional Kagome weave pattern. Each of the three topologies shown in Fig. 6.5a–c is efficient at supporting structural loads and especially the shear loads encountered in panel bending. In Lim et al. (2009) the mechanical behavior of sandwich panels with quasi-Kagome truss cores, fabricated by the expanded metal process is assessed; the mechanical strengths and failure mechanisms under compression and bending load are estimated based on elementary mechanics of materials and validated by their comparison with experimental results. Other lattice truss topologies have also been proposed based upon manufacturing considerations. In Hyun et al. (2003) a finite element method has been used to simulate the properties of panels with Kagome and tetragonal cores under compressive and shear loading.

The diamond textile structure (Fig. 6.5d) is made from layers of a plain weave metal fabric that are bonded to each other (Sypeck and Wadley 2001). In Lee et al. (2006) the compressive behavior of open-cell aluminum alloy and stainless steel woven textile diamond core materials have been investigated at three different deformation rate regimes; compressive tests were performed quasi-statically using a miniature loading frame, intermediate strain rate tests using a Kolsky bar apparatus, and high strain rates tests using a light gas gun. In Zok et al. (2003) metallic sandwich panels with textile cores have been analyzed subject to combined bending and shear and then designed for minimum weight; basic results for the weight benefits relative to solid plates are presented, with emphasis on optimizations that assure robustness (non-catastrophic failure); in addition selected numerical simulations were used to check the analytical results and to explore the role of strain hardening beyond failure initiation.

Various other micro-truss cellular topologies are shown in Fig. 6.6, namely: (a) octet-truss, (b) Wire-woven Bulk Kagome (WBK), (c) open cell diagonal, (d) 3D Warren truss, (e) lattice block of Zhou et al. (f) square-diagonal, (g) lattice block with hexagonal arrays, and (h) Wadley block. In Deshpande et al. (2001) the effective mechanical properties of the octet-truss lattice have been investigated both experimentally and theoretically; analytical and FE calculations of the elastic properties and plastic yielding collapse surfaces are reported; the intervention of elastic buckling of the struts is also analyzed and a good agreement is found between the predictions of the strength and experimental observations from tests on the octet-truss structures made by casting aluminum alloy. In Elsayed and Pasini (2010) the structural design of the microscopic architecture of lattice material with regular octet-truss cell topology and on the multistage design of an axially loaded member manufactured of this type of cellular solid is investigated. In Aboudi and Gilat (2005) multiphase lattice blocks with periodic structure are analyzed by a continuum-based micromechanical approach and effective stiffness tensors, global initial yield surfaces, global damage thresholds, effective inelastic stress–strain responses, and critical yielding temperatures of lattice blocks are derived.

The Wire-woven Bulk Kagome (WBK) is another truss type cellular metal topology fabricated by systematic assembling of helical wires in six directions. WBK looks promising with respect to morphology, fabrication cost, and raw materials. In Lee and Kang (2009a), analytical solutions for the calculation of the material properties and the maximum loads carried by a WBK core sandwich panel under bending are derived and the effect of the unit cell geometrical details such as the curved shape of the struts on the mechanical properties is assessed; optimization of the slenderness ratio of the WBK core for minimum weight of the sandwich panel is also performed. In Lee and Kang (2009b) compressive behaviors of WBK cores

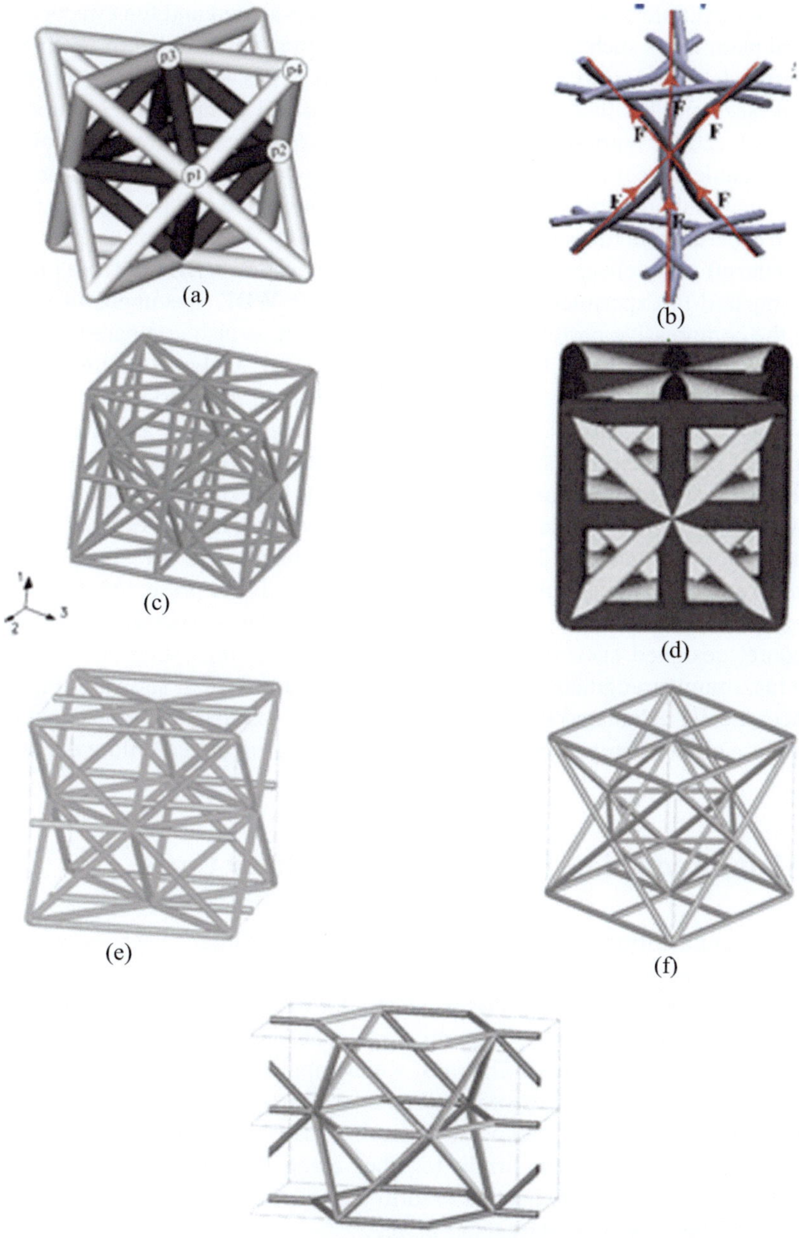

Fig. 6.6 Various other micro-truss lattice cellular topologies (**a**) octet-truss (Deshpande et al. 2001), (**b**) Wire-woven Bulk Kagome (WBK) (Hyun et al. 2009), (**c**) open cell diagonal (Aboudi and Gilat 2005), (**d**) 3D Warren truss (Doyoyo and Hu 2006), (**e**) lattice block of Zhou et al. (Aboudi and Gilat 2005) (**f**) square-diagonal (Aboudi and Gilat 2005), and (**g**) lattice block with hexagonal arrays (Aboudi and Gilat 2005)

fabricated of stainless wires are experimentally investigated and the effects of geometrical parameters such as wire diameter, strut length, and number of layers on the compressive behavior are studied; two different types of the specimen having pointed faced and flat faced geometry are analyzed and compared to each other. In Hyun et al. (2009) numerical simulation of the mechanical performance of WBK made of stainless steel wire and its sensitivity on stochastic imperfections on geometry and material properties under shear loads are presented and the results are compared to experimental measurements. In Lee et al. (2007) the compressive properties, such as the effective elastic modulus and the peak stress of WBK cores are predicted and compared to experimental measurements using WBK specimens of different sizes; the compressive performance of the WBK core is also compared with other widely used core types and the failure mechanisms are analyzed by finite element analysis. In Lim and Kang (2006) new approaches based on tri-axial weaving of wires to create ideal trusses, i.e., tetrahedral and Kagome truss are presented; the mechanical properties of respective sandwich panels in compression and bending loading are analyzed by elementary beam theory and experimental tests, from which relative density, stiffness, and strength of the sandwich panels are derived and compared. In Doyoyo and Hu (2006) the results of a parametric investigation on the failure of metallic lattices subjected to multi-axial loads are presented, focusing on parameters related to struts geometry, i.e., strut slenderness ratio and strengthening procedure, designed such that minimum-strengthening represents the octet-truss while maximum strengthening represents the 3D Warren truss, shown in Fig. 6.6d; the theoretically calculated failure envelopes include microscopic global and localized failure modes compare well with the numerically predicted failure data.

3 Main Production Technologies of Cellular Cores

Various manufacturing processes can be applied for cellular materials production depending mainly on the type of cellular material to be produced, the material and the size of the elementary cell. Prismatic cellular materials, usually of honeycomb type, can be created by three different ways each one of which presents certain advantages and disadvantages. The most widely applied is the expansion manufacturing process, in which adhesive is spread over sheets of material, usually of rectangular geometry and low thickness, in parallel strips. The sheets are then stacked so that the glue keeps them firmly together along the strips; consequently, the stacked layers are cut and expanded to form the cellular core. The schematic representation of this process is shown in Fig. 6.7.

The prismatic cellular materials can also be produced by the corrugation manufacturing process, in which grooved metal sheets are joined together by adhesives. The schematic representation of the process is shown in Fig. 6.8. A third option for manufacturing prismatic cellular materials is by the proper assembly of rectangular geometry slotted plates, known as the strip slotting method (Fig. 6.9).

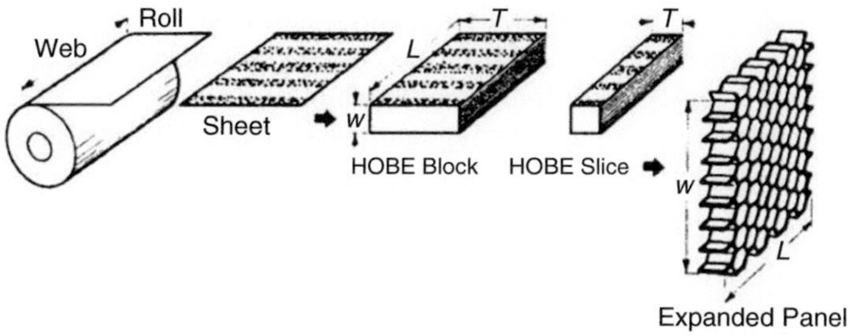

Fig. 6.7 Expansion manufacturing process for honeycomb cellular materials, after Rupani et al. (2017)

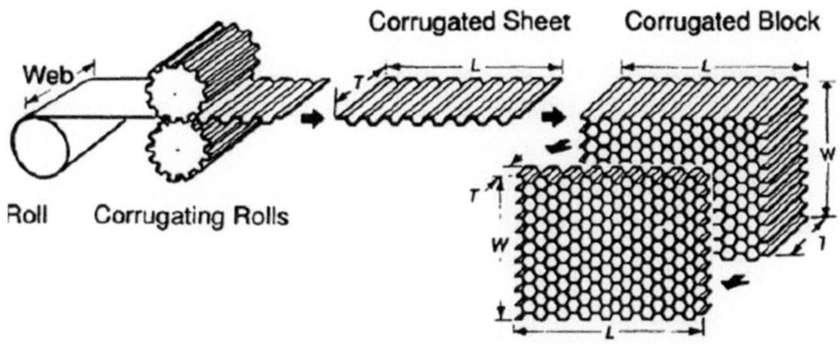

Fig. 6.8 Corrugation manufacturing process for honeycomb cores, after Rupani et al. (2017)

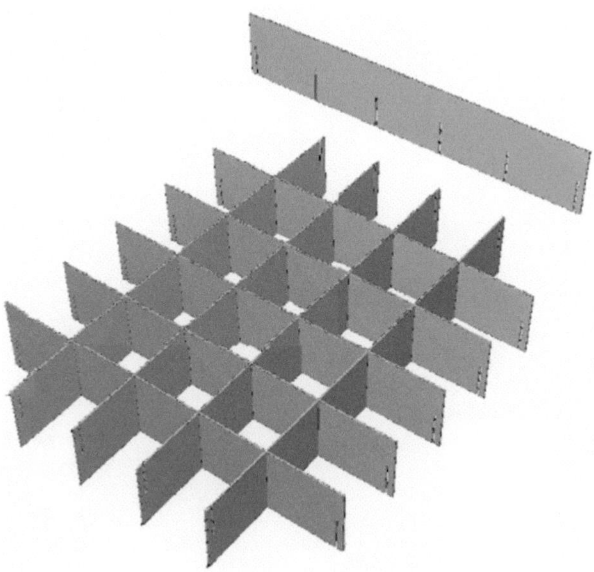

Fig. 6.9 Manufacturing prismatic cellular materials by the strip slotting method

The open lattice cellular materials are more attractive for novel aeronautical application due to their advantages over the closed cell conventional honeycombs, as mentioned previously. The conventional ways to manufacture open-cell strut based lattice cores are by sheet forming and perforated sheet folding/drawing, by wire/hollow tube lay-up, and by casting/investment casting methods (Wadley 2006). Schematic illustrations of the above manufacturing approaches are shown in Figs. 6.10, 6.11, and 6.12.

However, it should be stressed that in order to obtain the maximum effectiveness of a lattice core, its properties should be tailored as much as possible to the type and level of the applied loading; therefore, the core unit cell is required to have variable dimensions throughout its volume. The construction of such graded cores cannot be easily performed by the previously mentioned methodologies. A method that enables the production of functionally graded cores is the Additive Manufacturing (AM). Among the different variations of AM, Selective Laser Melting (SLM) has the ability to produce open-cell lattice cores of various geometry with high geometrical precision of up to 20 μm. In Fig. 6.13, the geometric precision and detail of a strut made by the SLM process is presented.

In the frame of SLM technique, the base material is metallic powder, either from specific Aluminum alloys, stainless steel (316L), or titanium (Ti64). In Fig. 6.14 typical powder grains are shown, ranging between 10 and 45 μm in diameter.

Fig. 6.10 Manufacturing of open-cell strut based lattice material by the sheet forming and perforated sheet folding/drawing process

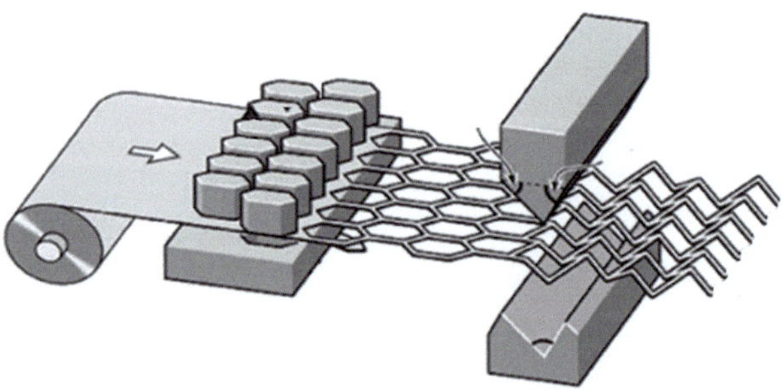

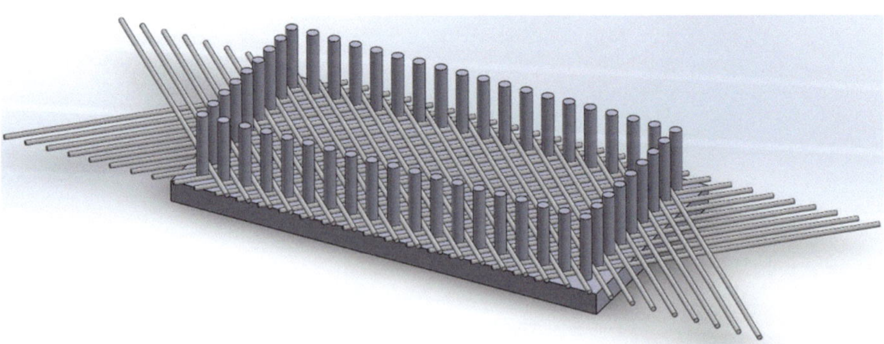

Fig. 6.11 Manufacturing of open-cell strut based lattice material by the wire/hollow tube lay-up process

Fig. 6.12 Manufacturing of open-cell strut based lattice material using the casting/investment casting methods

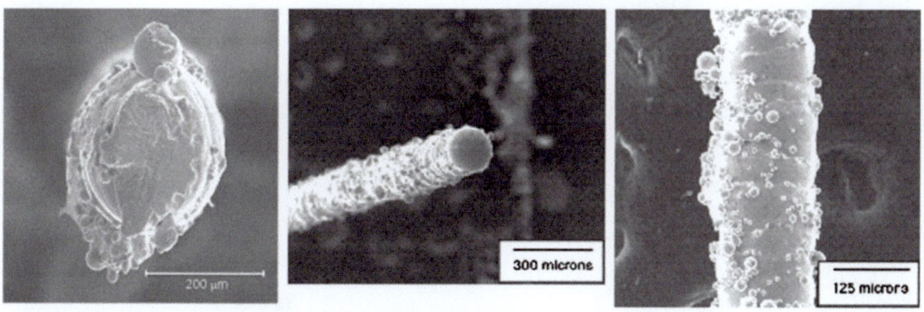

Fig. 6.13 Manufacturing of open-cell strut based lattice material using the Selective Laser Melting (SLM) additive manufacturing process, after Tsopanos et al. (2010)

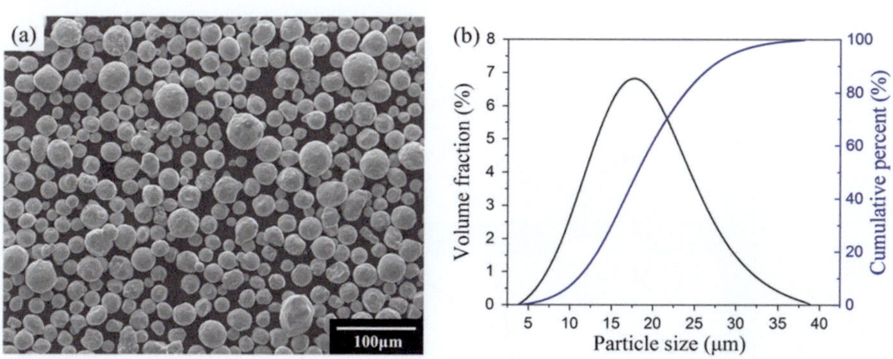

Fig. 6.14 Powder grains with variable particle size, after Wang et al. (2017) (**a**) Morphology of 316L powder; (**b**) particle size distribution of 316L powder.

The basic steps of SLM process are the following: initially a layer of metal powder is spread onto a metallic base that can be moved vertically. A laser beam is driven over the base surface at a preselected pattern and selectively melts the powder. Consequently the base moves down at a step equal to the average powder particle size and a new layer of powder is spread for the process to be repeated. A schematic illustration of the process is shown in Fig. 6.15. In Fig. 6.16 struts and lattice blocks created by the SLM method after non-fused metal powder removal are shown.

Many open lattice cellular topologies may be produced by SLM; however, limitations also exist. In particular, the SLM process is not capable of creating horizontal or almost horizontal struts. The smallest strut angle that can be created is about 25° to the horizontal. Another disadvantage of the SLM method for the production of cellular lattice materials is related to the long production time. In case of Fig. 6.16, each 50 μm layer deposit requires about 30 s, followed by the laser scan which requires also about the same time; provided that the production of 20 mm^3 blocks at a resolution of 50 μm requires 400 layers, the production process takes about 6 h. The diameter of the struts of the lattice core depends on the energy applied, which is controlled by the laser power and the exposure time. The density of the blocks increases when the laser energy increases. Further details on the SLM technique can be found in several publications and reports (e.g., Tsopanos et al. 2010; Brooks et al. 2005). Experimental results have shown that although the same amounts of laser energy can be obtained by different com-

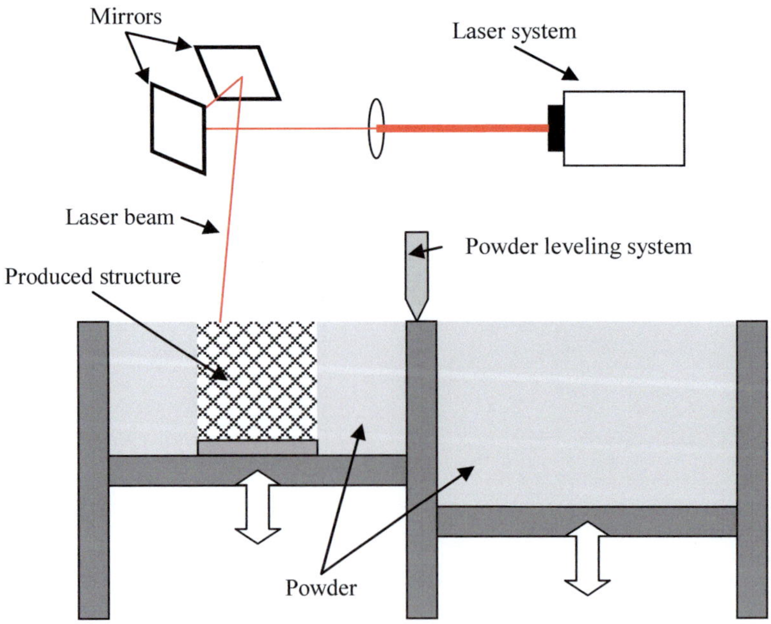

Fig. 6.15 Schematic illustration of the Selective Laser Melting process

Fig. 6.16 Struts and open lattice blocks of circular cross section, after Tsopanos et al. (2010)

binations of laser power and exposure time, cellular materials different of different properties are created, which indicates the requirement for the production process tailoring to the geometrical details of the unit cell.

4 Analysis and Simulation of Lattice Cellular Cores

An overview of analysis and simulation approaches of lattice cores is briefly presented in the present section and demonstrated in the case of a Body-Centered Cubic (BCC) unit cell. A respective lattice core composed of BCC unit cell is presented in Fig. 6.17. Based on this basic BCC unit cell, other unit cell geometrical variations can be developed, e.g., a BCC unit cell with additional pillar strands along the vertical edges of the unit cell (BCC-z), or the horizontal unit cell edges (BCC-h).

An analytical determination of the complete set of homogenized elastic constants, comprising elastic modulus, shear modulus, and Poisson's ratios in all the principal unit cell directions is available for many unit cell topologies of lattice cores. Such a set of homogenized elastic constants is presented in Ptochos and Lampeas (2012a) and Ptochos and Lampeas (2012b) for regular and cuboid BCC unit cells in the form of closed relationships depended on truss and unit cell dimensions.

An examination of the BCC unit cell reveals the existence of symmetry with respect to geometry, material, and loading in three-orthogonal planes, namely *xy*, *yz*, and *xz* (Fig. 6.18); therefore, the unit cell deformation will be symmetric in these planes, leading to the requirement of considering the one-eighth of the BCC unit

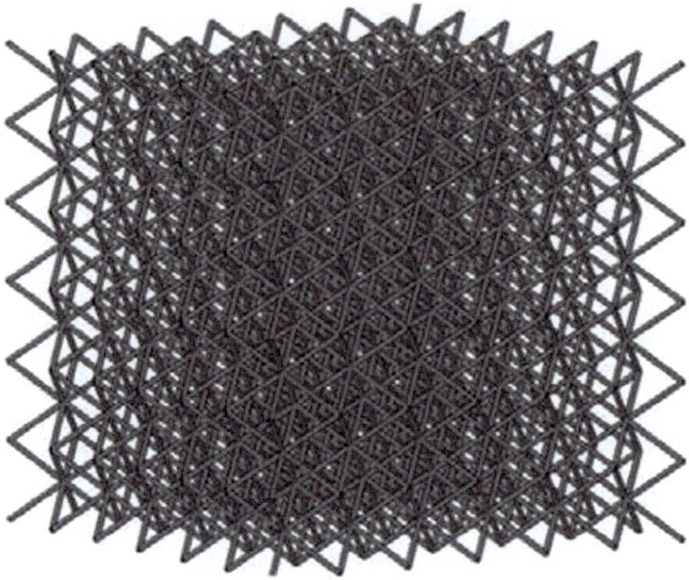

Fig. 6.17 Lattice core composed of BCC unit cells

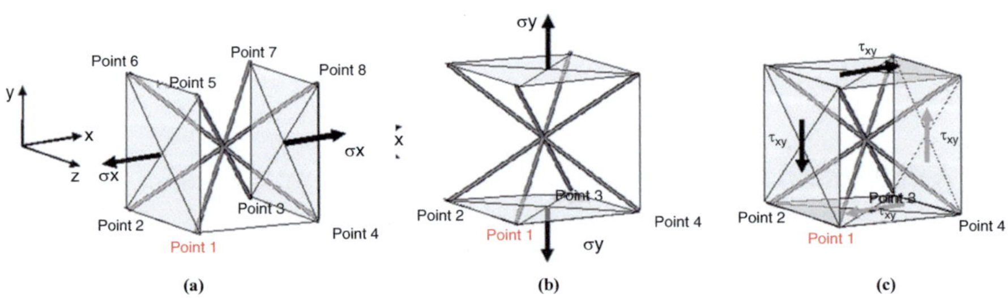

Fig. 6.18 Different loading cases used in the calculation of unit cell elastic constants and failure (**a**) Tension in the x-direction, (**b**) Tension in the y-direction and (**c**) shear in the xy-plane

cell, comprising only one strut along with appropriate boundary conditions. The unit cell boundaries are periodic planes; therefore, at these strut points periodic boundary conditions were applied, i.e., rotations in all directions are set to zero, while the other strut end is kept fully constrained.

In the general three-dimensional case, six loading cases related to normal and shear loads on the unit cell *xy*, *yz*, and *xz* planes (i.e., σ_x, σ_y, σ_z, τ_{xy}, τ_{yz}, and τ_{xz}) may be applicable. In Fig. 6.18, the loading cases related to σ_x, σ_y, and τ_{xy} stresses are presented. In the present analytical approach, the unit cell struts are considered as beams and closed form beam bending solutions are used to predict their failure conditions. The strut aspect ratio (*r/L*) is usually lower than 0.1, i.e., the limitations of classical beam bending theory are valid; otherwise, shear deflection terms in the analytical solutions are introduced.

Due to the complexity of the derived equations when all remote stress components are considered, the analytical failure prediction methodology is illustrated hereafter for the case of biaxial loading at the x–y plane, i.e., only σ_x and σ_y stresses are considered. In such a case the resulting loading system on each beam of the unit cell comprises one axial force ($F_{x'}$) and one shear force ($F_{y'}$), while boundary conditions include constrain of all rotations at the beam edges, as shown in Fig. 6.19. The solution of the above configuration leads to the determination of the stress field in the unit cell.

The loading forces ($F_{x'}$), ($F_{y'}$) on the beam end, which are calculated as functions of unit cell geometry and biaxial loading ($\sigma_x - \sigma_y$ in the present demonstration case), are indicatively presented in the following equations:

$$F_{x'} = \frac{abd\sigma_x}{4\sqrt{a^2+b^2+d^2}} + \frac{abd\sigma_y}{4\sqrt{a^2+b^2+d^2}} \tag{6.1}$$

$$F_{y'} = \sqrt{\frac{b^4d^2\sigma_x{}^2}{16\left(a^2+b^2\right)} + \left(\frac{abd^2\sigma_x}{4\sqrt{a^2+b^2}\sqrt{a^2+b^2+d^2}} - \frac{ab\sqrt{a^2+b^2}\sigma_y}{4\sqrt{a^2+b^2+d^2}}\right)^2} \tag{6.2}$$

In Eqs. (6.1) and (6.2), a, b, and d stand for the unit cell dimensions in x, y, and z Cartesian coordinates, respectively, and refer to the general case of a unit cell with rectangular parallelepiped shape. For the case of a cubic unit cell $a = b = d = l$, similar expressions for the loading system are derived for all possible combinations of unit cell loading, i.e., any combination of the remote stress tensor components (σ_x, σ_y, σ_z, τ_{xy}, τ_{yz}, τ_{xz}) at the unit cell boundaries, resulting to the derivation of unit cell stress field for any possible loading mode. Having derived the stress field expressions, failure criteria for the most important failure modes are formulated.

Analytical approaches are usually limited to linear elastic analyses, and thus only elastic constants can be determined. In order to enable determination of the nonlinear behavior of lattice cores, numerical simulation, usually based on Finite Element approaches are usually applied. The developed FE models comprise detailed modeling of the core unit cell geometry with beam or shell elements. Application of nonlinear analysis including geometrical nonlinearity and material nonlinear behavior make the prediction of various failure modes feasible, as well as, crush compaction and final collapse of the core. Core micro-struts material properties cannot be

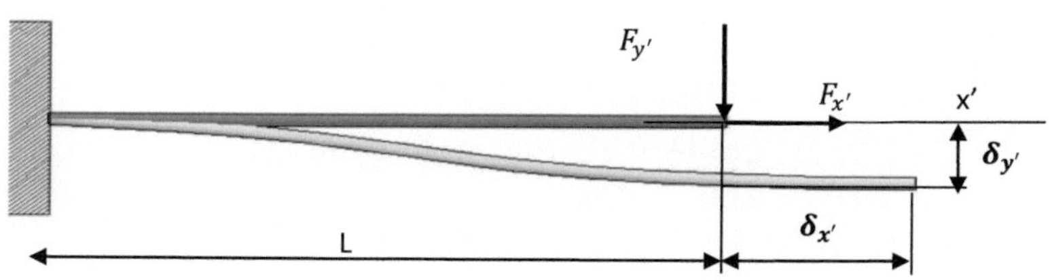

Fig. 6.19 The loading system of strut of the beam representing the BCC unit cell

obtained from the bulk material data sheets, but require specific mechanical tests like the ones presented in Tsopanos et al. (2010), in order to take into account the effects of manufacturing processes and scale. FE modeling experience has revealed that each micro-strut should be modeled with at least four beam or shell elements along its length; this results to an average element size in the range of 0.5–2 mm for typical lattice topologies, which is very small compared to the overall core dimensions, leading to very large computation times if used in the design of real sandwich structures.

Microscale core structural numerical simulation serves as a useful tool in order to select the optimum geometrical and materials parameters, which can provide tailored mechanical properties. An example of a study on the dependence of the macroscopic elastic modulus Ex on the micro-strut radius to length ratio of a regular BCC unit cell is presented in Fig. 6.20, using analytical and finite element analyses using two types of beam elements (Ptochos and Lampeas 2012b).

The calculated core unit cell linear elastic and nonlinear behavior may be consequently used in the development of homogenized material models of the core material from a macroscopic point of view. This is obtained by performing virtual tests on unit cells for different load combinations. In the homogenization approach a detailed microscale core model is used to derive equivalent material properties that can be used when the core is discretized by solid elements. Different types of homogenization principles are applicable in the case of cellular cores: simple techniques capable to calculate only the orthotropic effective elastic behavior of the core, or more sophisticated techniques which can calculate the nonlinear response and the different core failure modes. The calculation of homogenized elastic stiffness properties are usually derived based on the assumption that the average strain energy stored in a representative volume filled with the homogenized medium for specific boundary conditions and loading is equal to the average strain energy stored in the detailed micro-strut model which occupies the same volume and is subjected to the same loading and boundary conditions.

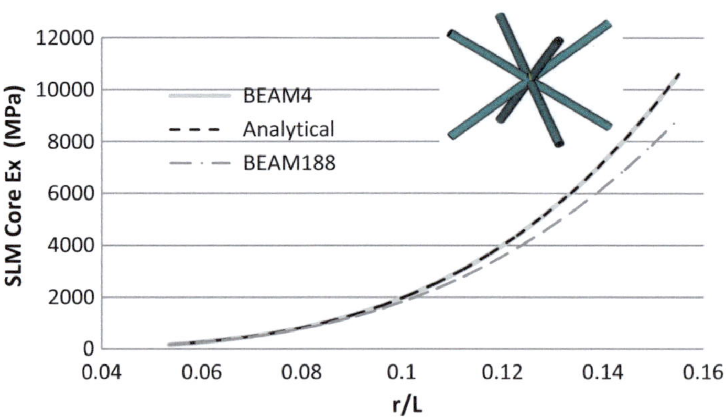

Fig. 6.20 Calculated macroscopic elastic modulus as a function of micro-strut radius to length ratio in a regular BCC unit cell

More advanced homogenization techniques can predict the nonlinear unit cell behavior by performing virtual tests under tension, compression, and shear loading (as well as their combinations, such as biaxial tension/compression, tension/shear etc.) of the unit cell model under periodic boundary conditions, in order to calculate stress–strain curves which are used as input for development of solid element material models. An example of the application of homogenization in the case of a BCC core block is presented in Fig. 6.21, after Ptochos and Lampeas (2012b).

Some limitations also exist however in the homogenization solution, the most important of which are the difficulty to compute very localized failure modes (micro-buckling, micro-plasticity), high stress gradients and edge effects. To take into account such effects, multiscale approaches are required, comprising detailed micro-strut modeling with beam/shell elements at the edge and the critical regions of the structure zones, while the remaining volumes can be homogenized and represented by solid elements with appropriately calculated material properties.

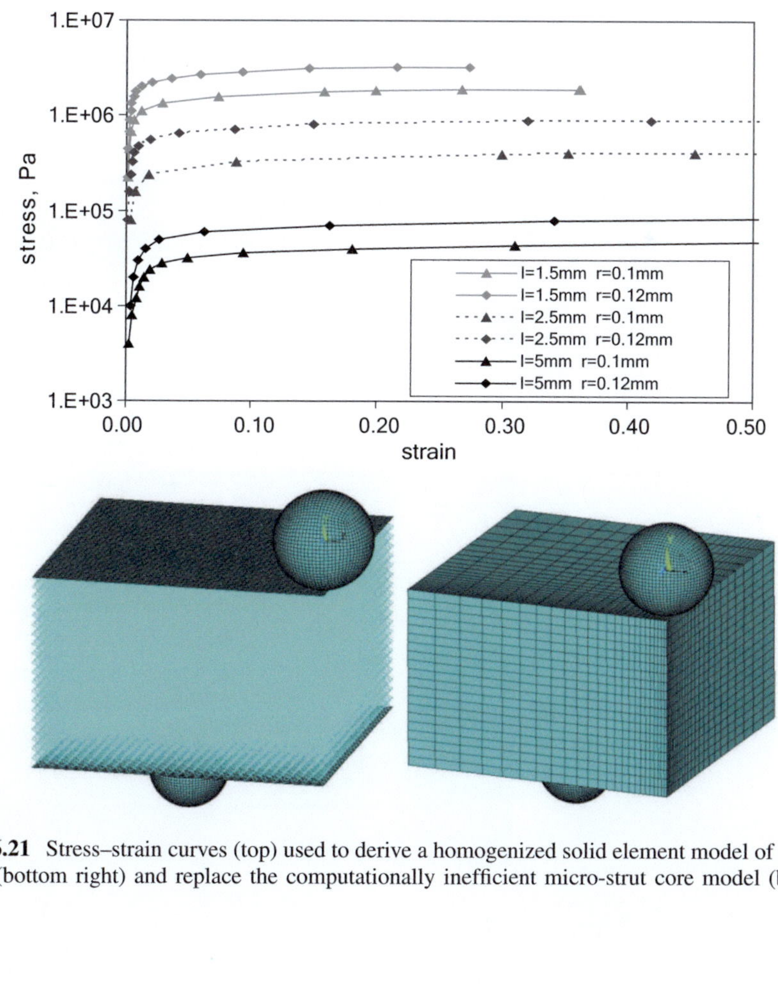

Fig. 6.21 Stress–strain curves (top) used to derive a homogenized solid element model of a BCC core (bottom right) and replace the computationally inefficient micro-strut core model (bottom left)

5 Analysis and Simulation of Sandwich Structures with Cellular Cores

The mechanical and functional properties of cellular materials make them irreplaceable as core materials in aerospace sandwich constructions, where core materials with low weight and high performance at reasonable cost are required. Such sandwich materials constructions are especially suited for lightweight constructions when the main loads are flexural or impact, e.g., Fig. 6.22.

As the lightweight core normally has a low shear modulus, the core shear deformation may result to local failure modes of the panel, which are not typical in conventional stiffened skin aircraft structures. There are several failure modes that might appear in a sandwich panel, as shown in Fig. 6.23. Among them, there are three local failure modes in edgewise compression, specifically face wrinkling, face dimpling, and shear crimping. Face wrinkling, as shown in Fig. 6.23c, d, occurs when sandwich panel skin buckles as a plate on elastic foundation. The skin buckling load and deflection direction depends on the core compression strength, the skin/core bond tensile strength, and initial eccentricities of the skin. Dimpling, or intracellular buckling, that usually occurs in honeycomb or lattice core panels with thin skins, refer to skin buckling between the core geometrical features as shown in

Fig. 6.22 Sandwich construction with cellular core loaded (**a**) in bending (Yang et al. 2013) and (**b**) impact (Labeas and Ptochos 2013)

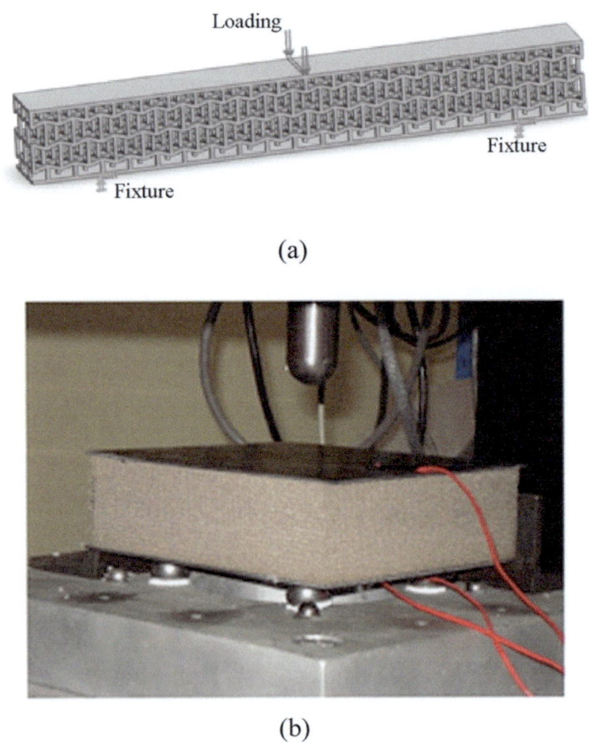

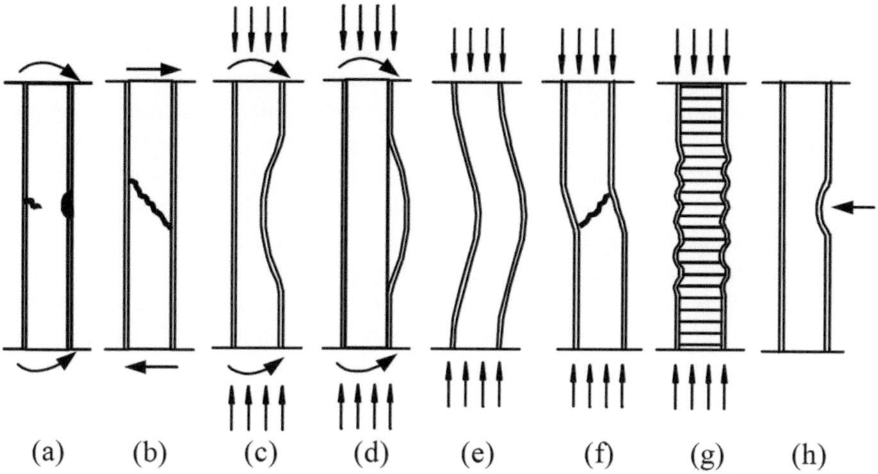

Fig. 6.23 Potential failure modes of sandwich panels, (**a**) Face yielding/fracture, (**b**) core shear failure, (**c**, **d**) face wrinkling, (**e**) general buckling, (**f**) shear crimping, (**g**) face dimpling and (**h**) local indentation (Kaminski et al. 2000)

Fig. 6.23g for the case of honeycomb core. Face dimpling may be severe enough so that permanent dimples remain after removal of load and dimples may grow across the core cell walls and result in a wrinkling of the facings. Shear crimping failure, Fig. 6.23f, appears to be a local mode of failure, but is caused by global buckling, Fig. 6.23e, combined with low core shear modulus leading to crimping of the sandwich with core failure in shear at the crimp. As soon as the crimp appears, the overall buckle may disappear. Therefore, although examination of the failed sandwich indicates crimping or shear instability, failure may initiate by overall buckling that finally causes crimping.

The behavior of a structure composed of sandwich type materials is usually determined utilizing FE analysis. There are two modeling approaches followed for the simulation of such structures: shell element models and shell combined to solid element models. The preferable approach depends on the structural behavior of the sandwich core and skins and the required level of accuracy of the analysis. In the shell element approach the sandwich structure is modeled using layered shell elements. The core is considered as a layer in the stack. Shell elements that take into account transverse shear stiffness are required for this type of analysis at the least a Mindlin–Reissner formulation for shell elements is required. This approach is suitable for global structural analyses of sandwich structures, where no local failures are expected in skins or core, as these failures cannot be predicted by the core shell element model.

In the shell combined to solid element approach the sandwich structure core is modeled using three-dimensional solid elements, while the skins are modeled using shell or layered shell elements depending on whether they consist of metallic or composite material. This approach generally leads to larger and computationally more expensive FE models, as the element aspect ratio limitations of the solid elements lead to relatively dense core meshes. On the other hand, the numerical results include more information about the structural behavior, including transverse compression and shear stresses which can be used for the prediction of skin and core failure modes.

In the case of composite skins, they are simulated using orthotropic material models, while core material behavior can be obtained using a homogenization process, as described above. Depending on the selected modeling approach, skins are modeled as the top and bottom plies of a layered composite shell element with appropriate stacking sequence, or as separate layered shells in the shell combined to solid modeling approach. These layered shell elements are attached to the outer faces of the solid core elements, as shown in Fig. 6.24. For simplicity reasons, modeling of the skins without a mid-plane offset, as presented in Fig. 6.24 is possible due to the its negligible contribution of the sandwich panel total stiffness due to the core relatively low mechanical properties. Thus, the distance between the upper and the lower nodes of the sandwich structure is equal to the distance between the skin mid-plane; when such an approach is followed, the skin and core surfaces share common nodes, and therefore, skin debonding cannot be considered as a potential failure mode. Contrary to solid element nodes, shell element nodes include rotational degrees of freedoms. Thus, in order to avoid unrealistic rotations at the shell/solid interface, equal mesh size should be defined for both skin and core and each shell element should be attached to a solid element.

In the shell element modeling approach, the skin in-plane stress and strain data are taken from representative skin layers, while core transverse shear stress information is taken only from the shell element core layer. For the shell combined to solid elements modeling approach, the skin in-plane stress and strain data are obtained from the skins shell elements, while the core transverse shear stress is obtained from maximum transverse shear stress of the core solid elements. In each case, the obtained stress and strain results are used to check whether specific design criteria such as stiffness requirements, damage tolerance, stability, etc. are fulfilled.

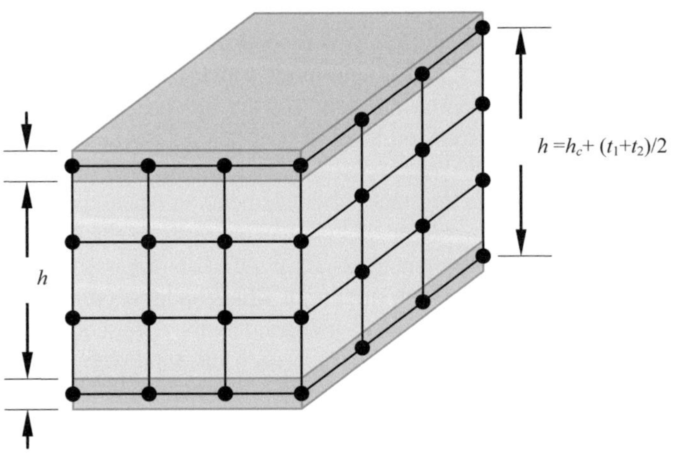

Fig. 6.24 FE modeling strategy for sandwich analysis using shell combined to solid elements approach (t_1 and t_2 are skin thicknesses)

When design under static loading is considered, the sandwich structure needs to fulfill criteria resulting from strength and stiffness requirements related to the skins and core materials and global and local stability. The structure needs to be able to undertake the expected ultimate load in all cases. The overall static strength at ultimate load level may have to be validated by the respective experimental testing.

Provided that sandwich structures are often used in aeronautical structures as impact energy absorption elements, in the fuselage, wing and tail components. Low velocity impact, i.e., stone impact or tool drop during fabrication/maintenance, may result to in-service damage of the structure. Thus, the damage tolerance of sandwich structures has also to be proven, as the ability to of a damaged part sustain a sufficient residual strength. Typical values of impact energies considered to create the initial damage vary from 35 to 90 J for external parts and from 35 to 50 J for internal parts. An example of low velocity impact testing and simulation of a sandwich panel of skins and BCC core is presented in Labeas and Ptochos (2013). In Figs. 6.25 and 6.26, representative results from this work are presented.

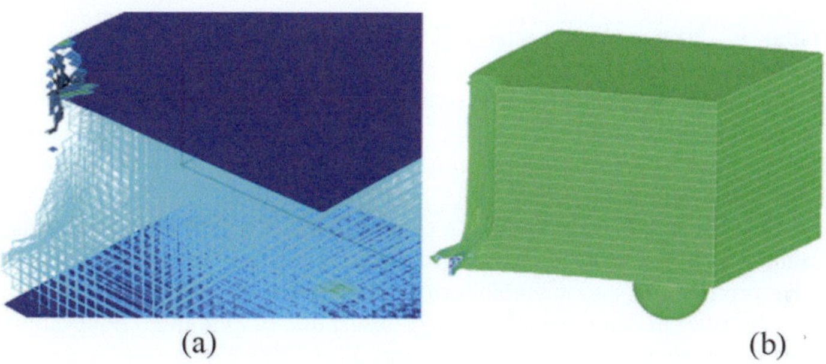

Fig. 6.25 Numerical simulation of low speed impact on open cellular sandwich material (**a**) detailed model and (**b**) homogenized core material model

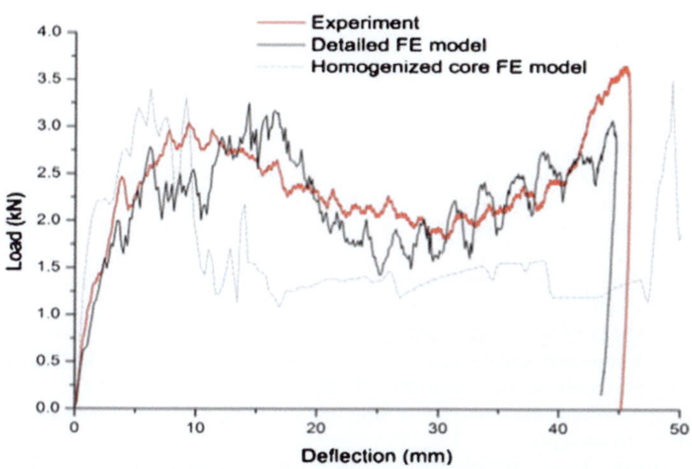

Fig. 6.26 Comparison of experimentally measured and numerically calculated load deflection curves of low speed impact on open cellular sandwich material

6 Conclusions

Sandwich constructions with cellular cores and metallic or composite skins are particularly efficient when the main loading is bending or impact. The most interesting category of cellular cores for the aerospace sector is the open lattice or open micro-truss cores, which offer specific advantages compared to the conventional honeycomb or foam cores. For this reason several types of regular open lattice cellular cores have been developed (as described in Sect. 2) combined with different appropriate techniques for their efficient manufacturing (Sect. 3). The analysis, simulation and design of cellular sandwich structures (Sects. 4 and 5) comprise very important areas of active research and development.

References

Aboudi J, Gilat R (2005) Micromechanical analysis of lattice blocks. Int J Solids Struct 42(15):4372–4392

Biagi R, Bart-Smith H (2007) Imperfection sensitivity of pyramidal core sandwich structures. Int J Solids Struct 44(14–15):4690–4706

Bouwhuis BA et al (2008) Edge effects in compression testing periodic cellular metal sandwich cores. J Mater Sci 43(9):3267–3273

Brooks W, Tsopanos S, Stamp R, Sutcliffe CJ, Cantwell WJ, Fox P, Todd J (2005) The production of open cellular lattice structures using selective laser melting. In: 6th national conference on rapid design, prototyping, and manufacturing, Buckinghamshire Chilterns University College

Caserta G et al (2010) Static and dynamic energy absorption of aluminum honeycombs and polymeric foams composites. Mech Adv Mater Struct 17(5):366–376

Chiras S et al (2002) The structural performance of near-optimized truss core panels. Int J Solids Struct 39(15):4093–4115

Cote F et al (2007) Structural response of pyramidal core sandwich columns. Int J Solids Struct 44(10):3533–3556

Deshpande VS, Fleck NA (2001) Collapse of truss core sandwich beams in 3-point bending. Int J Solids Struct 38(36–37):6275–6305

Deshpande VS et al (2001) Effective properties of the octet-truss lattice material. J Mech Phys Solids 49(8):1747–1769

Doyoyo M, Hu JW (2006) Multi-axial failure of metallic strut-lattice materials composed of short and slender struts. Int J Solids Struct 43(20):6115–6139

Elsayed MSA, Pasini D (2010) Multiscale structural design of columns made of regular octet-truss lattice material. Int J Solids Struct 47(14–15):1764–1774

Evans AG et al (2001) The topological design of multifunctional cellular metals. Prog Mater Sci 46(3–4):309–327

Flores-Johnson EA, Li QM (2010) Indentation into polymeric foams. Int J Solids Struct 47(16):1987–1995

Gibson L, Ashby M (1997) Cellular solids: structure and properties. Cambridge University Press

Goga VI (2010) Testing and application of new phenomenological material model for foam materials. Portál pre odborné publikovanie. ISSN 1338-0087

Hyun S et al (2003) Simulated properties of Kagomé and tetragonal truss core panels. Int J Solids Struct 40(25):6989–6998

Hyun S et al (2009) Effect of imperfections on the mechanical behavior of wire-woven bulk kagome truss PCMs under shear loading. J Mech Sci Technol 23(5):1270–1277

Jang W-Y, Kyriakides S (2009) On the crushing of aluminum open-cell foams: part II analysis. Int J Solids Struct 46(3–4):635–650

Kaminski ML, Amdahl J, Frieze P, Gordo JM (2000) Ultimate strength. In: Proceedings of the 14th international ship and offshore structures congress (ISSC), Nagasaki, 2–6 Oct 2000, vol I. pp 253–322

Kooistra GW et al (2004) Compressive behavior of age hardenable tetrahedral lattice truss structures made from aluminium. Acta Mater 52(14):4229–4237

Kooistra GW et al (2008) Shear behavior of aluminum lattice truss sandwich panel structures. Mater Sci Eng A 472:242–250

Labeas GN, Ptochos E (2013) Investigation of sandwich structures with innovative cellular metallic cores under low velocity impact loading. Plast Rubber Compos 42(5):194–202

Lee YH, Kang KJ (2009a) A wire-woven cellular metal: part-I, optimal design for applications as sandwich core. Mater Des 30(10):4434–4443

Lee BK, Kang KJ (2009b) A parametric study on compressive characteristics of Wire-woven bulk Kagome truss cores. Compos Struct 92(2):445–453

Lee S et al (2006) Deformation rate effects on failure modes of open-cell Al foams and textile cellular materials. Int J Solids Struct 43(1):53–73

Lee Y-H et al (2007) Wire-woven bulk Kagome truss cores. Acta Mater 55(18):6084–6094

Lim J-H, Kang K-J (2006) Mechanical behavior of sandwich panels with tetrahedral and Kagome truss cores fabricated from wires. Int J Solids Struct 43(17):5228–5246

Lim C-H et al (2009) A new type of sandwich panel with periodic cellular metal cores and its mechanical performances. Mater Des 30(8):3082–3093

Liu T et al (2006) Design optimization of truss-cored sandwiches with homogenization. Int J Solids Struct 43(25–26):7891–7918

Liu T et al (2007) Minimum weights of pressurized hollow sandwich cylinders with ultralight cellular cores. Int J Solids Struct 44(10):3231–3266

Lorato A et al (2009) The transverse elastic properties of chiral honeycombs. Compos Sci Technol 70(7):1057–1063

Pettermann HE, Hüsing J (2012) Modeling and simulation of relaxation in viscoelastic open cell materials and structures. Int J Solids Struct 49(19–20):2848–2853

Ptochos E, Lampeas G (2012a) Shear modulus determination of cuboid metallic open-lattice cellular structures by analytical, numerical and homogenisation methods. Strain 48(5):415–429

Ptochos E, Lampeas GN (2012b) Elastic modulus and Poisson's ratio determination of micro-lattice cellular structures by analytical, numerical and homogenisation methods. J Sandw Struct Mater 14(5):597–626

Queheillalt DT, Wadley HNG (2005a) Cellular material lattices with hollow trusses. Acta Mater 53(2):303–313

Queheillalt DT, Wadley HNG (2005b) Pyramidal lattice truss structures with hollow trusses. Mater Sci Eng A 397(1–2):132–137

Queheillalt DT et al (2008) Mechanical properties of an extruded pyramidal lattice truss sandwich structure. Scr Mater 58(1):76–79

Rathbun HJ et al (2004) Measurement and simulation of the performance of a lightweight metallic sandwich structure with a tetrahedral truss core. J Appl Mech 71(3):368–374

Rathbun HJ et al (2005) Strength optimization of metallic sandwich panels subject to bending. Int J Solids Struct 42(26):6643–6661

Rupani S, Jani SS, Acharya G (2017) Design, modelling and manufacturing aspects of honeycomb sandwich structures: a review. Int J Sci Eng Dev Res 2:526–532

Sugimura Y (2004) Mechanical response of single-layer tetrahedral trusses under shear loading. Mech Mater 36(8):715–721

Sypeck DJ, Wadley HNG (2001) Multifunctional microtruss laminates: textile synthesis and properties. J Mater Res 16(3):890–897

Tsopanos S et al (2010) The influence of processing parameters on the mechanical properties of selectively laser melted stainless steel microlattice structures. J Manuf Sci Eng 132(4):041011

Wadley HNG (2006) Multifunctional periodic cellular metals. Philos Trans R Soc A Math Phys Eng Sci 364(1838):31–68
Wadley HNG et al (2003) Fabrication and structural performance of periodic cellular metal sandwich structures. Compos Sci Technol 63(16):2331–2343
Wang J et al (2003) On the performance of truss panels with Kagome cores. Int J Solids Struct 40(25):6981–6988
Wang S, Liu Y et al (2017) Research on high layer thickness fabricated of 316L by selective laser melting. Materials 10(9):1055
Xiong J et al (2010) Fabrication and crushing behavior of low density carbon fiber composite pyramidal truss structures. Compos Struct 92(11):2695–2702
Xue Z, Hutchinson JW (2003) Preliminary assessment of sandwich plates subject to blast loads. Int J Mech Sci 45(4):687–705
Yang L, Harrysson O, West H, Cormier D (2013) A comparison of bending properties for cellular core sandwich panels. Mater Sci Appl 4:471–477
Zhang X et al (2017) Thermo-fluidic comparison between sandwich panels with tetrahedral lattice cores fabricated by casting and metal sheet folding. Energies 10(7):906
Zhu F et al (2010) Some theoretical considerations on the dynamic response of sandwich structures under impulsive loading. Int J Impact Eng 37(6):625–637
Zok FW et al (2003) Design of metallic textile core sandwich panels. Int J Solids Struct 40(21):5707–5722

Chapter 7
Integral, Disruptive, and Multifunctional Aircraft Structures

Ronny Sachse, Daniel Fernandez, Yves Klett, and Peter Middendorf

1 Introduction

It is widely acknowledged that through the application of fiber-reinforced composites in aircraft structures, significant weight and cost savings can be achieved. However, by simply replacing the material in existing metal designs (commonly known as "black metal design"), the potential of composite structures can only be partially exploited. Completely new and disruptive design approaches must therefore be found, which account for the inherent design freedom associated with composite materials.

Figure 7.1 summarizes design requirements for a pressurized fuselage and highlights the potential for additional functional integration from structures and materials, as well as manufacturing point of view. The structure must therefore on the one hand comply with a wide range of loading scenarios defined by the certification authorities. Following a damage-tolerant design approach, residual strength of a structure must be demonstrated after a defined number of load cycles with realistic damages introduced to the structure. Additionally, crash and fire damage resistance as well as sufficient thermal and acoustic isolation must be provided. On the other hand, production and in-service costs define the competitiveness of the final product. They are affected, among other factors, by repairability and maintainability as well as part and functional integration (Nicolais et al. 2011).

A rigorous integral design philosophy of composite structures with additional functional integration is an efficient way to address most of the aforementioned requirements. Functional integration can be broadly classified as structural (e.g., strength, durability) and nonstructural (e.g., thermal isolation, electrical conductivity) functional integration. The integration itself can be achieved at the material (e.g., nanoparticles) and structural (e.g., integral design) level. In this chapter, three

R. Sachse (✉) · D. Fernandez · Y. Klett · P. Middendorf
Institute of Aircraft Design, Stuttgart, Germany
e-mail: sachse@ifb.uni-stutgart.de; daniel.fernandez@ifb.uni-stuttgart.de

S. Pantelakis, K. Tserpes (eds.), *Revolutionizing Aircraft Materials and Processes*, https://doi.org/10.1007/978-3-030-35346-9_7

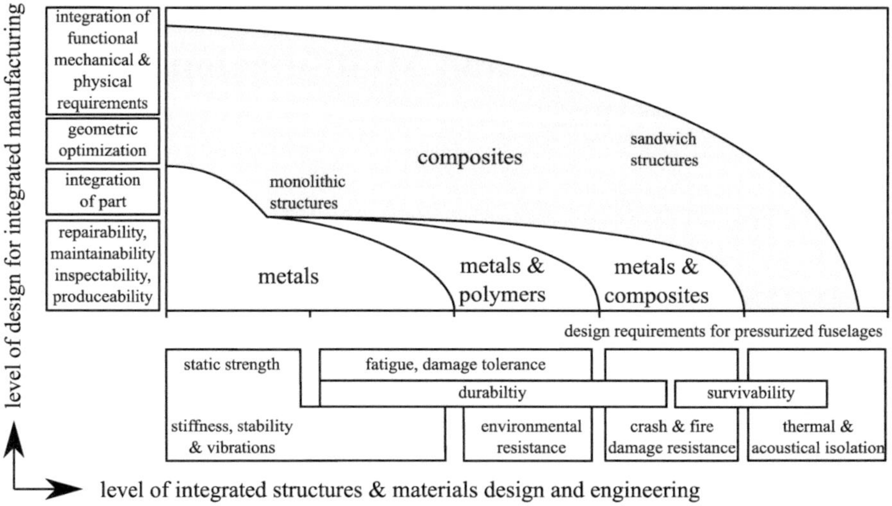

Fig. 7.1 Potential for functional integration in fuselage structures. (Adapted from Nicolais et al. (2011))

concepts shall be presented, which address a particularly wide range of functional integration. When executed correctly, the final product can be highly competitive compared to conventional designs (Gibson 2010; Jagath Narayana and Burela 2018).

Section 2 presents the concept of disbond-arrest features (DAF) for the damage-tolerant design of bonded joints as an enabling technology for an integral design of composite structures.

Section 3 discusses the concept of selected stitching in integral stringer-skin structures. Selective reinforcement through stitching is applied to dry textile preforms to improve impact resistance and damage tolerance of the joint while minimizing adverse effects on intralaminar stiffness and strength.

Section 4 introduces the concept of multifunctional sandwich structures and highlights the potential of functional integration through folded core structures.

2 Damage-Tolerant Design of Bonded Joints: Concept of Disbond-Arrest Features

Bonding of composite structures has the potential of significant cost and weight savings as demonstrated by Lockheed Martin during the Advanced Composite Cargo Challenge (Erripis 2009). The rigorous integral design philosophy applied to the composite fuselage and vertical tail of the X-55A was enabled by bonding technologies. Compared to conventional designs, the structural part count was reduced by 90% while cutting down development time and costs by half.

Despite its great advantages, a commercial implementation of this concept is not possible today. Certification regulations prohibit the use of purely bonded composite joints in primary aircraft structures. This is partially attributed to the widely acknowledged sensitivity of adhesive joints to manufacturing parameters and environmental aging. They can result in so-called weak bonds, which can lead to a premature failure of the joint unless timely detected. Consequently, a damage-tolerant design approach is requested by certification authorities.

2.1 Damage-Tolerant Design Approach

The certification guideline AMC 20-29 (European Aviation Safety Agency n.d.) addresses adhesive joints in primary composite airframe structures. Besides a sufficient surface treatment and process stability, it requires the damage-tolerant design of bonded joints. Following CS 23.573(a), limit load capability must be substantiated through one of the following methods (also applicable to CS 25, CS 27, CS 29):

1. Limitation of disbond size: Disbond sizes must be limited to a maximum disbond size by design features to ensure load-carrying capability according to CS 23.573(a)(3).
2. Proof testing: Critical limit design load must be applied to each critical bonded joint of each production article.
3. Nondestructive testing: The strength of each joint must be ensured through nondestructive inspection techniques that are proven to be repeatable and reliable.

Closely examining the available methods, it becomes obvious that proof testing (2) is not an option for most applications. Furthermore, no nondestructive inspection technique (3) has yet been established to reliably detect weak bonds (European Aviation Safety Agency n.d.).

The limitation of disbonds using design features beyond a critical disbond size (1) therefore remains the only viable option in most application. Starting from local weak bonds, disbond growth must be addressed in either of the following ways (European Aviation Safety Agency n.d.):

(a) No-growth approach: The joint must be designed to prohibit any disbond growth, and limit load capability must be sustained.
(b) Slow-growth approach: Disbond growth rate can be shown to be slow, stable, and predictable. Inspection intervals must be established to ensure that the damage can be detected and repaired while limit load capability is still ensured.
(c) Arrested-growth approach: It must be shown that through the use of design features, any disbond growth can be mechanically arrested before it becomes critical.

The current certification strategy by industry standards is based on the limitation of disbond sizes (Kruse et al. 2018). Mechanical joints are installed to ensure no-growth capability, leading to significant additional weight and expensive manufacturing processes.

2.1.1 The Concept of Disbond-Arrest Features

The arrested-growth certification approach holds many advantages over the conventional no-growth approach. It relies in the application of disbond-arrest features (DAF), which shall be defined as any design element according to AMC 20-29 (European Aviation Safety Agency n.d.) whose primary purpose is the mechanical arrest of disbond growth within the adhesive bondline.

Figure 7.2 illustrates the concept of DAF for a single lap joint of a fuselage skin with an initial weak bond. The edges of adhesive joints are highly loaded, while stresses on the inside are comparably low. Consequently, a local and growing disbond does not necessarily lead to a direct failure of the joint but rather to a redistribution of load transfer. This in turn provides additional design freedom for the application of spatially distributed DAF. Compared to the no-growth approach, where primary load transfer is achieved through mechanical joints, DAF must be considered as a local support of the adhesive joint, while the primary load transfer is still ensured by the adhesive joint itself. The application of dedicated DAF in bonded composite joints has been systematically investigated for the first time within the project "Boltless Assembling of Primary Aerospace Composite Structures" (BOPACS) (BOPACS 2016) funded by the European Union as part of the 7th Framework Programme.

With the ultimate goal of a road map to certification acknowledged by certification authorities, a thorough understanding of fatigue crack growth in bonded joints has to be proven. Disbond-arrest mechanisms must be identified, and dedicated DAF must be developed and characterized accordingly, paying special attention to the system's robustness and quality control. An essential next step is the development of design tools, which allow the prediction of disbond growth under fatigue loading in the presence of DAF. Finally, the concept of DAF must be proven on representative test specimen following a building block approach (Fig. 7.12).

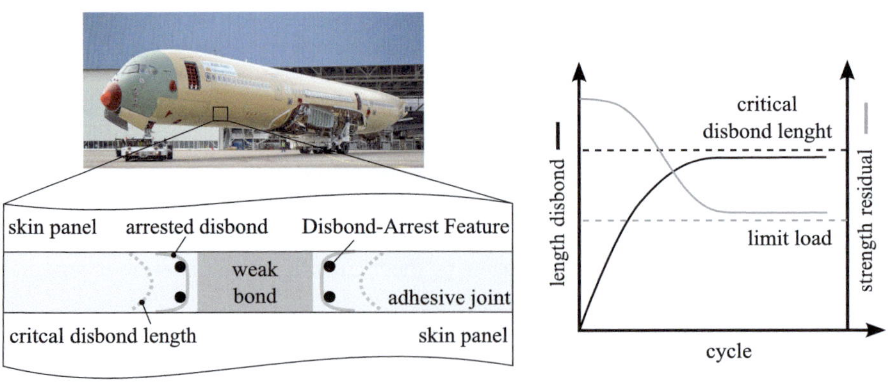

Fig. 7.2 Schematic illustration of the concept of disbond-arrest features and arrested growth. (Photograph taken from Brooklyn Decker (2017))

2.2 *Design and Functionality of Disbond-Arrest Features*

2.2.1 Disbond-Arrest Mechanisms in Bonded Joints

The concept of arrested growth or slow growth is well established in metal structures. When transferring the knowledge to adhesively bonded joints, fundamental differences must be acknowledged. While crack growth in thin metal structures can often be treated as one dimensional, disbonds or cracks in bonded composite joints grow in a two-dimensional or even tree-dimensional manner constrained by the adherents. Further, an adhesive joint cannot be reduced simply to the adhesive bondline, but in fact is designed as a chain connecting the adherent and adhesive through interfaces. Consequently, the failure of a joint is driven by the weakest link (Habenicht 2009). The most significant disbond-arrest mechanisms in bonded composite joints are (BOPACS 2016; Sachse et al. 2017):

Reduction of crack tip loading: Fig. 7.3 depicts crack growth rate da/dN within the adhesive bondline as a function of strain energy release rate range ΔG acting at the crack tip. Already a small reduction in loading leads to a significant reduction in crack growth rate. Limiting loading to the "threshold" region, a technical disbond-arrest is possible.

Suppression of peeling load: Due to the constrained crack growth within an adhesive layer, the macroscopic crack growth rate strongly depends on the type of external loading. By far the most critical loading is peeling (Mode I). This loading mode results in a smooth crack surface, parallel to the interface. On the contrary, shear loading (Mode II) leads to small cracks in the bondline at an angle to the interface. These cracks eventually coalesce resulting in a reduced macroscopic crack growth rate. Figure 7.3 shows the Paris' law for Mode I, Mode II, and shear-

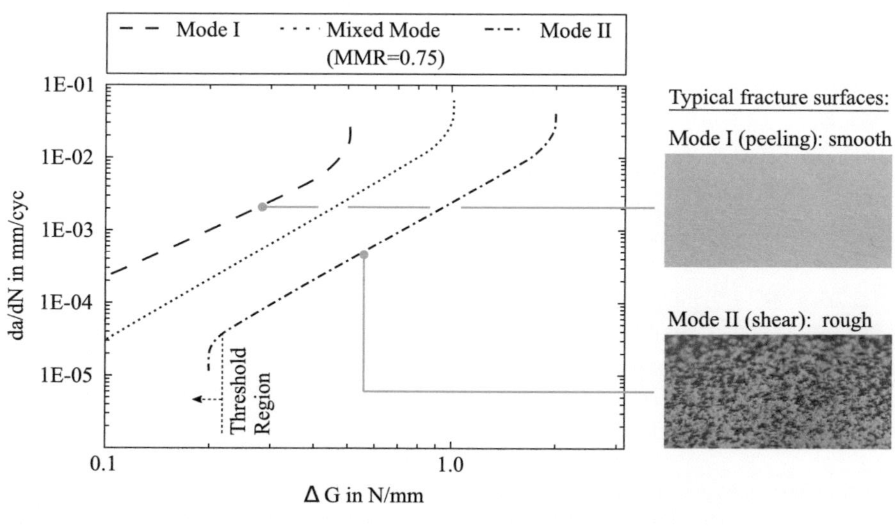

Fig. 7.3 Typical crack growth rate (left) and Paris' law for different loading types

dominant mixed-mode loading, as well as representative fracture surfaces for a toughened epoxy-based adhesive. It can be found that already small peeling loads significantly increase crack growth rate.

Compressive loading: The effect of compressive loading on fatigue crack growth is twofold: First, compressive stresses lead to an additional load path through friction at the crack surfaces, which reduces the absolute loading at the crack tip. Second, the stress state in the adhesive is affected leading to a further inclination of micro-cracks nucleating ahead of the crack tip. This in turn reduces the macroscopic driving force for crack growth parallel to the bondline mid-surface.

Extension of crack front length: Load transfer in cracked bonded joints is spatially distributed over the crack front. Since the adhesive layer is only highly stressed at its edges, it is possible to have cracks growing toward the center of the joints without overloading it. Through a dedicated design, the crack front can be actively extended leading to an overall decrease of loading at the crack front (see also Fig. 7.4).

Material selection: The fatigue crack growth resistance strongly depends on the fracture toughness of the adhesive material and can be tailored chemically by applying, e.g., carbon nanotubes (Liu and Bae 2011) or additional particles (Barbosa

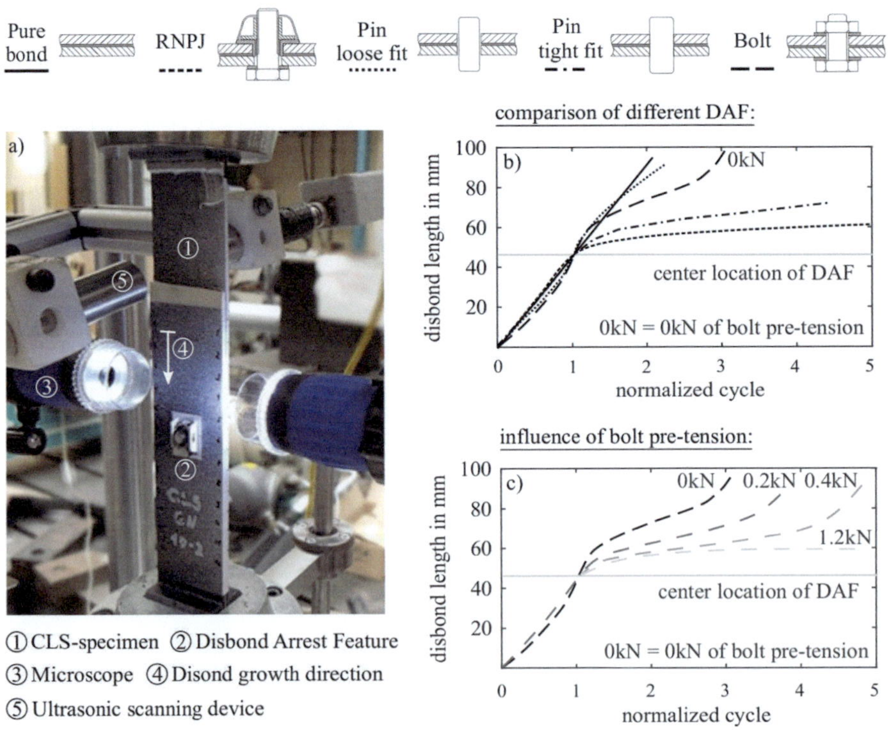

Fig. 7.4 Test setup and principle effect of different through-the-thickness features on disbond growth in cracked-lap shear (CLS) specimen measured by means of microscopes at the specimen edge (Sachse et al. 2017)

et al. 2012). These particles then act as internal crack stoppers, following similar principles to those discussed in this section.

2.2.2 Disbond-Arrest Feature for Bonded Joints

Following the discussion of disbond-arrest mechanisms, a range of DAF, e.g., corrugation (Tserpes et al. 2016), laser bondline stripping (Kruse et al. 2015), and separation of adhesive zones (BOPACS 2016), were developed and characterized during the BOPACS project. From these, the following two concepts have shown the most promising results.

Through-the-Thickness Reinforcement

The logical first step, and also the most widely studied disbond-arrest concept, is to introduce through-the-thickness reinforcement (TTTR) features like rivets, bolts, or pins. On the one hand, they are already at a high technology readiness level (TRL), fostering industrial acceptance. On the other hand, the concept is not limited to the bondline itself but reinforces the laminate over the entire thickness. An overloading of the composite adherent resulting in crack growth into the laminate can therefore be addressed with the same feature.

Sachse et al. (2015, 2016, 2017) compared three different through-the-thickness reinforcement features in bonded cracked-lap shear (CLS) specimens. Pins were selected for their enhanced design freedom compared to conventional rivets (e.g., *z*-pinning (Pegorin et al. 2015) or staple-like pins (Löbel et al. 2013)). The rivetless nut plate joint (RNPJ) (Sachse et al. 2015, 2016, 2017) was adopted for its additional functional integration to perform bondline thickness control. Finally, bolts were selected to demonstrate the effect of pre-tension.

Figure 7.4 presents the test setup (a) and summarizes the principle effect of through-the-thickness reinforcement features installed at a disbond length of 45 mm (b). The rivetless nut plate joint continuously reduces disbond growth rate up to a factor of more than 100 once the disbond passes the DAF. Pins installed with a 0:01 mm clearance rather adversely affected crack growth, leading to an initial acceleration of disbond growth ahead of the pins location. The efficiency can be improved by installing the pins with a tight fit resulting in a crack growth reduction factor of about 5–10. Bolts installed without pre-tension only marginally affect fatigue crack growth. Of particular concern is the reacceleration of crack growth rate once the crack has grown past the bolt, violating the requirement of robustness. Additional pre-tension (c), however, significantly improves the bolts' disbond-arrest efficiency, resulting in a disbond growth reduction similar to the rivetless nut plate joint for pre-tension loads higher than 1.2 kN.

Figure 7.5 exemplary shows the two-dimensional crack growth measured by means of an air-coupled ultrasonic scanning device. A common observation made for all DAF relying on through-the-thickness reinforcement is the spatially varying

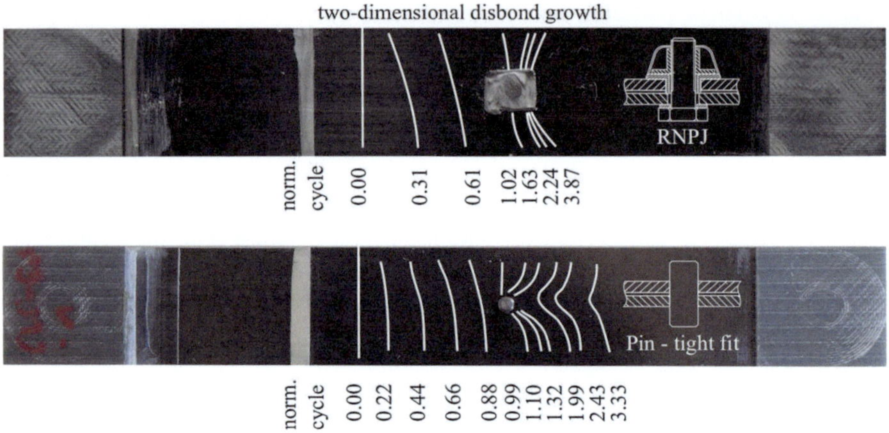

Fig. 7.5 Principle effect of different disbond-arrest feature configurations on disbond growth in CLS specimen measured by means of air-coupled ultrasonic scanning device

effect on disbond growth. This is attributed to two main reasons: (1) The effect of peeling load suppression is reduced as the bending of the laminate increases; (2) bearing loads are transferred only locally, followed by a gradual load distribution in the specimen width direction through shear. As a consequence, the initial straight crack front becomes curved, which leads to an extension of the crack front length and, consequently, to a reduction of disbond growth rate.

Hybrid Bondline

The selection of adhesive material is governed by several requirements including sufficient thermal and moisture resistance, good creep properties, and damage tolerance. Therefore, not necessarily the adhesive with the best fatigue resistance is chosen. Nevertheless, it can be argued that a local application of an adhesive with improved fatigue resistance can be beneficial to arrest disbond growth until the joint is being repaired. Löbel (2016) developed the hybrid bondline concept. This combines epoxy-based adhesive bonding with local thermoplastic welding. Strips of polyvinylidene fluoride (PVDF) are locally integrated into the prepreg laminate layup before curing, representing the bonding surface ply. Opposing strips of PVDF are then welded by means of autoclave or ultrasonic welding, and the epoxy-based adhesive is cured.

Figure 7.6 shows the principle effect of the hybrid bondline in a CLS specimen. The disbond growth is effectively arrested ahead of the PVDF region. At elevated load levels, additional delamination can occur within the adherent, which however does not extend beyond the PVDF region. Löbel (2016) attributes the disbond-arrest capability to a blunting of the growing crack tip in the vicinity of the ductile thermoplast and thus to a reduction of crack tip loading as well as to a suppression of peeling load.

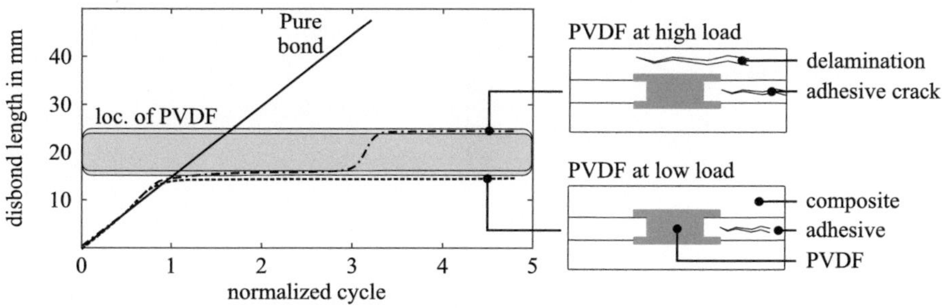

Fig. 7.6 Principle effect of PVDF disbond-arrest feature on disbond growth in CLS specimen measured by means of microscopes at the specimen edge. (Adapted from Löbel (2016))

2.3 Design Methodology

The arrested-growth approach to damage-tolerant bonded joints using DAF provides the designer with an additional design freedom that cannot be treated with classical analytical or stress-based design methods. It requires the assessment of critical disbond sizes, shapes, and locations with respect to disbond growth and disbond arrest under fatigue loading. Furthermore, the residual strength of the structure must be proven. The finite element method is an ideal tool to analyze this complexity.

Sachse et al. (2017) proposed a method specifically designed to simulate disbond growth under high-cycle fatigue loading in the presence of DAF. It is based on the concept of cohesive zone modeling first proposed by Barenblatt (1962), in which stresses acting on a cohesive surface are related to the surface separation. The model is extended to represent fatigue disbond growth, calculating fatigue damage rate by means of Paris' law. A load envelope approach is adopted, in which a constant fatigue load is applied to the model and loading cycles are represented by an internal fatigue frequency correlating simulation time to fatigue cycles. Key features of the material model are summarized as follows:

- Trilinear traction-separation law: to model adhesive plasticity
- Mode-dependent fracture toughness: to model static disbond growth
- Mode-dependent Paris' law: to model fatigue disbond growth
- Crack tip degradation approach: to model complex arbitrary crack shapes
- Adaptive numerical frequency: to efficiently model high-cycle fatigue

Figure 7.7 compares numerical and experimental results of fatigue disbond growth in a CLS specimen with a rivetless nut plate joint installed. The model closely predicts disbond arrest, as well as the two-dimensional disbond shape. By analyzing the adhesive load history during fatigue damage, it is possible to visualize and quantify the localization effect associated with the DAF. The results closely correlate to the disbond shape (Fig. 7.5). Similar plots were created for other DAF and can be used in a design handbook for arrested-growth design of damage-tolerant bonded joints.

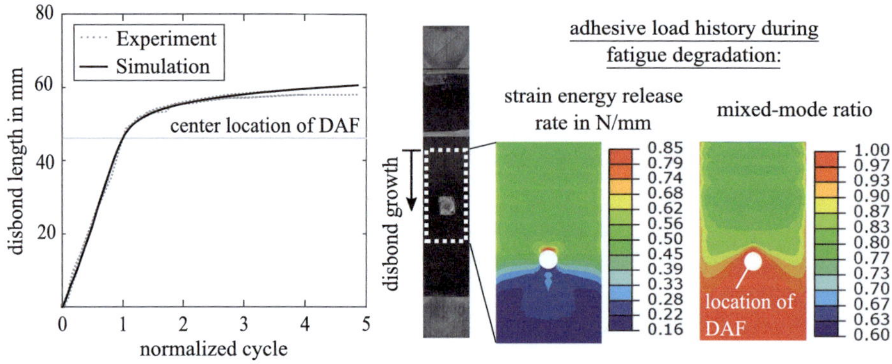

Fig. 7.7 Exemplary simulation results of fatigue disbond growth in CLS specimen with a rivetless nut plate joint installed: one-dimensional disbond growth (left); adhesive load history (right). (Adapted from Sachse et al. (2017))

2.4 *Application to High-Load Transfer Joints*

While the CLS specimen represents an ideal joint geometry for the characterization of DAF, only low-load transfer found in stringer-skin connections can be represented correctly. High-load transfer joints, e.g., fuselage skin connections, represent the more interesting business case for the arrested-growth design of damage-tolerant bonded joints.

Among others, Kruse et al. (2018) therefore developed a representative joint configuration, named wide single lap shear (WSLS) specimen. The test setup is shown in Fig. 7.8 and relates to the illustration of the concept of DAF presented in Fig. 7.2. Each bonded panel had a width and length of 500 mm with an overlap length of 60 mm. An initial disbond was introduced by means of a square PTFE strip at the center of the joint. Two types of DAF were investigated: (1) lock bolts placed next to the initial disbond and (2) thermoplastic strips located along the joint's edge. The typical disbond growth behavior for the configurations investigated is depicted in Fig. 7.8. The authors used the disbond area as metric to compare two-dimensional disbond growth. The disbond starts to grow from the corners of the initial disbond. It then progresses in a curved shape until final rupture. Lock bolts are able to significantly slow down disbond growth and eventually arrest it in the area between the lock bolts. Thermoplastic strips are also able to slow down disbond growth rate significantly; however, the disbond shape seems unaffected, and no full arrest can be observed. Reaching a critical disbond area, disbond growth accelerates until final rupture of the joint.

The authors (Kruse et al. 2018) remark that the number of tested specimen is still too low to draw a final conclusion, but test results indicate that lock bolts can be efficiently applied as disbond-arrest features in high-load transfer joints.

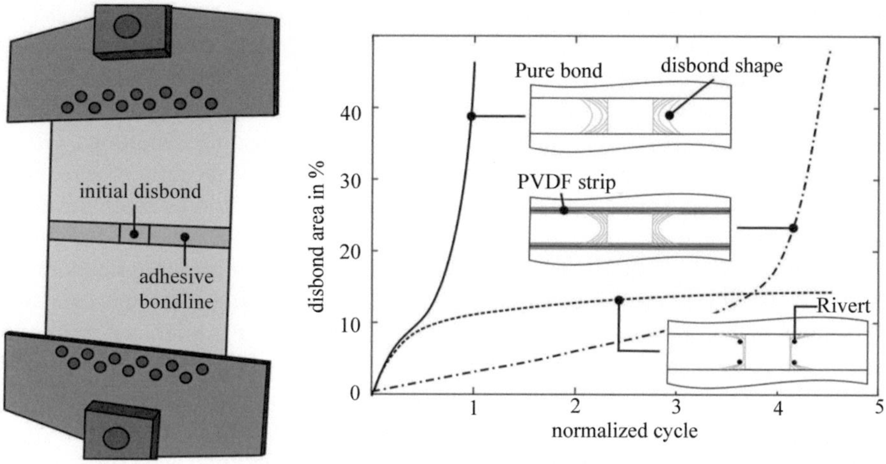

Fig. 7.8 Test setup for wide single lap shear (WSLS) specimen (left); principle effect of different disbond-arrest feature configurations on disbond growth in WSLS specimen measured by means of air-coupled ultrasonic scanning device. (Adapted from Kruse et al. (2018)) (right)

2.5 Concluding Remarks

Adhesive bonding can be considered an enabling technology for the integral design of composite aircraft structures. To ensure damage tolerance, the concept of disbond-arrest features was developed and validated on laboratory scale as well as structural level. Out of many different configurations, through-the-thickness reinforcement joints, as well as the hybrid bondline concept, have shown the greatest potential to securely arrest disbond growth. A design methodology was proposed to efficiently integrate the concept into an industrial development environment. By combining adhesive bonding and dedicated DAF, a multifunctional design solution can be obtained which potentially reduces part count significantly and leads to a faster and more efficient manufacturing processes.

3 Damage-Tolerant Design of Integral Stringer-Skin Structures: Concept of Selective Stitching

Composite materials offer the possibility of designing and manufacturing integral structures, reducing the number of parts, and enabling cost and weight savings. However, complex integral structures necessarily involve interfaces between subparts which are subjected to peak loads and stress concentrations. On a representative aerospace fuselage structure, such interfaces appear, for example, between the fuselage skin and the reinforcing stringers and frames. In a state-of-the-art carbon fiber reinforced plastic (CFRP) structure, the interface between the said subparts is typically a layer of the epoxy matrix, equivalent to the interface between any other two layers of carbon fiber.

Without additional reinforcement, stress concentrations in subpart interfaces generate cracks that grow and propagate under nominal loading conditions. Given the strict crack-growth-related regulations involved in the certification of aerospace structures, the design of integral CFRP structures requires additional mechanisms to achieve compliance.

Structural stitching offers the possibility of introducing reinforcing fibers in the thickness direction to connect laminate layers. The reinforcing stitching seams not only show disbond-arrest features (DAF) but also increase the tolerance to impact damage of a laminate. Furthermore, the selective stitching approach focuses on the application of structural stitching only to areas with extraordinary stress configurations, such as the aforementioned interfaces between subparts of an integral composite structure, thereby reducing the overall manufacturing time and cost.

As an additional technology that introduces DAF (see Sect. 2.2), structural stitching aspires to forward the certification efforts that involve the no-growth approach. This section provides a selection of research that advances the understanding of selective stitching and its success as a viable design approach for integral composite structures.

3.1 Stitching Technologies Within the Composite Manufacturing Process

With the rise of out-of-autoclave composite manufacturing processes and the use of the preforming manufacturing route over prepreg solutions, multiple textile processes have been adapted to the composite production plans. Stitching is one of these technologies imported from the textile industry, and composite manufacturing has benefited from the extensive knowledge base available.

In the context of an FRP laminate that has been produced by stacking dry fiber textile preforms and applying one of the available liquid composite molding (LCM) methods to add the matrix material, stitching offers the possibility of introducing through-the-thickness reinforcement (TTTR) fibers that connect the plies of the laminate. As such, stitching represents an additional step in the production chain. In addition, the stitching techniques used for composite structures need to meet the industry's requirements. Apart from fulfilling the appropriate compliance and certification requirements, stitching of lightweight structures must take into consideration their 3D character and allow seams to be placed along complex 3D paths, often on geometries with reduced accessibility.

In order to address those challenges, one-sided stitching techniques were developed. These, unlike the conventional double-lock stitching technique, only require access to one side of the stitched preform and are designed as stitching heads which can be mounted as end effectors on a robot arm. The three most common one-sided techniques are shown in Fig. 7.9, along with their seam geometries.

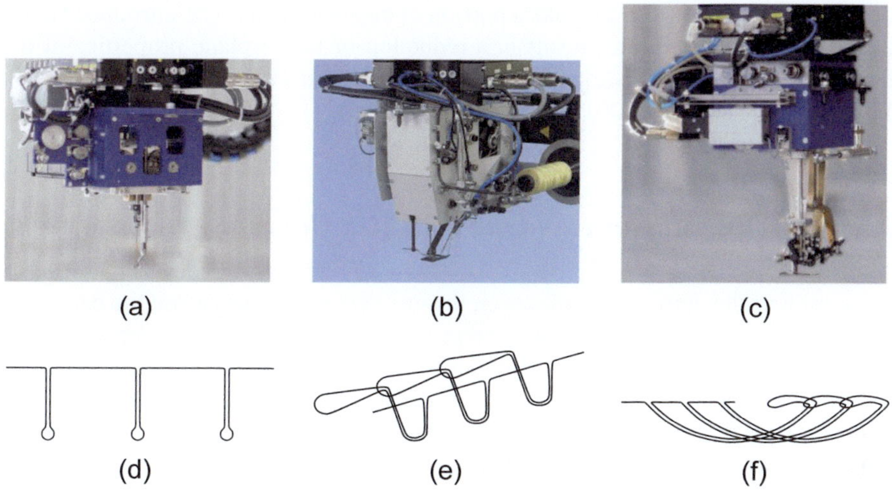

Fig. 7.9 Stitching techniques and their corresponding seam geometries. (**a**) Tufting head (Tufting Head Datasheet 2016). (**b**) Two-needle head (Two Needle Head RS 530 2016). (**c**) Blind-stitching head (RS 510 Blind-Stitch head 2016). (**d**) Tufting seam. (**e**) Two-needle seam. (**f**) Blind-stitching seam

Tufting is a very simple technique that merely introduces a double thread into the preform. This technique requires no thread tension during the process and ensures the fixation of the thread through friction with the stitched textile. Both the two-needle and the blind-stitching techniques require thread tension to form a seam with heavy interlacing. The former introduces the thread in the material with the oblique needle and pulls it back up with the vertical hook. The latter uses a curved needle that introduces the thread in the material and brings it back up to the top surface. The blind-stitching head then uses a catcher lever to hold the loop in place and interlace it with the next stitch.

3.2 The Influence of Stitching Reinforcement on the Mechanical Properties of a Laminate

What these stitching techniques have in common is that they introduce local reinforcement that connects the layers of the textile preform. In doing so, stitching reinforcement greatly improves the interlaminar or out-of-plane properties of a laminate. Representative parameters like the interlaminar strength, or the delaminated area after impact, are dominated in unstitched laminates by the properties of the matrix and are noticeably enhanced by the introduction of TTTR.

Nevertheless, the manipulation of the preform, the repeated introduction of the stitching needle, and the tension on the stitching thread (when applicable) are factors that damage the textile fiber preform. For every stitching cycle, the needle displaces fiber bundles and breaks a portion of their filaments. The introduced damage and the displaced tows result in a reduction of the in-plane properties of the laminate, as reported by various authors (Jain and Mai 1997; Karahan et al. 2010; Weimer and Mitschang 2001; Yudhanto et al. 2015).

3.3 *Selective Stitching and the PRSEUS Concept*

As a result of the detrimental effects on the in-plane performance and the beneficial increase of the out-of-plane mechanical properties, the stitching of composite laminates poses an optimization problem that is discussed under the concept "selective stitching." A selective stitching configuration therefore aims at maximizing the benefits of stitching in terms of increased damage tolerance and interlaminar strength while keeping the drawback of in-plane properties to a minimum.

The NASA-led project PRSEUS investigated the effects of structural stitching at panel level. A full-scale fuselage panel was manufactured and tested, in which crack-arresting and crack-turning features had been implemented. Figure 7.10 shows the elements and techniques that have been integrated in the PRSEUS concept.

The stringers in this panel consist of pre-cured pultruded rods, and the frames have a foam core over which the carbon fiber stacks are draped. These frames also contain a keyhole opening at the sections where they are connected to the stringers.

Both the stringers and the frames are stitched to the skin using the two-needle technique, and the positions of the stitching seams are marked in Fig. 7.10 with red dashed arrows.

During the final testing phase, a 200 mm-long notch was mechanized across the central stiffener and through the complete thickness. The damaged panel was loaded axially until failure while monitoring crack propagation stemming from the central notch (see Fig. 7.11).

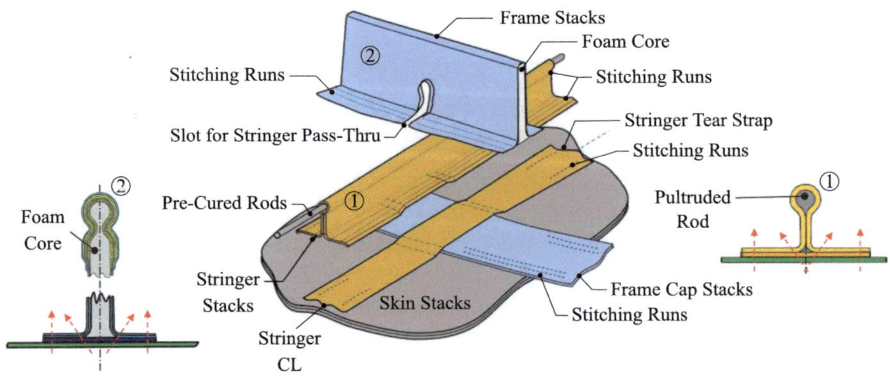

Fig. 7.10 Exploded view of the PRSEUS concept. (Adapted from Bergan et al. (2014) and Velicki and Jegley (2011))

Fig. 7.11 Damage propagation in the PRSEUS panel (Velicki and Jegley 2011)

The first image shows an early stage of the axial loading, where initial cracks propagate from the slot corners, at 45° to the load direction. At an intermediate loading stage, these cracks have reached the reinforcing seams along the stringer feet and have changed direction, continuing to propagate in the direction of the stringer foot edge. As the cracks continue to grow in the loading direction, they eventually reach the feet of the 90 frames. At this point, crack growth is halted again before final failure.

PRSEUS is therefore an example of how structural selective stitching can contain damage growth and alter the failure mechanisms involved in a full-scale fuselage panel.

3.4 Damage-Tolerance Enhancement

The PRSEUS concept described in the previous section is a useful and ambitious proof of concept for the capabilities of selective stitching. The large size of the tested panels and their complex geometry, however, involve high manufacturing and testing costs. As a consequence, the number of selective stitching configurations that can be tested has to be drastically reduced, which in turn limits the understanding of the mechanisms involved.

In order to overcome this, tests can be carried out along the so-called testing pyramid (see Fig. 7.12). In it, test specimens are classified according to their size and geometrical complexity. Tests at the bottom of the pyramid are simple to manufacture and test, allowing a wide range of stitching configurations to be tested at an affordable cost. As the complexity increases, the knowledge gained in the lower levels of the pyramid can be used to shed sub-optimal stitching configurations and reduce the number of tested specimens.

Using a systematic approach to the optimization of selective stitching configurations, the effects of a broad range of them on the damage tolerance of composite

Fig. 7.12 Testing pyramid. (Partially adapted from Herzog (2016))

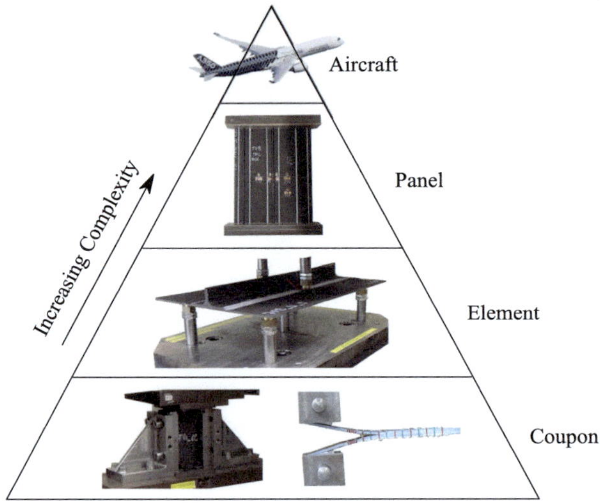

laminates can be studied at coupon level. The efficiency of stitching to improve damage tolerance of composite structures, especially for barely visible impact damage (BVID), has been widely documented in the corresponding literature (Dell'Anno 2007; Erber et al. 2009; Scarponi et al. 2007; Tan et al. 2011).

The compression after impact (CAI) testing procedure, described in the Airbus Norm AITM1-0010 (Airbus Industrie Test Method 2005), enables the evaluation and comparison of techniques designed to enhance the tolerance to impact damage of a laminate. This procedure takes a flat 150 mm/100 mm coupon and measures its residual compression strength after being subjected to impact damage. During the compression test, the vertical sides of the specimen are simply supported to avoid buckling failure modes (see Fig. 7.13a).

Figure 7.13f shows, for example, the effect of tufting on the residual strength of CAI specimens.[1] Figure 7.13b–e show the front and back sides of a tufted and a blind-stitched CAI specimen.

The residual strength of stitched specimens is normalized with the average residual strength of unstitched specimens and represented over the reinforcement density (R_D) used in each stitching configuration. R_D is defined as the weight of structural reinforcement thread introduced in the thickness direction per areal unit.

The design parameter R_D enables a straightforward comparison of selective stitching configurations that involve different stitching techniques and reinforcement materials. It allows, for example, to fairly compare the effects of a carbon fiber

[1] Where "U" stands for unstitched, "T" for tufting, "BS" for blind stitching, and "L", "M," and "H" for low, medium, and high stitching densities, respectively. The stitching thread materials used are represented by the letters "A" (aramid) and "C" (carbon).

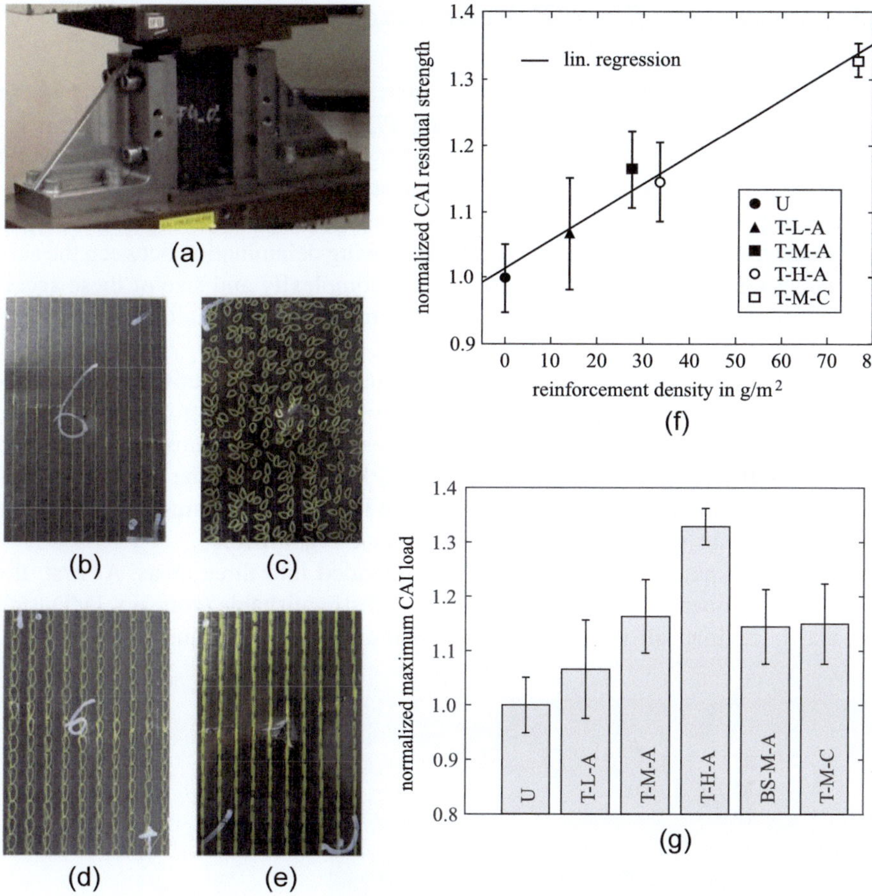

Fig. 7.13 CAI testing. (**a**) CAI testing setup. (**b**) T-L-A: front. (**c**) T-L-A: back. (**d**) BS-M-A: front. (**e**) BS-M-A: back. (**f**) CAI residual strength for tufting configurations. (**g**) Normalized CAI loads

and an aramid thread together in Fig. 7.13f even if the linear densities of these off-the-shelf threads are not exactly the same. Thus, it is possible to plot these results together and extract useful design guidelines, such as the remarkable linear approximation that can be inferred between the normalized CAI residual strength and the R_D.

Furthermore, efforts have been made to evaluate the efficiency of different parameter combinations, including additional stitching techniques, stitching densities, and thread materials. Figure 7.13g shows the normalized CAI loads for different stitching configurations.

3.5 Application: Seven-Point Bending (7PB) Test

A novel testing procedure was developed to evaluate the integrity of stringer-skin composite joints: the seven-point bending (7PB) test. In this test, a stringer-reinforced skin panel is placed on five supporting points while two load introduction points press on the skin, detaching it from the stringer foot. See Fig. 7.14a.

Selective stitching configurations can be studied using this test method in order to evaluate their efficiency in containing the growing delamination between the skin and the stringer foot. The greater geometrical complexity and size of these specimens place them one level above the CAI specimens described in the previous section, i.e., the structural element level (see Fig. 7.12).

Figure 7.14d shows the load-displacement curves of two selective stitching configurations and a reference unstitched category (7PB U). A tufting configuration using a carbon fiber thread (7PB T-M-C) contains two tufting seams along each side of the stringer foot (see Fig. 7.14b). Category 7PB 2N-M-A, on the other hand, uses the two-needle technique and an aramid thread to stitch two seams connecting the skin and the T-stringer under the stringer web (see Fig. 7.14c).

The load-displacement curves can be subdivided into three areas. At first, the load increases linearly, and all specimens are indistinguishable (zone no. 1). During this elastic regime, all the load introduced into the skin is transmitted onto the

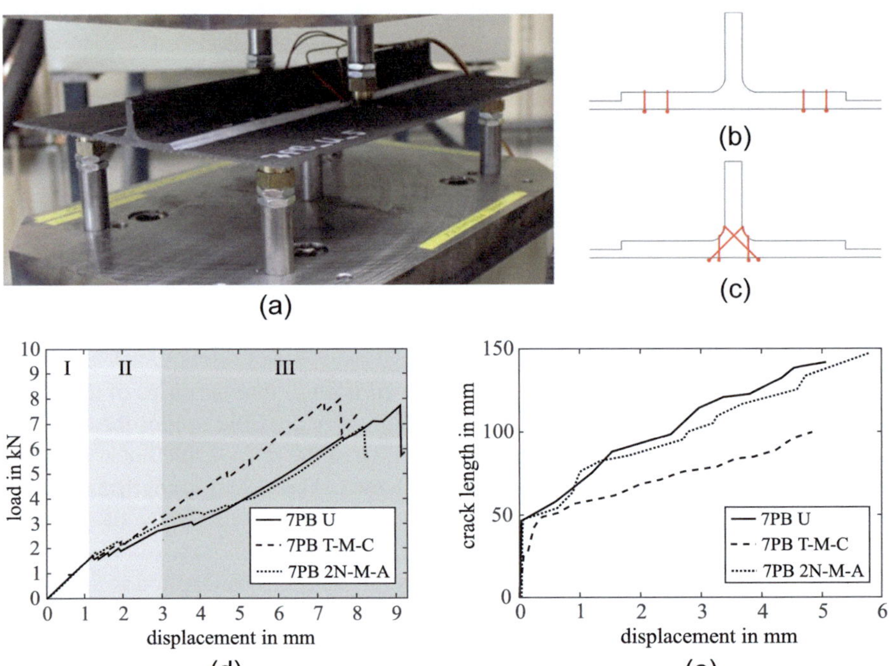

Fig. 7.14 Seven-point bending testing. (**a**) 7PB testing setup. (**b**) T-M-C stitching configuration. (**c**) 2N-M-A stitching configuration. (**d**) 7PB load-displacement curves. (**e**) Crack growth along the stringer foot

stringer through the stringer-skin interface. In zone no. 2, the curves become jagged but remain roughly together. In this section of the plot, micro-cracks begin to appear on the edge of the stringer foot and begin to coalesce, forming visible cracks that start to propagate. The final zone no. 3 begins with the separation of the 7PB T-M-C curves from those belonging to the two other categories. The load increase rate remains constant for the former, while there is a significant reduction of it for the latter. This can be explained by the fact that the tufting seams along the stringer foot edge are able to contain the crack growth toward the middle of the stringer.

The crack growth along the stringer foot edge was also monitored during the test, and the corresponding results are plotted in Fig. 7.14e. Both the unstitched and the 2N-M-A configurations exhibit very similar behaviors, signaling that stringer foot edge lies outside the sphere of influence of the central 2N-M-A seams. The T-M-C configuration, on the other hand, is capable of reducing the growth rate and the extent of the crack.

The 7PB test can therefore be successfully used, similarly to the PRSEUS testing procedure, to assess the efficiency of a selective stitching configuration to contain damage growth in a stringer-skin interface. Beyond the structural detail level of the testing pyramid, additional testing campaigns have been carried out at panel level comparable to PRSEUS (see Sect. 3.3 and Fig. 7.12). In doing so, the investigation and selective reinforcement of the stringer-skin interfaces of multiple-stringer panels have shown that the compression and shear residual strengths can be improved. Furthermore, the buckling modes of said panels have been successfully modified through targeted selective stitching configurations (Fernandez et al. 2017).

3.6 Concluding Remarks

Selective stitching is an efficient method to introduce through-the-thickness reinforcement in a composite laminate. Using these techniques, promising results have been obtained in terms of damage-tolerance improvement and crack-arresting capabilities. Furthermore, proof-of-concept experiments, such as PRSEUS, have been carried out in order to validate the selective stitching approach at a complex structural level. As a result of this empirical research, useful design guidelines have been established, and a deeper understanding of the reinforcing mechanisms has been gained. This knowledge base can now be implemented in computational models which enable an even more refined optimization of selective stitching configurations.

4 Multifunctional Sandwich Structures

Sandwich structures enable the design and construction of predominantly laminar aircraft components with unmatched lightweight performance. Compared to monolithic concepts, sandwich parts can help to save weight by providing panels with high specific strengths and stiffnesses.

By nature of its multilayered, nonuniform structure, manufacturing, integration, and maintenance are usually more complex than for monolithic counterparts. In contrast to general aviation, where sandwich is commonly used in all parts of smaller aircraft, this complexity and the associated overhead for testing and certification currently limit the otherwise desirable use of sandwich in commercial aircraft to quite specific areas. By far the most use of sandwich is seen in aircraft interiors, where nearly all larger surfaces in the cabin, including floors, linings, and monuments, are executed as sandwich. Here, the sandwich concept enables lightweight structures under fairly controlled ambient conditions and mostly in noncritical parts. On the structural side, sandwich parts are applied in fairings, flaps, rudders, and other non-primary parts. Again, this limitation is imposed because due to their more complex structure, sandwich components using state-of-the-art material concepts seem to be more difficult to maintain and check for integrity in mission-critical parts. More difficult does not mean impossible, though, and especially in helicopter blades, sandwich sees widespread application in absolutely mission-critical parts, meeting highest requirements for performance and reliability. There are also some notable exceptions to the application of sandwich for primary structures in smaller business aircraft: the Beechcraft Starship features both fuselage and wing parts as sandwich structures, which seem to stand up formidably against fatigue and aging (Tomblin et al. 2018), and the much more recent HondaJet uses sandwich panels for front and aft fuselage sections (Fujino 2011).

In addition to their structural performance potential, sandwich constructions can incorporate additional primary or secondary functionalities. This is the upside of the combination of different materials with drastically different densities and different physical and mechanical properties. Figures 7.15 and 7.16 demonstrate the potential of the application of sandwich in comparison with state-of-the-art monolithic/differential designs for the case of a sandwich fuselage. Design drivers for this concept included a drastic reduction of the number of frames and sections and the elimination of stringers altogether. These were enabled by the much higher, uniform bending stiffness of sandwich panels and a considerable increase of thermal insulation in comparison with standard aluminum panels. These drivers not only reduced the mass of the fuselage system but also lowered manufacturing cost due to a drastically reduced number of parts and especially the reduced overhead associated with assembly, riveting, sealing, and bonding steps.

4.1 State of the Art in Multifunctional Sandwich Structures

Sandwich parts are currently predominantly chosen for their lightweight potential but usually fulfill more than just a purely mechanical function. A very straightforward passive functionality is heat insulation, which is inherent to most sandwich core materials due to their low density, which typically ranges from about 25 to 120 kg/m^3 (Hexcel Composites 2011), and together with materials with low heat conductivity (e.g., foams or aramid composites) results in highly functional insulators. For civil

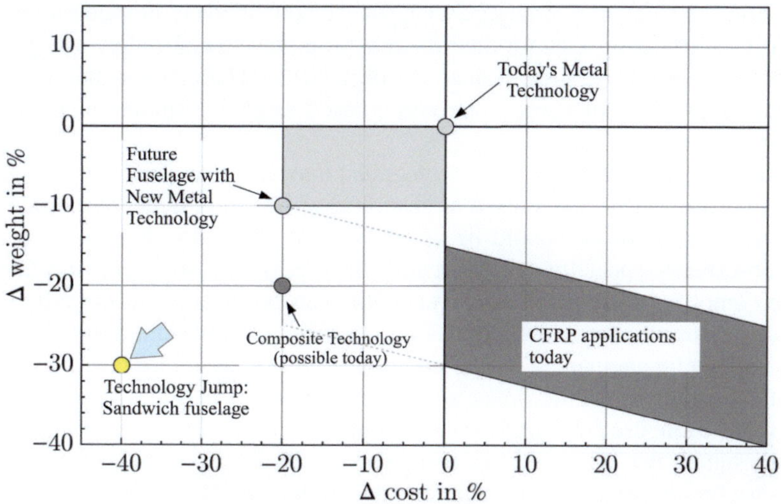

Fig. 7.15 Comparison of weight and cost saving potential for current metal and CFRP technology vs. the Ventable Shear Core (VeSCo) sandwich concept. For the structure, a 30% weight and 40% cost saving potential were projected (Kolesnikow and Herbeck 2004)

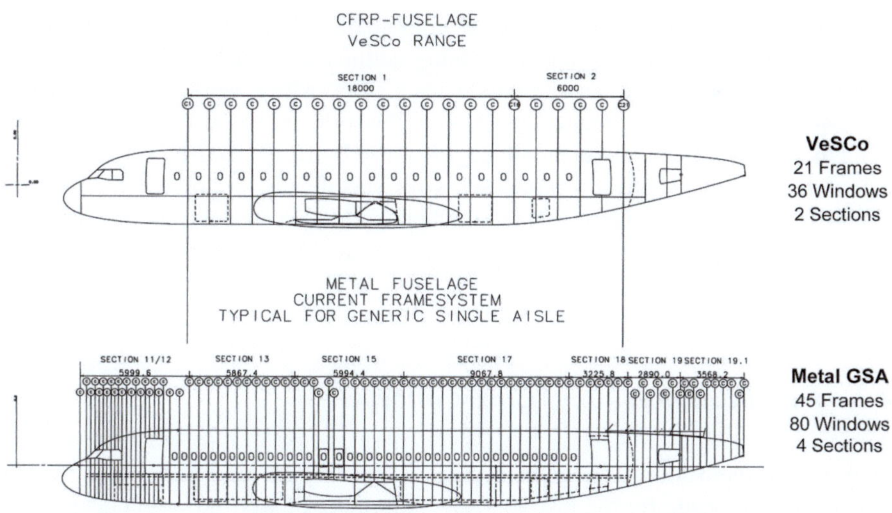

Fig. 7.16 Blueprint of the Ventable Shear Core (VeSCo) concepts in comparison with a standard metal fuselage; use of sandwich panels drastically reduces the number of sections and frames (Klett et al. 2007)

engineering and logistics, some sandwich panel type with foam- or vacuum-insulated cores is even primarily used for their thermal properties to realize low-energy buildings or efficient refrigeration storage (Fischer 2011). Thermal insulation is also a welcome side effect of using honeycomb-based sandwich linings in the aircraft interior.

Another functionality that can be integrated into sandwich parts concerns acoustics. Sandwich panels can be used to absorb acoustic noise, either by porous or by resonance absorption. It should be noted that not all sandwich panels are prime noise absorbers: especially very light and stiff sandwich components can conduct, or even amplify, noise quite efficiently. This can be the case in aircraft linings, which by themselves are excellent studies in efficient lightweight design but sometimes need to be augmented by the addition of heavy elastomer layers to provide acceptable acoustics in the noisy aircraft environment. In some cases, sandwich panels are even used as loudspeaker components, amplifying and transducing acoustic signals that are introduced using so-called exciters, replacing the otherwise necessary frame membrane structure of standard loudspeakers.

In general, if lightweight or stiffness performance is not the primary driver of a sandwich application, sound insulation can be reached by using comparably soft and/or porous foam cores with good damping properties, possibly in conjunction with more specifically sound-absorbing (e.g., textile or elastomer) layers.

For a number of special applications though, sandwich can provide both excellent specific structural performance in combination with finely tuned acoustic behavior. This is usually achieved by using cellular core materials, in which the enclosed cell volumes act as Helmholtz resonators, which work well for usually narrow-banded, specific frequency ranges. This principle is used extensively in modern aircraft nacelles to attenuate engine noise by using acoustic liners consisting of honeycomb sandwich structures with specifically perforated face sheets (Barbosa et al. 2012; Kempton 2011). Another application area of resonance absorbers is interior design, where the acoustic properties of rooms, offices, and other venues are tailored.

4.2 *New Multifunctional Core Concepts*

In terms of innovation, sandwich design—especially in aerospace—has overall remained fairly static for the last 30 years. A number of well-proven concepts and materials and their combination have been around for a long time, and the conservative nature of aerospace material development together with the demanding requirements on new materials did not overly encourage innovation. Progress was mostly incremental and focused on the processing side to reduce manufacturing costs.

One project that prompted renewed efforts to provide truly innovative sandwich concepts was the development of the Ventable Shear Core (VeSCo) next-generation single-aisle aircraft sandwich fuselage as a key component to save both weight and cost (see Figs. 7.15 and 7.16) (Ilcewicz et al. 1997; Klett et al. 2007; Kolax 2004;

Yuan et al. 2012). Available and aerospace-proven core materials like honeycombs and polymethacrylimide (PMI) foams did not meet the required specifications like avoidance of long-term moisture accumulation and impact resistance that were deemed essential for the application of sandwich in the fuselage (Heimbs et al. 2008).

During the search for alternatives, folded core structures (also called foldcores) were identified as a novel material with potential for complex, multifunctional applications. Foldcores can be manufactured efficiently from flat material sheets and result in a cellular structure that bear some resemblance with honeycombs but provide a larger design freedom in terms of cell geometry and material, as shown in Fig. 7.17, which demonstrates the flexibility to generate cores with drastically different properties. Figure 7.18 shows the principle and features some hardware developed by the Institute of Aircraft Design, Stuttgart (Klett 2013).

Compared to honeycomb or foam cores, foldcores met the requirements for the fuselage shell concept. Water accumulation caused by diffusion or micro-crack seepage can be easily removed with properly designed unit cell geometries (Fig. 7.19), and near-net shaping of cylindrical panels is easily achieved by tailored folding patterns. Figure 7.20 shows a demonstrator section of the VeSCo fuselage.

Foldcore provides excellent shear performance. Figure 7.21 shows a recent comparison of three core types with a density of 32 kg/m^3. In addition to the highest shear stiffness, the tested foldcore provides a more isotropic behavior when compared to honeycombs and shows promise for new applications that require high specific bending stiffnesses of sandwich components (Grzeschik et al. 2018). In addition, specifically tailored foldcore structures can be used to control failure modes in crash-relevant components (Sturm et al. 2014).

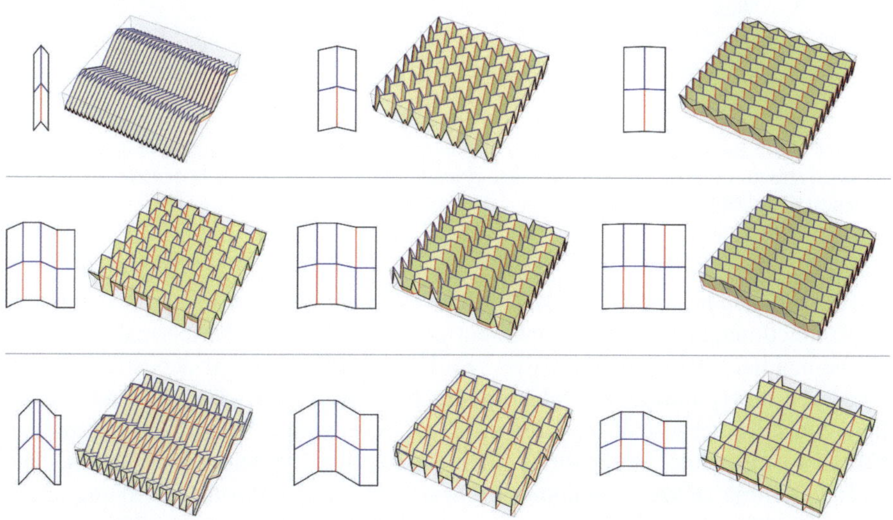

Fig. 7.17 Foldcore unit cell shape variations. All cores feature the same density and overall dimensions but differ drastically in terms of their mechanical and acoustic properties (Klett and Drechsler 2011)

Fig. 7.18 Left: Foldcore manufacturing principle. Right: Aerospace-grade aramid composite foldcore (Klett 2013)

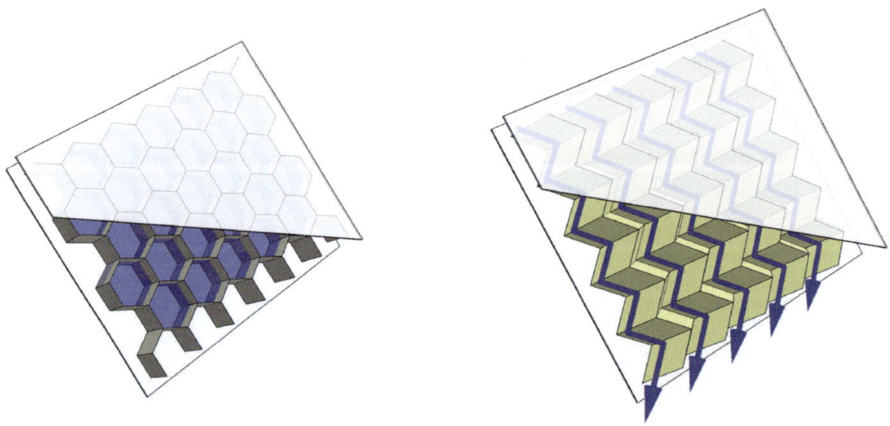

Fig. 7.19 Left: Long-term exposition to moisture together results in accumulation in honeycomb sandwich structures. Foldcore structures can provide inherent open channels that can be used for drainage (Klett 2013)

The next step in sandwich design will be to incorporate not only passive but also active functionalities. Currently inherent properties like passive thermal insulation can be augmented to provide active ventilation and thermal or acoustic management. The newly introduced foldcore structures lend themselves to a number of these concepts, because of the large available, open volume (Fig. 7.19) (Klett et al. 2017a, b). Possible applications include structural heating, ventilation, and air-conditioning (HVAC) components, heat exchanger elements, morphing structures, and tailored aerodynamic and acoustic components (Boermans 2006).

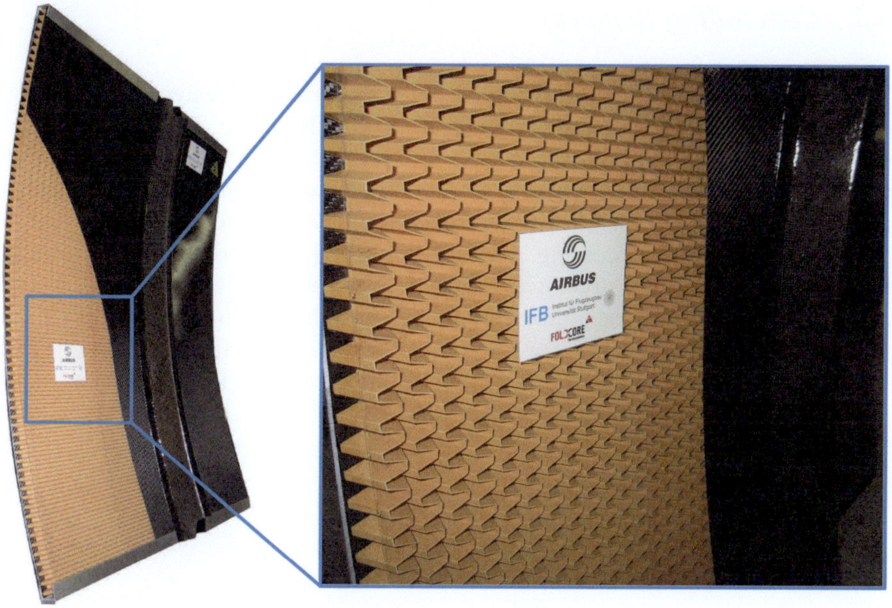

Fig. 7.20 Demonstrator section of the VeSCo shell: CFRP face sheets are combined with an open near-net shape foldcore (Klett 2013)

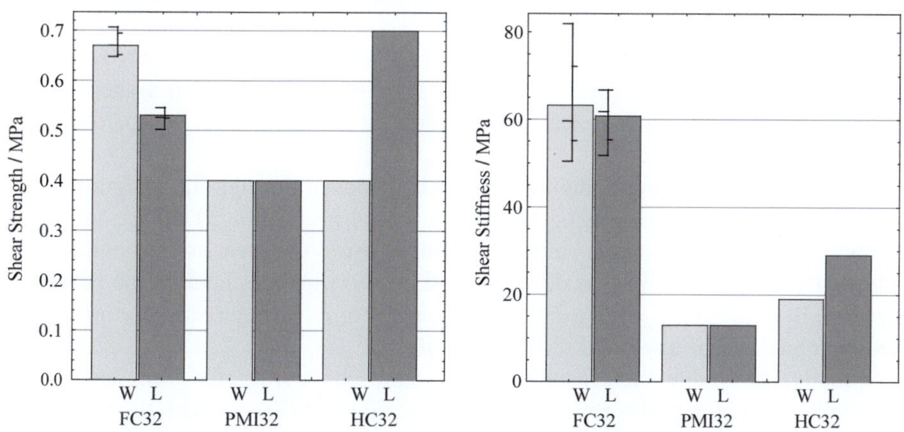

Fig. 7.21 Comparison of transversal strengths and stiffnesses for foldcore (FC32), foam (PMI32), and honeycomb (HC32) cores, measured in two perpendicular directions (W&L). The new foldcore material shows significant potential for shear-centered applications (Grzeschik et al. 2018)

4.3 Concluding Remarks

New development in the area of sandwich cores and sandwich design paradigms can open up new application areas for advanced lightweight design. The introduction of these new materials and concepts in aerospace represents a long-term project which requires close cooperation between academia and industry and also a subsequent adaption of design guidelines and certification standards to incorporate newly gained knowledge on processing, manufacturing, and long-term behavior and reliability of new sandwich structures. Foldcore materials enable the construction of multifunctional, highly integrated components which help to save weight on the system level. The case study for the VeSCo sandwich fuselage concept showed compelling cost and weight saving potential compared to standard metal and composite technologies, and a number of successful follow-up research projects have significantly advanced the state of the art in this field.

5 Conclusions

Rigorous changes to the conventional design philosophy of metallic aircraft structures are necessary to fully exploit the potential of composite materials. In this chapter, three highly promising integral and multifunctional design concepts have been presented, which represent different levels of design maturity and functional integration in composite aircraft structures.

The concept of disbond-arrest features in bonded joints is still closely related to the conventional CFRP fuselage design, addressing certification issues of the bonding process. Stringers as well as skin sections are manufactured separately and joint together thereafter.

A greater level of functional integration is achieved through the integral design of stringer-skin structures using composite textile preforming. Damage-tolerance requirements of the impact-sensitive stringer-skin interface are accounted for through selective stitching, as demonstrated on a component level in the PRSEUS project. By adhesively joining integrally manufactured stringer-skin sections, both technologies can be exploited most efficiently.

Finally, the double-shell fuselage design represents a truly multifunctional solution with great potential for weight and cost savings. In this concept, the part count for the fuselage is reduced drastically, and additional functional integration like active ventilation and thermal or acoustic management is possible by use of new core materials. While sandwich fuselages have already been successfully implemented in general aviation, the realization of an airliner fuselage still represents an extremely ambitious project on many different levels, because the concepts differ drastically from the current state of the art, and like all radically new design paradigms comes with considerable risks and benefits.

References

Airbus Industrie Test Method (2005) Determination of compression strength after impact; AITM1-0010, October 2005

Barbosa AQ, da Silva LFM, Ochsner A, Abenojar J, del Real JC (2012) Influence of the size and amount of cork particles on the impact toughness of a structural adhesive. J Adhes 88(4–6):452–470

Barenblatt GI (1962) The mathematical theory of equilibrium cracks in brittle fracture. In: Dryden HL, von Karman T (eds) Advances in applied mechanics, vol 7 of Advances in applied mechanics. Academic Press, New York, pp 55–129

Bergan A, Bakuckas J, Awerbuch J, Tan T-M (2014) Assessment of damage containment features of a full-scale PRSEUS fuselage panel. Compos Struct 113:174–185

Boermans L (2006) Research on sailplane aerodynamics at delft university of technology. Recent and present developments. In: Netherlands Association of Aeronautical Engineers (NVvL)

BOPACS (2016) Boltless assembling of primary aerospace composite structures: final report

Brooklyn Decker (2017) Airbus A350 XWB fuselage to final assembly. https://amam92.blogspot.com/2017/11/airbus-a350-xwb-fuselage-to-final.html

Dell'Anno G (2007) Effect of tufting on the mechanical behaviour of carbon fabric/epoxy composites. PhD thesis, Cranfield University, Cranfield

Erber A, Witzel V, Drechsler K (2009) Damage tolerance of structural stitched CFRP laminates. In: Deutscher Luft- und Raumfahrtkongress

Erripis IK (2009) X-55 (acca)

European Aviation Safety Agency (n.d.) Amc 20-29: composite aircraft structure

Fernandez D, Begemann B, Middendorf P, Horst P (2017) Untersuchung des Beul- und Tragverhaltens von selektiv genahten, versteiften FKV-Paneelen unter mehraxialer Belastung. In DLRK, Munich

Fischer S (2011) Rechnerische Ermittlung der mechanischen Eigenschaften von Faltkernen. PhD thesis, Universitt Stuttgart

Fujino M (2011) Case study 4 hondajet. Fundamentals of aircraft and airship 616 design, vol 2, pp 615–647

Gibson RF (2010) A review of recent research on mechanics of multifunctional composite materials and structures. Compos Struct 92(12):2793–2810

Grzeschik M, Klett Y, Middendorf P (2018) Reality check—mechanical potential of tessellation-based foldcore materials. In Proceedings of the 7th international meeting on origami in science, mathematics and education

Habenicht G (2009) Kleben: Grundlagen, Technologien, Anwendungen. VDI-Buch. Springer, Berlin, 6., aktualisierte aufl. edition

Heimbs S, Middendorf P, Hampf C, Hähnel F, Wolf K (2008) Aircraft sandwich structures with folded core under impact load. In Proceedings of the 8th international conference on sandwich structures (ICSS 8), vol 3, 05

Herzog J (2016) Airbus A350-941, https://de.m.wikipedia.org/wiki/Datei: Airbus_A350-941_F-WWCF_MSN002_ILA_Berlin_2016_20.jpg

Hexcel Composites (2011) Hexweb acousti-core product data. Technical report

Ilcewicz L, Smith PJ, Hanson CT, Walker TH, Metschan SL, Mabson GE, Wilden KS (1997) Advanced technology composite fuselage: program overview. Technical report NASA-CR-4734, NASA

Jagath Narayana K, Burela RG (2018) A review of recent research on multifunctional composite materials and structures with their applications. Mater Today Proc 5(2):5580–5590

Jain LK, Mai YW (1997) Recent work on stitching of laminated composites—theoretical analysis and experiments. In: Scott ML (ed) Eleventh international conference on composite materials, Gold Coast, Queensland, Australia, 14–18th July 1997, proceedings. Australian Composite Structures Society, Melbourne

Karahan M, Ulcay Y, Eren R, Karahan N, Kaynak G (2010) Investigation into the tensile properties of stitched and unstitched woven aramid/vinyl ester composites. Text Res J 80(10):880–891
Kempton A (2011) Acoustic liners for modern aero-engines. In: 15th CEAS-ASC workshop and 1st scientific workshop of X-noise EV Lausanne, Switzerland
Klett Y (2013) Auslegung multifunktionaler isometrischer Faltstrukturen für den technischen Einsatz. Number ISBN 978-3-8439-1025-5. Institut für Flugzeugbau, Universität Stuttgart
Klett Y, Drechsler K (2011) Designing technical tessellations. In: Wang-Iverson P, Lang RJ, Yim M (eds) Origami[5], number ISBN 1568817142, pp 305–322. CRC Press
Klett Y, Drechsler K, Wentzel H, Kolax M (2007) Design of multifunctional folded core structures for aerospace sandwich applications. In: 1st CEAS European air and space conference
Klett Y, Middendorf P, Sobek W, Haase W, Heidingsfeld M (2017a) Potential of origami-based shell elements as next-generation envelope components. In 2017 IEEE international conference on advanced intelligent mechatronics (AIM), pp 916–920
Klett Y, Zeger C, Middendorf P (2017b) Experimental characterization of pressure loss caused by flow through foldcore sandwich structures. In: ASME 2017 international design engineering technical conference, vol 5B
Kolax M (2004) Concept and technology: advanced composite fuselage structures. JEC Composites, 6/7
Kolesnikow B, Herbeck L (2004) Carbon fiber composite airplane fuselage: concept and analysis. Technical report, Deutsches Zentrums fr Luft- und Raumfahrt (DLR)
Kruse T, Koerwien T, Heckner S, Geistbeck M (2015) Bonding of CFRP primary aerospace structures—crack stopping in composite bonded joints under fatigue. In 20th international conference on composite materials, Copenhagen
Kruse T, Koerwien T, Meer T, Geistbeck M (2018) Certification by means of disbond arrest features and results (EU-FP7 project BOPACS). In: Use of bonded joints in military applications
Liu Z, Bae DHO (2011) A study on joining technology of aluminum alloy sheet using nano-adhesives. Int J Mod Phys B 25(31):4265–4268
Löbel T (2016) Disbond stopping concept: hybrid bondline. PhD thesis, Technische Universität Braunschweig
Löbel T, Kolesnikov B, Scheffler S, Stahl A, Hühne C (2013) Enhanced tensile strength of composite joints by using staple-like pins: working principles and experimental validation. Compos Struct 106:453–460
Nicolais L, Meo M, Milella E (eds) (2011) Composite materials: a vision for the future. Springer, London
Pegorin F, Pingkarawat K, Mouritz AP (2015) Comparative study of the mode i and mode ii delamination fatigue properties of z-pinned aircraft composites. Mater Des 65:139–146
RS 510 Blind-Stitch head (2016). www.ksl-lorsch.de
Sachse R, Pickett AK, Adebahr W, Klein M, Käß M, Middendorf P (2015) Experimental investigation of mechanical fasteners regarding their influence on crack growth in adhesively bonded CFRP-joints subjected to fatigue loading. In 20th international conference on composite materials, Copenhagen
Sachse R, Pickett AK, Middendorf P (2016) Numerical simulation of fatigue crack growth in the adhesive bondline of wide single lap shear specimen. In BOPACS symposium, Hamburg
Sachse R, Pickett AK, Essig W, Middendorf P (2017) Experimental and numerical investigation of the influence of rivetless nut plate joints on fatigue crack growth in adhesively bonded composite joints. Int J Fatigue 105:262–275
Scarponi C, Perillo AM, Cutillo L, Foglio C (2007) Advanced TTT composite materials for aeronautical purposes: compression after impact (CAI) behaviour. Compos Part B 38(2):258–264
Sturm R, Klett Y, Kindervater C, Voggenreiter H (2014) Failure of CFRP airframe sandwich panels under crash-relevant loading conditions. Comp Struct 112:11–21
Tan KT, Watanabe N, Iwahori Y (2011) X-ray radiography and micro-computed tomography examination of damage characteristics in stitched composites subjected to impact loading. Compos Part B 42(4):874–884

Tomblin J, Salah L, Davies C (2018) Aging effects evaluation of a Beechcraft starship main wing. In FAA Joint Advanced Materials & Structures (JAMS) Center of Excellence 6th annual technical review meeting May 19–20, 2010, 08

Tserpes KI, Peikert G, Floros IS (2016) Crack stopping in composite adhesively bonded joints through corrugation. Theor Appl Fract Mech 83:152–157

Tufting Head Datasheet (2016). www.ksl-lorsch.de

Two Needle Head RS 530 (2016). www.ksl-lorsch.de

Velicki A, Jegley D (2011) PRSEUS development for the hybrid wing body aircraft: 100 years of achievement and progress. AIAA Centennial of Naval Aviation Forum, pp 1–11

Weimer C, Mitschang P (2001) Aspects of the stitch formation process on the quality of sewn multi-textile-preforms. Compos A: Appl Sci Manuf 32(10):1477–1484

Yuan C, Bergsma O, Koussios S, Zu L, Beukers A (2012) Optimization of sandwich composites fuselages under flight loads. Appl Compos Mater 19(1):47–64

Yudhanto A, Lubineau G, Ventura IA, Watanabe N, Iwahori Y, Hoshi H (2015) Damage characteristics in 3D stitched composites with various stitch parameters under in-plane tension. Compos A: Appl Sci Manuf 71:17–31

Chapter 8
Nano-enabled Multifunctional Materials: Mechanical Behavior and Multi-scale Modeling

Konstantinos Tserpes and Spiros Pantelakis

1 Introduction

Multifunctional materials (MM) are materials with multiple functions inspired by the human body which consists of organs with multiple functions that respond to the environment. A very good example is the human skin which contains hairs for sensing and protection; nerves for sensing and motion; glands for excreting oils; epidermis for structure and protection; veins and arteries for healing, thermal management, and nutrition; and sweat glands and ports for thermal management.

Structural MM beyond the basic strength and stiffness that typically drive the design of structures possess also other functionalities such as enhanced electrical, magnetic, optical, locomotive, power generative properties. A description of the general multifunctional concept is given in Fig. 8.1. Structural MM are considered as the future materials for aeronautics as they are expected to have a tremendous impact on structural performance by reducing size, weight, cost, power consumption, and complexity while improving efficiency, safety, and versatility (Nemat-Nasser et al. 2005).

Introduction of MM into aircraft structures will, and in some cases already do, allow savings in the number of parts, thus reducing the number of assemblies. Moreover, the effective integration of MM will eliminate the traditional electronic boards, connectors, and bulky cables, thus yielding to major weight and volume savings and increasing the system-level efficiency. The potential advantages of the use of MM are already evident in the unmanned air vehicles (UAVs) for which a considerable increase of flight endurance time is achieved through the combination of structure and batteries realized through the use multifunctional material systems (James et al. 2002).

K. Tserpes (✉) · S. Pantelakis
Laboratory of Technology and Strength of Materials, Department of Mechanical Engineering and Aeronautics, University of Patras, Patras, Greece
e-mail: kitserpes@upatras.gr

S. Pantelakis, K. Tserpes (eds.), *Revolutionizing Aircraft Materials and Processes*, https://doi.org/10.1007/978-3-030-35346-9_8

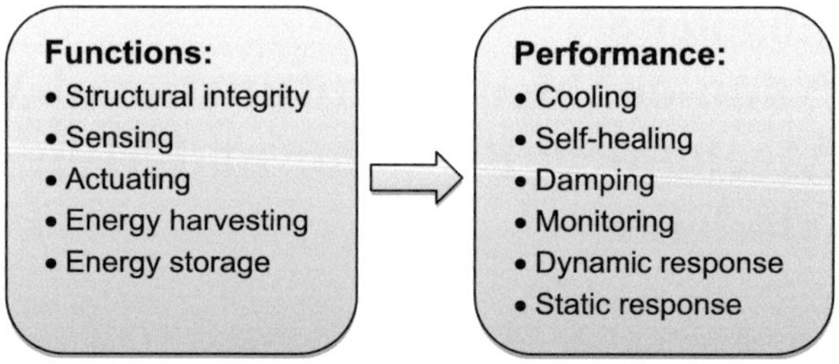

Fig. 8.1 The multifunctional concept

Although a remarkable progress has been achieved in the development and implementation of MM in the last two decades, for most concepts, the technology readiness level remains low (between 3 and 4). Research in the area is intense and is being conducted mainly through manufacturing trials and tests. In this endeavor, models in the simulation-driven design tools and virtual testing and optimization tools can make a significant contribution by reducing the development time and cost.

This chapter is divided into three parts. In the first part, an introduction on MM is given with a focus on aircraft applications. In the second part, the experimentally characterized mechanical behavior of nano-enabled MM is discussed, and in the third part, multi-scale models developed to predict the properties of nano-enabled MM as functions of material and processing parameters are presented.

2 Types of MM

There are many types of MM and many ways to categorize them. According to Ferreira et al. (2016), the most important types are the following:

- Carbon nanomaterials and nano-enabled materials
- Functionally graded materials (FGMs)
- Piezoelectric materials (actuation materials, sensing materials, energy-harvesting/generating materials)
- Shape memory materials
- Others

In the next sections, a brief description of the morphology, the functions, and the performance of the different types of MM are given.

2.1 Carbon Nanomaterials

Carbon nanomaterials are different spatial configurations of the covalent C-C bond. These are fullerenes, carbon nanotubes (CNTs), graphene, graphite, and diamond. Carbon nanomaterials possess excellent mechanical, electrical, and thermal properties owing to the covalent C-C bond (Papanikos et al. 2005; Tserpes et al. 2006;

Tserpes and Papanikos 2007; Tserpes 2011). These properties make them perfect candidates to be used as fillers in order to enhance the properties of matrix materials. Especially, CNTs due to their fiber-like structure have attracted the interest of the scientific community for the last 15 years. In hundreds of works, CNTs have been embedded into polymers to enhance their strength and assign to them electrical conductivity. The first results were not promising due to manufacturing problems. The formation of CNT agglomerates due to van der Waals forces and the creation of a weak interphase between the CNTs and the polymer are two parameters that have been found to dominate the effectiveness of the reinforcement. Similar investigations have been performed on graphene and other carbon nanomaterials such as carbon nanofibers. Details on the mechanical behavior and the multi-scale modeling of carbon nanomaterial-based MM will be given in the chapter's Sects. 3 and 4, respectively.

2.2 *Functionally Graded Materials*

Functionally graded materials (FGMs) are characterized by the variation in composition and structure gradually over volume, resulting in corresponding changes in the properties of the material (Fig. 8.2). The materials can be designed for specific function and applications. FGMs are composite materials at which there isn't a clear interface between the different constituent materials because the change from one material to the other is gradual or graded. For example, a ceramic/metal FGM possesses a high fracture toughness and thermal conductivity on the metal side and a high hardness and thermal stability on the ceramic side (Jha et al. 2013).

FGMs were originally designed for use as thermal barriers in fusion reactors and aerospace structural applications and later on in structural parts at very high operation temperatures. At present, the research on FGMs is focused on the development of FGMs for thermal and mechanical loads, modeling and numerical simulation of FGM parts, fracture properties and crack propagation/analysis, the use of piezoelectric materials and CNTs along with FGMs, and improving and discovering better manufacturing processes. However, the main challenges for FGMs are related to manufacturing. Jha et al. (2013) have reviewed the current state of the art on manufacturing

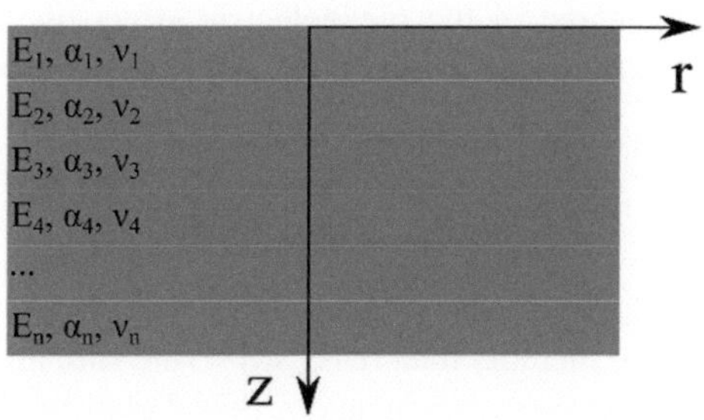

Fig. 8.2 Sketch of a functionally graded material

processes, the current models to estimate the FGM properties, and a compilation of recent research on FG plates and presented the need to focus more effort on improving the current most promising manufacturing method which is solid freeform.

2.3 Piezoelectric Materials

Piezoelectric materials are materials that generate an electric potential when loaded by mechanical stress. This effect, which is called the direct piezoelectric effect, can be used for sensing or energy harvesting. On the other hand, when an electric field is applied, the piezoelectric material will either change shape or generate mechanical stress. This is effect, which is called the inverse piezoelectric effect, which can be used for actuation and shape control. In practice, this means that if we apply a force to a piezoelectric material, it will develop a positive charge on one end and a negative charge on the other end. If we connect the two ends, current flows.

Despite the existence of several active materials such as shape memory alloys, magnetorheological fluids, and electrostrictives, piezoelectric materials remain currently the most widely used active material. This is because they are capable of actuating or sensing at frequencies from about 1 Hz to the MHz range with relatively linear behavior. Moreover, piezoelectric materials have high stiffness which provides them with strong voltage-dependent actuation.

The most basic applications of piezoelectric materials are actuation, sensing, and energy harvesting/generation. Using the inverse piezoelectric effect, it is possible to use piezoelectric materials for actuation. They are commercially available and are currently used for this purpose in a variety of fields including industrial, automotive, medical, aviation, aerospace, and consumer electronics applications. Piezoelectric actuators present one of the few applications of MM in aircrafts. More details about this application will be given in the following section. By using the direct piezoelectric effect, it is possible to use piezoelectric materials as sensors. Piezoelectric sensors are extensively used on structural health monitoring systems, both for sensing vibrations and static strain. So far, most applications have been in the civil engineering field, but there is an increasing interest in aircraft applications. Energy harvesting refers to harnessing energy from the environment and converting it into useful energy. Even though the ambient energy can take many forms, the one usually concerned when dealing with piezoelectric materials is vibration or strain, which takes advantage of the piezoelectric effect of these materials. The use of piezoelectric materials in this field has been a topic extensively investigated.

2.4 Shape Memory Materials

Shape memory materials (SMM) are stimulus-responsive materials that respond to thermal, pressure, and magnetic field. For example, thermo-responsive SMM are responsive to thermal changes. When thermo-responsive shape memory alloys (SMA) are heated above their transformation temperature, their martensitic crystal structure undergoes a change transforming into austenite, and the materials return to their original shape. The most used SMM are SMA and shape memory polymers (SMP). These materials find several applications, but they are mainly used in shape-morphing structures.

2.5 *Others*

Besides the MM that fit into four main categories, there have been proposed many other interesting MM. Table 8.1 lists some of them.

2.5.1 Self-Healing Materials

Self-healing materials (SHM) are materials that are capable of self-repairing damage without human intervention. SHM are classified by two ways: the type of triggering the healing mechanism and whether the healing mechanism is part or not of the material. If the triggering of the healing mechanism needs external stimulus, the SHM are called non-autonomic, and if not, they are called autonomic. On the other hand, when the self-healing process is applied by embedded materials such as microcapsules, the self-healing process is called extrinsic, and if the material itself has the ability to be healed, the self-healing process is called intrinsic. Review papers on SHM have been published by Hager et al. (2010) and Wu et al. (2008), and a relative book has been edited by Ghosh (2009).

Among the SHM that have been studied extensively are the SHM with a glass fiber repair mechanism (Trask et al. 2007; Hayes et al. 2007) (Fig. 8.3), the SHM with a microencapsulated healing agent (Andersson et al. 2007; Guadagno et al.

Table 8.1 Some MM and their functionalities (Data from Ferreira et al. 2016)

Multifunctional material	Functionality
Aptamer based	Use the presence of targeted molecules as a stimulus to trigger further changes to the system
Carbon fiber-reinforced composites for energy storage and sensing	Energy storage
Cement-based and conducting elastomer	Strain sensing
Commercial 3D printing materials	Composites with dynamically controllable surface texture
Copper nano-ink and NiF_2 on PET	Energy harvesting and storing
Foams, fiber-based and sandwich structures	Enhanced mechanical properties
Metallic-intermetallic laminate composites	Tailored composition, physical and mechanical properties
Polyaniline and porous silicon	Conduction of electric current without needing electrically conducting fillers

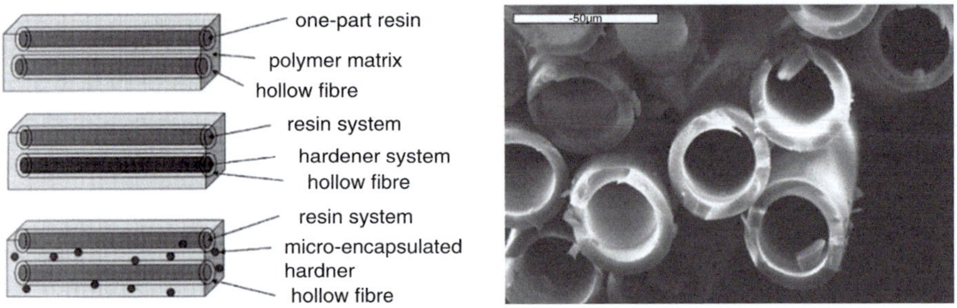

Fig. 8.3 (**a**) Schematic of different hollow fiber self-healing approaches (Hayes et al. 2007); (**b**) Typical hollow glass fiber (35 μm external diameter with 55% hollowness fraction) (Trask et al. 2007)

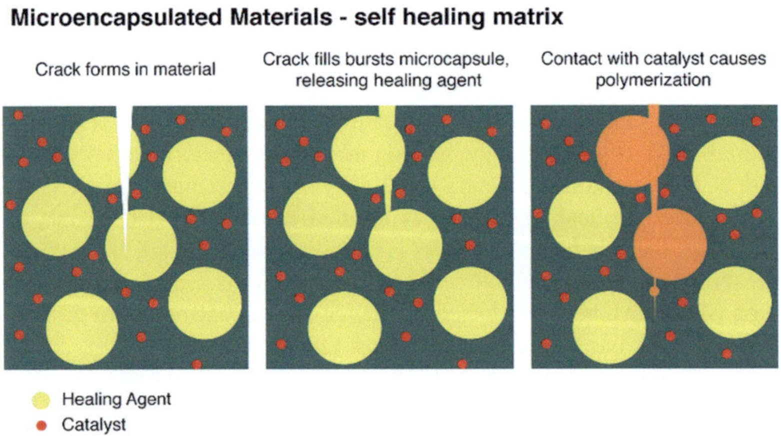

Fig. 8.4 Microencapsulated materials. (From https://bouncingideas.wordpress.com/2012/02/01/self-healing-materials/)

2016, 2017, 2014a, b) (Fig. 8.4), the SHM containing a microvascular network (Toohey et al. 2007), the SHM with a hollow fiber approach (Kousourakis and Mouritz 2010), and the SHM with thermoplastic additives (Varley et al. 2013). Perhaps the most popular technique is the use of microencapsulated healing agents. The mechanical behavior of aeronautical composites containing self-healing microcapsules is discussed in Sect. 5 of the chapter.

3 Aircraft Applications

At present, there are few applications of MM in aircraft structures, most of them at an experimental stage. The applications can be categorized into four technologies (Inman and Johnson 2013):

- The multifunctional spars
- The electronic damping
- The shape-morphing parts
- Others

3.1 Multifunctional Spars

The basic desired functions for multifunctional spars are the energy harvesting to run small electronics, the integrated energy storage, the active gust control, and the active damping. In Wang and Inman (2013), the concept and design of a multifunctional composite sandwich structure for simultaneous energy harvesting and vibration control for small UAV application were examined. In particular, the wing itself is able to harvest energy from normal vibration or sunlight, sense the wind disturbances, and alleviate wind gust by the application of reduced energy control (REC) laws, which is supplied by the available energy harvested from ambient vibration. The proposed multifunctional wing spar is shown in Fig. 8.5.

For the energy harvesting, also characteristic is the work of Thomas and Qidwai on the development of "structure + power" systems for unmanned vehicle applications. The authors have embedded lithium polymer batteries into structural components to carry load and have tested the structures under mechanical load to determine if batteries remain functional. A demonstrator of this technology is the AeroVironment Wasp MAV aerovehicle. In a similar work, Pereira et al. (2006) have embedded novel thin-film lithium batteries into structural composites and tested it under tensile loading. Charge/discharge tests have shown no damage to the cells under loading. Inman et al. (2011) have developed a self-charging structure concept based on the integration of piezoelectric materials and a thin-film battery. Novel thin-film batteries, which are part of the structure, are charged by piezo layers. Flexible batteries are used because they allow load-bearing capability. The aim is to integrate

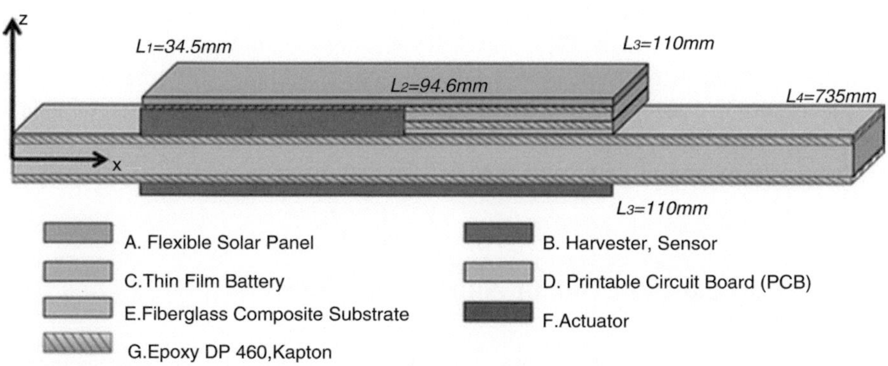

Fig. 8.5 Multifunctional wing spar design showing various functionalities including self-sensing, self-harvesting, self-storage, and self-control (Wang and Inman 2013)

the harvester into the design of the host structure and to eliminate some of the existing components. The concept has been tested with success in a wing spar for UAV applications.

The flight of UAVs because they are small and lightweight can be severely affected by gusts. Systems for gust alleviation in UAVs contain energy harvesting from air flow, energy storage, control, and sensing. The components of such a system include a minimum energy controller and a flexible solar array layer consisting of a harvesting piezoelectric ceramic material, an electronics layer, a battery layer, an actuation piezoelectric ceramic material, and a sensing layer. Research on gust alleviation systems via MM is still conducted in a laboratory environment.

3.2 *Electronic Damping*

Damping behavior of aircraft structures is critical from the skin to engine blades. Electronic damping control via MM aims to compensate for temperature and frequency variations which affect the damping behavior of materials. The developed systems consist of an embedded controller to keep the settling time fixed as temperature and frequency vary. The controllers used are based on flexible electronics. Research is mainly conducted on cantilever beams in a laboratory environment; however, the readiness level of this technology is relatively high. Chen et al. (2006) have successfully installed a piezoelectric composite-based damping system for a vertical fin of the F/A-18 fighter aircraft.

3.3 *Morphing Aerosurfaces*

Shape morphing is one of the most common applications of MM in aircraft structures. Morphing refers to shape changing of aircraft parts which aims to improve flight control, performance, and maneuverability. The MM used in shape-morphing applications are smart SMA, PZT, and SMP. Most of the reported applications concern the morphing wing. Sofla et al. (2010) presented a concise classification on the recent activity in conceptual design, prototype fabrication, and evaluation of shape-morphing wing. The basic functions include wing planform alternation, out-of-plane transformation of wing, and airfoil profile adjustment.

4 Mechanical Behavior of Nanocomposites

Until today, there are no applications of nano-enabled MM on aircraft structures. Only recently, engineers in the UK unveiled the first graphene-skinned aircraft, the "Juno". The main material used in Juno is the graphene-doped prepreg composite material which offers reduced weight, protection against lightning strike, and ice buildup.

Despite the absence of aircraft applications, in Europe, in the last decade, a significant progress has been made on the development of nano-enabled MM through the implementation of several research projects. In this section, a part of that has been performed on the mechanical behavior, and multi-scale modeling of nano-enabled MM in the Laboratory of Technology and Strength of Materials of University of Patras in the frame of several European Union-funded research projects is described.

4.1 Tensile Behavior of MWCNT/PP Nanocomposite

The Plasticyl™ PP2001 material is a commercial nanocomposite consisting of polypropylene reinforced by MWCNTs. It is ideal for standard injection molding and extrusion processes, while it possesses enhanced electrical conductivity at low-loading, steady, and/or improved key mechanical properties and easier processing. In Chanteli and Tserpes (2014), the mechanical behavior of the Plasticyl PP2001 MM has been experimentally characterized using tension tests and SEM tests. The work has been performed in the frame of the SARISTU project. The aim was to correlate the mechanical behavior with the morphological characteristics of the nanocomposite.

The experimental engineering tensile stress-strain curves of the tested materials are depicted in Fig. 8.6. The reference PP material exhibits a nonlinear behavior from the start, and the respective curves show a small scatter. In the nanocomposite, a clear enhancement in the Young's modulus and maximum stress is obtained.

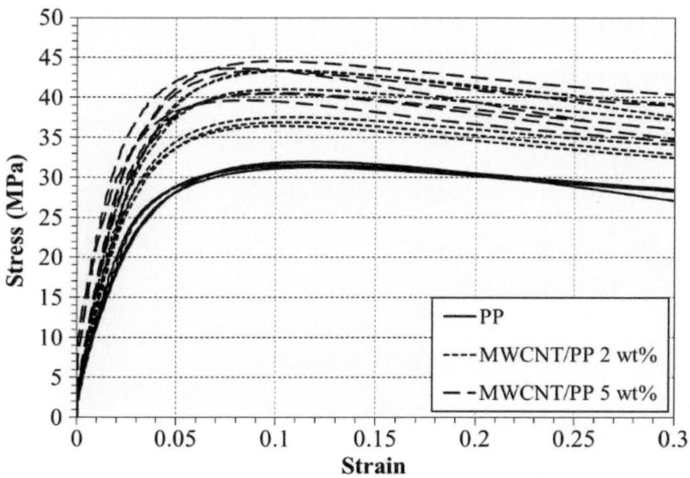

Fig. 8.6 The stress-strain curves measured by the ARAMIS system for all tests

Differentiation of the curves of the MWCNT/PP 2 wt% and 5 wt% materials is not obvious due to the large scatter. Table 8.2 lists the Young's moduli and maximum stress for each specimen. The comparison between the average values of the properties reveals a significant increase in both the Young's modulus and maximum stress of the nanofilled material. Moreover, the comparison between the 2 wt% and 5 wt% materials reveals clearly that the larger content of MWCNTs leads to a larger increase in the properties of the PP material.

The specimens of which the mechanical properties deviate a lot from the average values have been tested by SEM. The SEM images are shown in Fig. 8.7. In Fig. 8.7,

Table 8.2 Measured mechanical properties of the PP and nanocomposite

Specimen	Young's modulus (MPa)	Maximum stress (MPa)	Average Young's modulus (standard deviation)	Average maximum stress (standard deviation)
PP_01	1629.2	–[a]	1641.33 (85.75)	31.55 (0.33)
PP_02	1515.6	–[a]		
PP_03	1592.0	31.22		
PP_04	1767.6	31.40		
PP_05	1687.9	31.98		
PP_06	1657.5	31.61		
MWCNT 2 wt%_01	1885.2	41.06	1974.73 (146.7)	39.80 (3.22)
MWCNT/PP 2 wt%_02	2092.1	43.36		
MWCNT/PP 2 wt%_03	2057.3	43.41		
MWCNT/PP 2 wt%_04	2007.8	36.43		
MWCNT/PP 2 wt%_05	2086.8	36.94		
MWCNT/PP 2 wt%_06	1719.2	37.57		
MWCNT/PP 5 wt%_01	2175.3	44.61	2055.22 (121.82)	42.06 (2.06)
MWCNT/PP 5 wt%_02	1965.8	40.54		
MWCNT/PP 5 wt%_03	1912.4	40.49		
MWCNT/PP 5 wt%_04	2182.9	43.38		
MWCNT/PP 5 wt%_05	2039.7	43.68		
MWCNT/PP 5 wt%_06	2075.2[b]	39.68		

[a]No data are available from the ARAMIS optical system
[b]Estimated from the ARAMIS optical system due to strain gauge failure

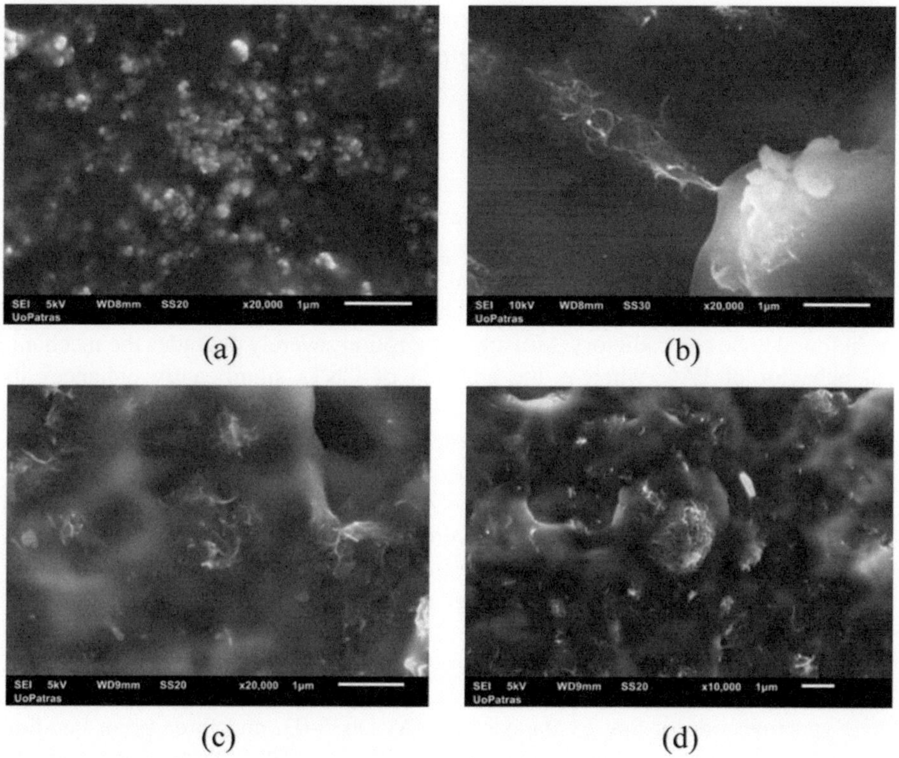

Fig. 8.7 SEM photos from (**a**) the MWCNT/PP 2 wt%_03 specimen (×20,000), (**b**) the MWCNT/PP 2 wt%_04 specimen (×20,000), (**c**) the MWCNT/PP 5 wt%_01 specimen (×20,000), and (**d**) the MWCNT/PP 5 wt%_03 specimen (×10,000)

a good dispersion and wetting of the MWCNTs is observed. These are signs of efficient reinforcement as indicated by the large properties measured for the MWCNT/PP 2 wt%_03 specimen. For the specimen MWCNT/PP 2 wt%_04 (Fig. 8.7), a sparse dispersion of MWCNTs and formation of agglomerations were observed. This observation explains the smaller properties of the specimen MWCNT/PP 2 wt%_04 compared to the MWCNT/PP 2 wt%_03 specimen. In the case of the MWCNT 5 wt% specimens, despite the good dispersion of the MWCNTs (Fig. 8.7), formation of agglomerations is more intense due to the larger MWCNT content. In conclusion, the properties of the specimens are determined by the size of agglomerations. In cases where agglomerations of small size were developed, such as the MWCNT/PP 5 wt%_01 specimen, the measured properties are above average values, while in cases where agglomerations of large size were developed, such as the MWCNT/PP 5 wt%_03 specimen, the measured properties are below the average values.

4.2 *Effect of Hygrothermal Aging on the Tensile Behavior of MWCNT/PA6 Nanocomposite*

4.2.1 Experimental

Polyamide 6 (PA6) is a thermoplastic material with good mechanical strength, high impact strength, good fatigue strength, very good wear resistance, and good sliding properties. However, these properties might be counterbalanced by water absorption as PA6 is a superabsorbent material. The effects of CNTs and hygrothermal aging on the mechanical behavior of PA6 have been studied independently and have been found to be contradictory. Moisture absorption severely degrades the mechanical behavior of PA6, whereas the addition of CNTs significantly enhances its mechanical properties. In Tserpes et al. (2017a, b), the effect of hygrothermal aging on multiwalled CNT (MWCNT)-reinforced PA6 has been experimentally investigated by means of hygrothermal tests, mechanical tests, and SEM tests.

The polymer used in the study is the Akulon® F132-C1 PA6. The MWCNTs were the NC7000 of Nanocyl®. They were used without any further purification. PA6 and MWCNT/PA6 specimens were subjected to the hygrothermal conditions: 25 °C/85%RH (medium temperature, high humidity) and 40 °C/85%RH (high temperature, high humidity) until saturation. In order to study the effect of hygrothermal aging on the mechanical behavior of the MWCNT/PA6 material, tension tests and three-point bending tests were conducted. Tension tests were conducted on dog-bone specimens according to standard ASTM D638-03, and three-point bending tests were conducted on bar specimens according to ASTM D790-03 standard. In addition, the morphology of the MWCNT/PA6 material was characterized by SEM tests conducted after the mechanical tests on selected specimens.

The weight of the specimens subjected to aging was measured periodically until the saturation point. As a measure of the absorbed humidity, the percentage normalized weight gain $M(t)$ was used. Figure 8.8 plots $M(t)$ with regard to exposure time for the different materials. It is observed that the presence of MWCNTs reduces the absorbability of PA6 material. Moreover, the absorbing rate increases significantly with the increase of temperature from 25 to 40 °C for both materials.

4.2.2 Results

The measured Young's modulus of the unaged and aged, neat PA6, and MWCNT/PA6 materials are compared in the diagram of Fig. 8.9. For the neat PA6 material, the standard deviation is rather small, and it is larger for the reinforced material, which is probably due to the variation of reinforcement quality. At RT conditions, the results show an extraordinary increase of Young's modulus of the PA6 material due to the addition of MWCNTs which reaches up to 180%. Aging is causing a decrease in the Young's modulus of both materials. The rate of decrease is larger for the MWCNT/PA6 material. In any case, the presence of MWCNTs compensates the

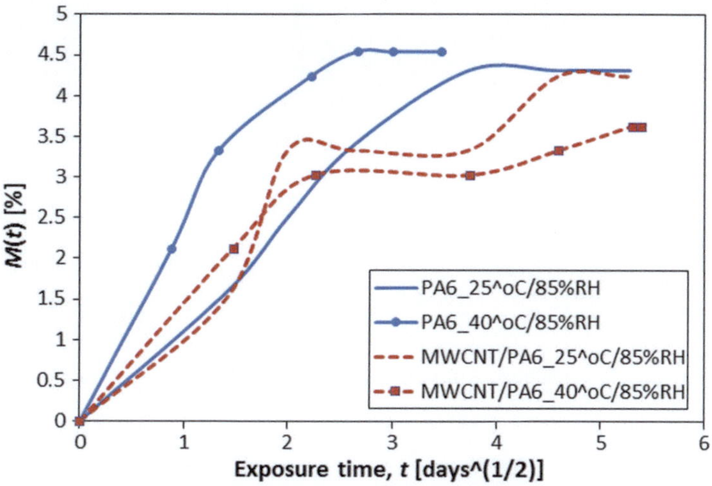

Fig. 8.8 Normalized weight gain $M(t)$ with regard to exposure time t

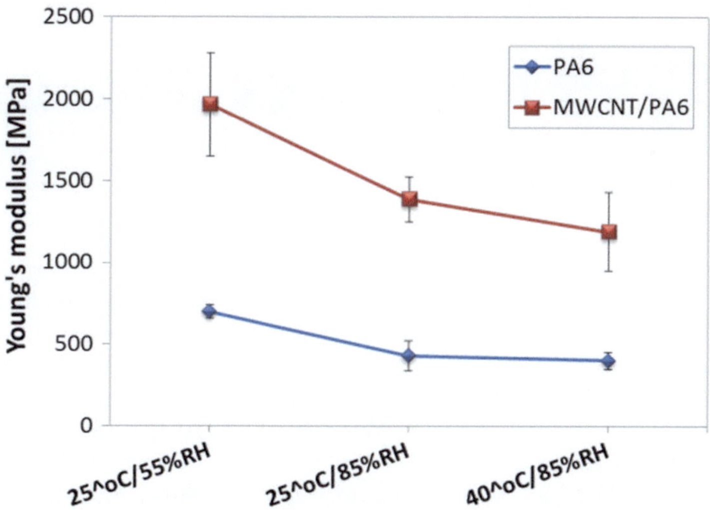

Fig. 8.9 Young's modulus for the different materials under different conditions

negative effect of aging since the Young's modulus of the aged MWCNT/PA6 material is not only higher than that of the aged PA6 material but also even higher than that of the unaged PA6 material.

The flexural modulus and flexural strength are compared for the different materials and environmental conditions in Fig. 8.10, respectively. The flexural modulus and flexural strength of PA6 material at RT conditions increase due to the addition of MWCNTs

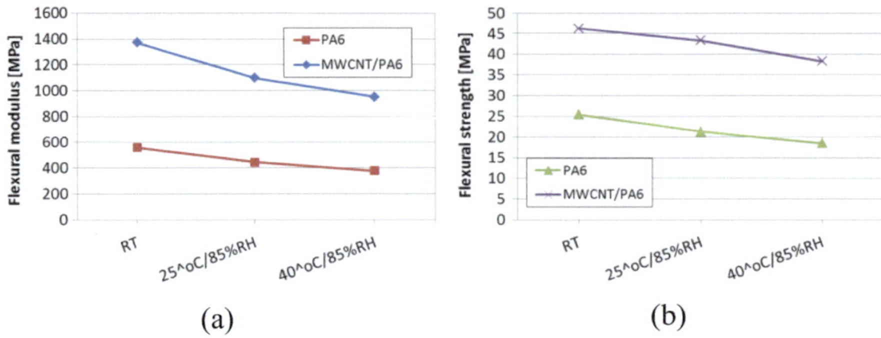

Fig. 8.10 (**a**) Flexural modulus at different conditions and (**b**) flexural strength at different conditions

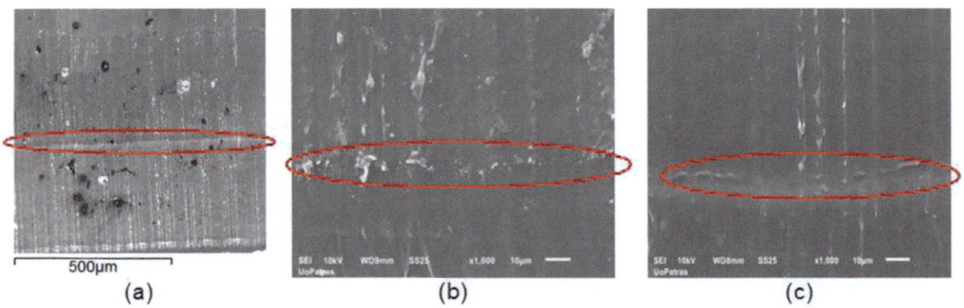

Fig. 8.11 SEM images of PA6 material: (**a**) RT, (**b**) 25 °C/85%RH, and (**c**) 40 °C/85%RH

by 146% and 82%, respectively. Aging causes a degradation of both properties for both materials. The degradation rate is larger for the flexural modulus and almost the same between the two materials. Again, the addition of MWCNTs compensates the negative effect of aging on the flexural properties of the PA6 material since both properties of the aged MWCNT/PA6 material are higher than those of the unaged PA6 material.

Figure 8.11 shows representative SEM images of the PA6 material for the different environmental conditions. The comparison of the images reveals a swelling (indicated areas in Fig. 8.11) of the material due to water absorption. SEM images of the MWCNT/PA6 material, illustrated in Fig. 8.12, show a uniform and dense dispersion of MWCNTs within the PA6 material, which is an indication of good reinforcement quality, and the division of material's surface into two areas of different roughness. The uniform and dense dispersion of MWCNTs is not influenced by water absorption. A closer look at the SEM images of the unaged and aged MWCNT/PA6 specimens, shown in Fig. 8.13, reveals that the pullout length of MWCNTs in the aged sample is significantly larger than in the unaged sample, which is attributed to the reduction of cohesion strength between MWCNT and PA6 due to the absorbed moisture.

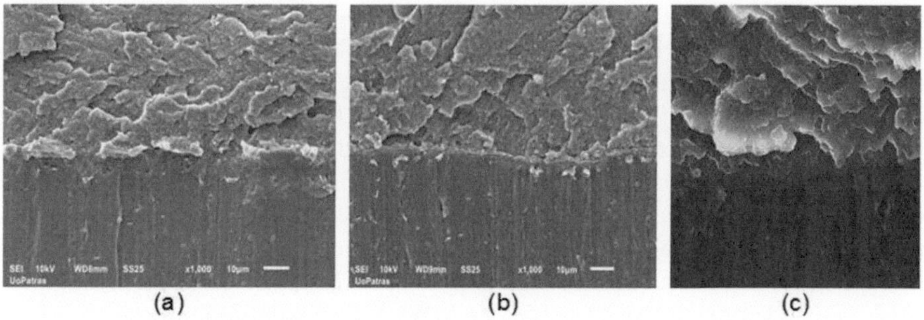

Fig. 8.12 SEM images of MWCNT/PA6 material: (**a**) RT, (**b**) 25 °C/85%RH, and (**c**) 40 °C/85%RH

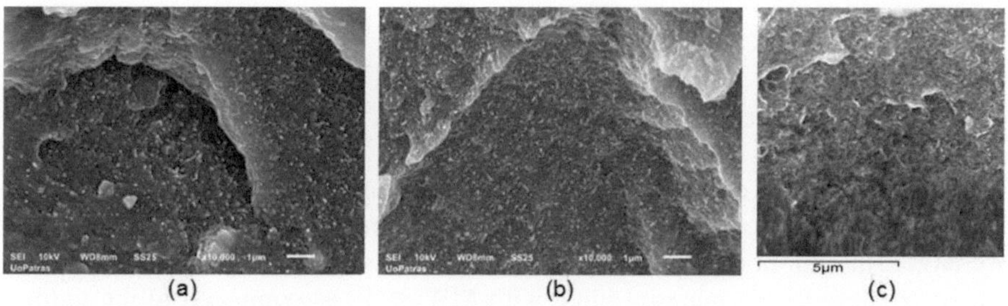

Fig. 8.13 SEM images of MWCNT/PA6 material: (**a**) RT, (**b**) 25 °C/85%RH, and (**c**) 40 °C/85%RH

4.3 MWCNT/GPOSS/RTM6-2 Nanocomposite

In Polydoropoulou et al. (2016), the synergistic influence of MWCNTs and the flame-retardant GPOSS on the mechanical behavior of the polymer RTM6-2 has been evaluated experimentally. The aim was to evaluate whether the optimized electrical conductivity and flame resistance have, as a side effect, a negative impact on the mechanical behavior of the nanocomposite. The epoxy polymers were subjected to tension, compression, flexural, and fracture toughness tests. SEM as well as energy dispersive spectroscopy (EDS) analyses were carried out to support a better understanding of the results obtained from mechanical tests.

4.3.1 Experimental

The material investigated is an epoxy polymer enhanced with 0.5 wt% MWCNTs and 5 wt% viscous liquid GPOSS functionalized with oxirane rings. The epoxy matrix formulation is based on a tetrafunctional epoxy precursor (TGMDA) under the commercial name RTM6-2, which is a two-component resin designed to fulfill the requirements of the aerospace industry. The CNTs used were the NANOCYL NC3100 series. The selected concentration of the MWCNTs has been proved to offer sufficient electrical conductivity to composite aerostructures resulting to an effective dissipation of lightning currents during flight (Guadagno et al. 2015, 2014a, b), while the mixture is characterized by good dynamic mechanical behavior (Raimondo et al. 2015). The epoxy resin was also filled with POSS compounds in order to improve the flame resistance. The GPOSS compound was dispersed at a percentage of 5 wt% in the epoxy matrix. Consequently, the final formulations of the epoxy matrix were as follows: epoxy RTM6-2, epoxy RTM6-2 + 0.5 wt% MWCNT, and epoxy RTM6-2 + 0.5 wt% MWCNT +5 wt% GPOSS.

The tension tests were conducted according to the specification ASTM D638-01, the compression tests according to ASTM D695-02a, the flexural tests according to the specification ASTM D790-03, and the mode-I fracture toughness according to the specification ASTM D5045-99.

4.3.2 Tensile Properties

The influence of the considered additives on the tensile properties is displayed in Fig. 8.14. For both tensile strength and Young's modulus, the observed deviations are large. An enormous increase over 120% (p-value = 0.004) is observed for the tensile strength of the material filled with MWCNTs as compared to the unfilled one. The addition of GPOSS into the polymer has led to a tensile strength reduction as compared to the polymer filled with MWCNTs. Nevertheless, the tensile strength of the material which is filled with both additives remains increased by more than 50% as compared to the unfilled one. The results on Young's modulus show an appreciable deviation, and no significant differences (p-value >0.05) were found.

To understand the material behavior presented above, SEM and EDS analyses were made. The results obtained from the SEM analysis revealed the presence of MWCNT agglomerates and GPOSS aggregates. As shown in Fig. 8.15, MWCNT agglomerates were extensively identified. Figure 8.15 refers to a specimen filled with MWCNTs while Fig. 8.15 to a specimen filled with MWCNTs and GPOSS. Furthermore, some indications of incomplete dissolution of GPOSS in the resin were observed; some residuals.

4.3.3 Compressive Strength

As shown in Fig. 8.16, a 7.80% decrease of compressive strength in the case of the material filled with MWCNTs and GPOSS, which is significantly lower (p-value = 0.004) as compared to the reference one, was observed. CNTs are less efficient to effectively reinforce the polymeric material under compression. As their aspect ratio increases, the buckling behavior of CNTs under compressive stresses becomes more pronounced diminishing their ability to further carry loads. Moreover, the presence of large particles of MWCNT agglomerates as well as GPOSS aggre-

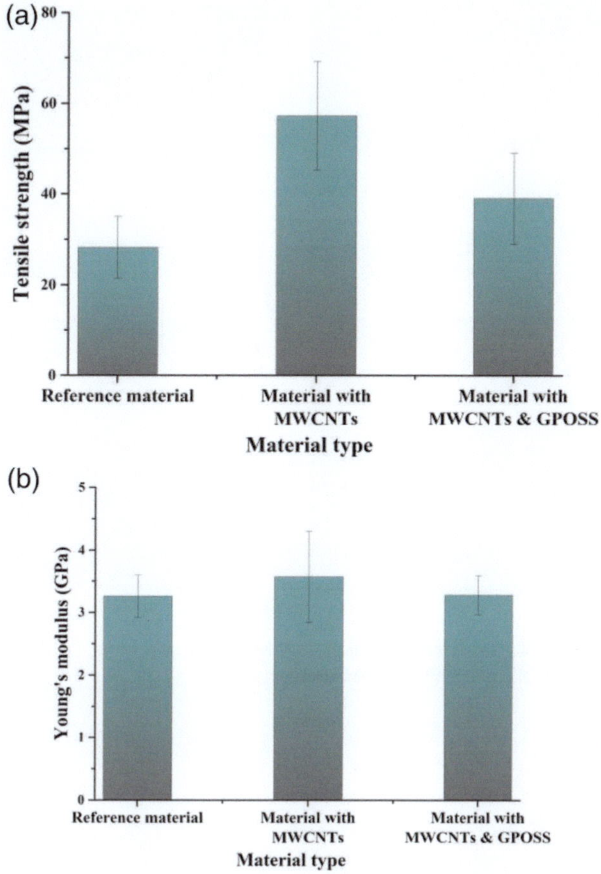

Fig. 8.14 Tensile strength (**a**) and Young's modulus (**b**) of reference, filled with MWCNTs as well as filled with MWCNT material

gates leads to local high stresses and, consequently, to local failures. To this end, the observed decrease in the compressive strength due to the combined effect of MWCNTs and GPOSS can be justified considering on one hand the ineffective load-bearing capability of the MWCNTs and on the other hand the presence of the inclusions, which have been revealed from the SEM and EDS analysis.

4.3.4 Flexural Properties

Flexural strength and flexural modulus values are displayed in Fig. 8.17. As shown, an insignificant decrease (p-value >0.05) in the flexural modulus is exhibited for both filled materials as compared to the reference one. As far as the flexural strength is concerned, the material filled with MWCNTs and GPOSS shows a higher reduc-

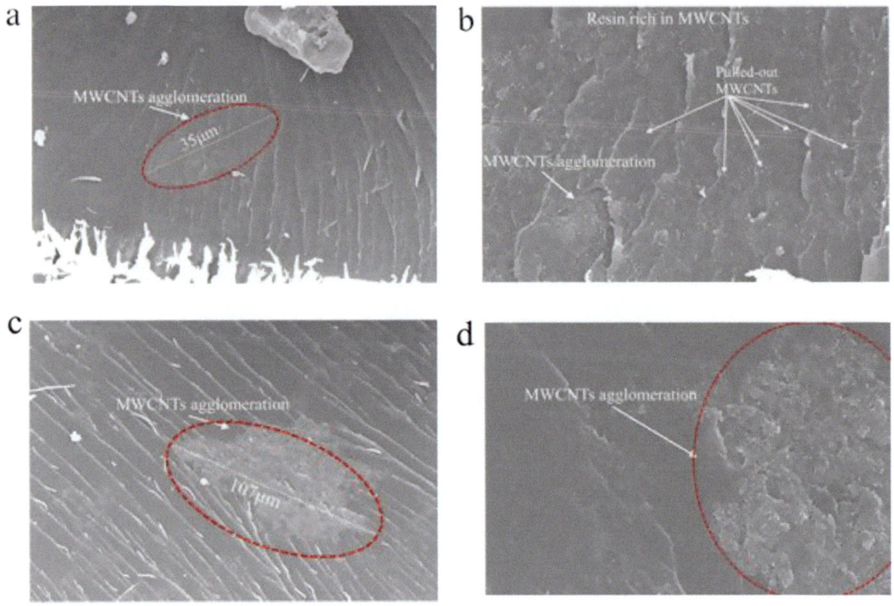

Fig. 8.15 (**a**, **b**) MWCNT agglomerates filled with MWCNT specimen; (**c**, **d**) MWCNT agglomerates filled with MWCNT and GPOSS specimen

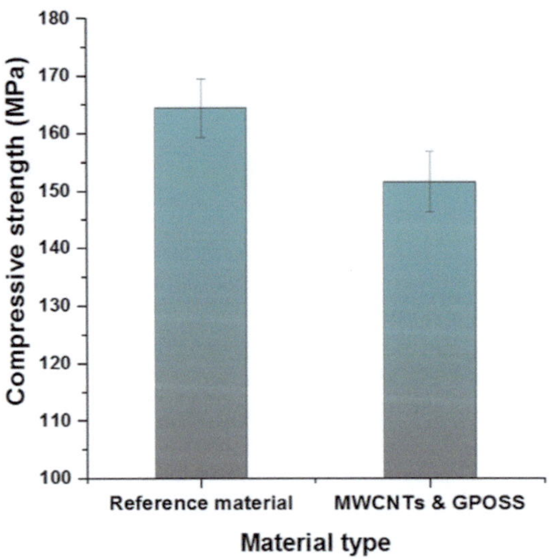

Fig. 8.16 Compressive strength

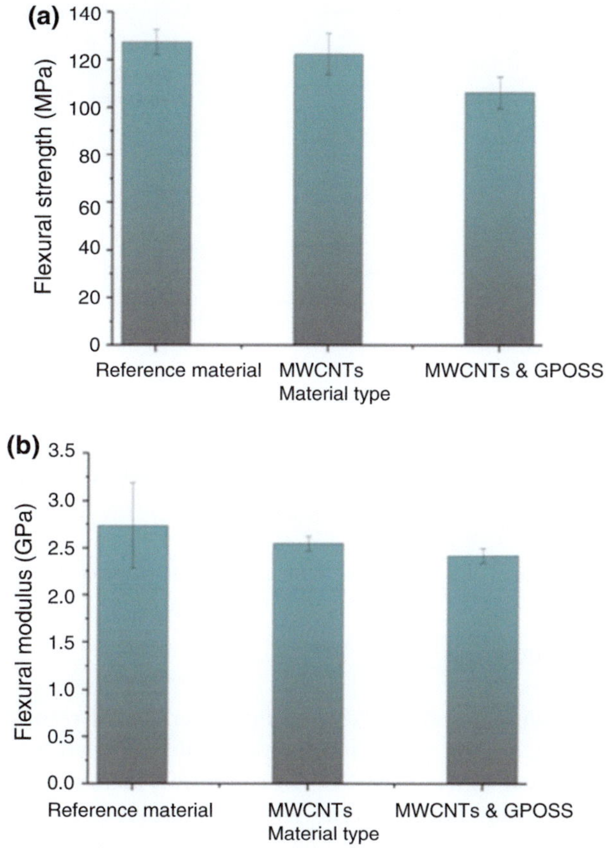

Fig. 8.17 Flexural properties

tion (p-value = 0.003) than the material filled only with MWCNTs (p-value = 0.016) as compared to the reference material. Under three-point bending, the upper part of the specimen is subjected to compression while the bottom part of the specimen to tension. As discussed, MWCNTs cause a significant increase in the case of tension, overriding the effect of the inclusions. On the other hand, as their aspect ratio increases, the buckling behavior of CNTs under compressive stresses becomes more pronounced diminishing their ability to further carry loads. The observed values of the flexural properties are the interacting effect of the abovementioned mechanisms as well as the deteriorating effect of the presence of the MWCNT and GPOSS agglomerates. MWCNT agglomerations seem to be the reason for the degradation in the flexural behavior by further increase of the concentration of the MWCNTs into the material. A more deteriorated flexural behavior is obtained as a cause of the synergistic effect of the MWCNTs and GPOSS into the polymeric resin, as the number of the inclusions is getting increased by the incorporation of the GPOSS into the polymeric resin.

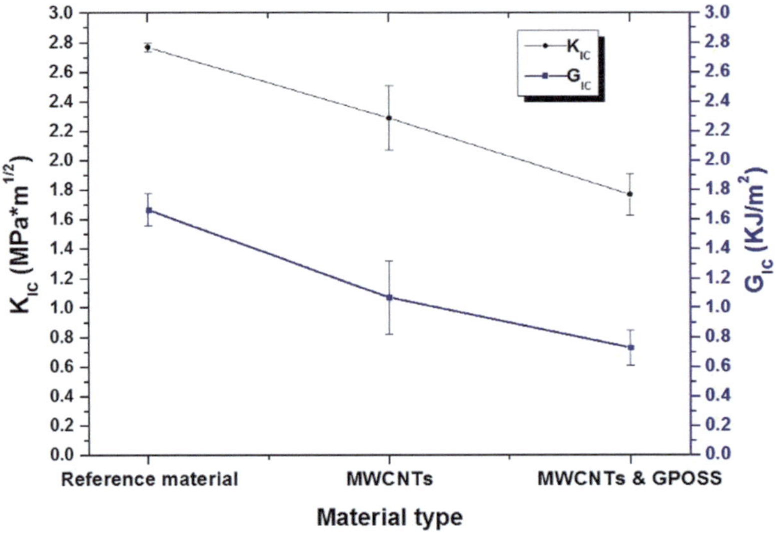

Fig. 8.18 Fracture toughness K_{IC} and critical strain energy release rate G_{IC}

4.3.5 Fracture Toughness Tests

The values of K_{IC} and G_{IC} calculated from the experimental results are shown in Fig. 8.18. As shown, the lowest K_{IC} and G_{IC} values are observed for the case of the material which is filled with both MWCNT and GPOSS materials. Specifically, G_{IC} fracture toughness of the filled material with both additives is appreciably lower (p-value = 0.002) than the reference one. The mechanisms involved in fracture toughness testing are more complex as compared to the uniaxial loading cases investigated so far in this study. With regard to the specimen geometry, three-point bending is expected to cause essential tensile stresses at the plane along the crack tip. Therefore, one should expect that well-dispersed CNTs would increase the G_{IC} values due to the strengthening effect of CNTs under tensile loading and possible crack bridging effects as the propagating crack may meet the CNTs lying transverse to the crack propagation plane. On the other side, the material investigated in this study was found to include appreciable amounts of MWCNT agglomerations as well as incompletely diluted GPOSS which act as defects causing a multiaxial stress state which facilitates the material failure. The results observed indicate that this deteriorating effect is overriding the beneficial effects of MWCNTs.

4.4 MWCNT/GPOSS/CFRP Composite

In Polydoropoulou et al. (2016), the synergistic effect of multiwalled carbon nanotubes and glycidyl polyhedral oligomeric silsesquioxanes on the compression after impact behavior of multifunctional CFRP plates has been investigated.

4.4.1 Experimental

The materials used for the study are CFRP plates enhanced with 0.5% (by wt) MWCNTs and 5% (by wt) viscous liquid GPOSS functionalized with oxirane rings. The matrix formulation is based on a tetrafunctional epoxy precursor (TGMDA) under the commercial name RTM6-2, which is a two-component resin designed to fulfill the requirements of the aerospace industry. The operation service temperatures range from −60 °C to 120 °C. However, the resin used in this study differs from the commercial one since only one type of hardener is used, instead of a mixture of hardeners used in the commercial resin. The epoxy matrix of the composite has been prepared by mixing a tetraglycidylmethylenedianiline (TGMDA) with epoxy reactive diluents 1,4-butanediol diglycidyl ether (BDE) at a concentration of 80%:20% (by wt) combined with 4, 4′-diaminodiphenylsulfone (DDS), as hardener.

The impact tests were conducted according to the specification ASTM D7136. This test method determines the damage resistance of composite plates subjected to a drop-weight impact event. The specimens have been subjected to low energy impact level, ranging from 25 to 30 J. Prior to impact, all specimens were subjected to C-scan analysis. Following to impact testing, the specimens were subjected to compression to determine the residual strength. For the determination of the residual strength of the impacted composite plates, the equipment used and the procedure followed meet the specifications according to the specification ASTM D7137.

4.4.2 Impact Behavior

The mean contact force-time and energy-time curves recorded during the impact test of all specimens are illustrated in Fig. 8.19. The graphs show a typical impact behavior of composite plates. In Fig. 8.19, almost the same values are observed for the first discontinuity in contact force as well as the peak contact force of both materials. However, the filled material shows systematically longer contact duration than the reference material. This duration discrepancy is likely due to creation of more damaged areas in the case of the filled material. After the peak force, a more rapid load drop is observed in the case of the reference material as compared to the filled one. Concerning the energy versus time results, as shown in Fig. 8.19, slight differences are observed. After the maximum impact energy value is reached, a slight drop leading to a constant energy is observed for both materials, reference and filled. The constant energy observed coincides with the absorbed energy by the specimens.

The internal damage was detected through C-scan analysis; representative results are illustrated in Fig. 8.20. The graph refers to a reference (Fig. 8.20), as well as an enriched MWCNT and GPOSS specimen (Fig. 8.20) before and after impact test. As it can be seen in Fig. 8.5, the red central circle indicates the damage caused by the impactor for both specimens. Yet, in the case of the enriched material, the damage also expands out of the boundaries of the impacted area. A distinct increase in damaged area is shown in the case of the enriched MWCNT and GPOSS material. According to C-scan graphs, the damaged area, which was calculated, was 640 ± 106 mm^2 and

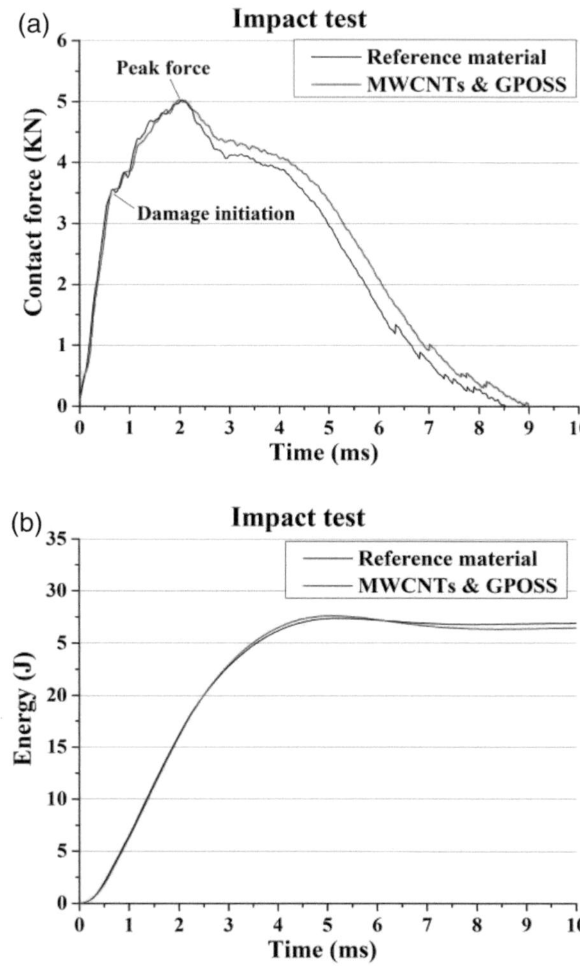

Fig. 8.19 (**a**) Force-time and (**b**) energy-time for both materials, reference and filled

1090 ± 174 mm^2 for the reference and the filled material, respectively. Based on a statistical analysis using the *t*-test's statistical significance with 95% confidence level, the calculated *p*-value was 0.000 which indicates a significant increase for the measured damaged area of the filled material as compared to the unfilled one.

4.4.3 CAI Behavior

The load-displacement curves of both reference and filled material are shown in Fig. 8.21. As can be seen in Fig. 8.22, concerning the reference material, a linear load-displacement curve until the maximum load is obtained, while in the case of the filled material (Fig. 8.8b), not all specimens demonstrate a linear curve, and the stan-

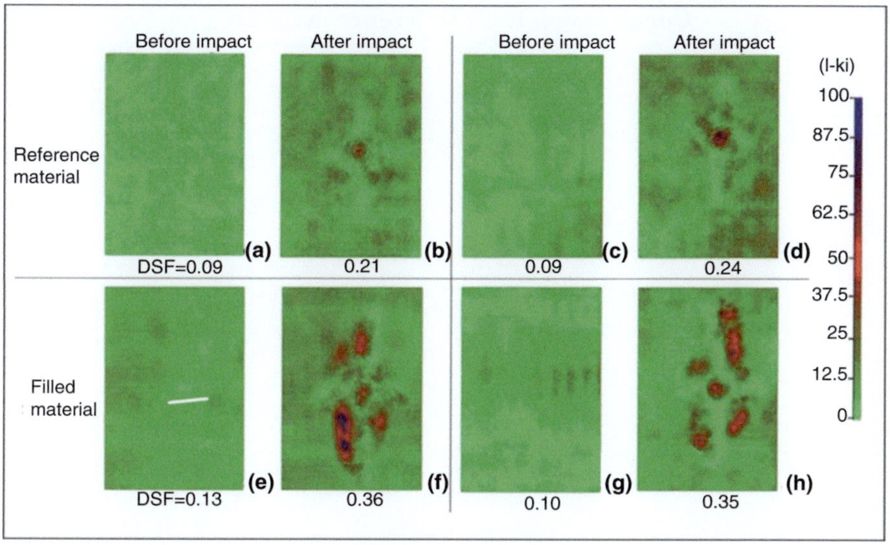

Fig. 8.20 C-scan graphs and corresponding DSFs of reference material before and after impact (**a**–**d**) as well as filled material before and after impact (**e**–**h**)

dard deviation is larger. In total, the compressive residual strength was appreciably reduced (p-value = 0.005) in comparison with the reference material, based on the same statistical analysis (t-test), as described above. The compressive strength was 158 ± 8.5 MPa and 133 ± 19 MPa for the reference and the filled material, respectively.

5 Mechanical Behavior of Self-Healing Materials

In Polydoropoulou et al. (2018), the effect of embedded self-healing microcapsules on the ILSS behavior of carbon fiber-reinforced composite materials has been studied. The results of the mechanical tests were discussed supported by scanning electron microscope (SEM) analysis.

5.1 Experimental

The self-healing material investigated is an epoxy polymer enhanced with 18 wt% microcapsules and 5 wt% HGI catalyst. The epoxy matrix is based on bisphenol A diglycidyl ether with a reactive diluent 1,4-butanediol diglycidyl ether. The cyclic olefins that act as healing agents are the 5-ethylidene-2-norbornene (ENB) and dicyclopentadiene (DCPD) at a fraction of 95:5%, respectively. The outer shell of the microcapsules is composed of poly(urea-formaldehyde) and the inner shell of

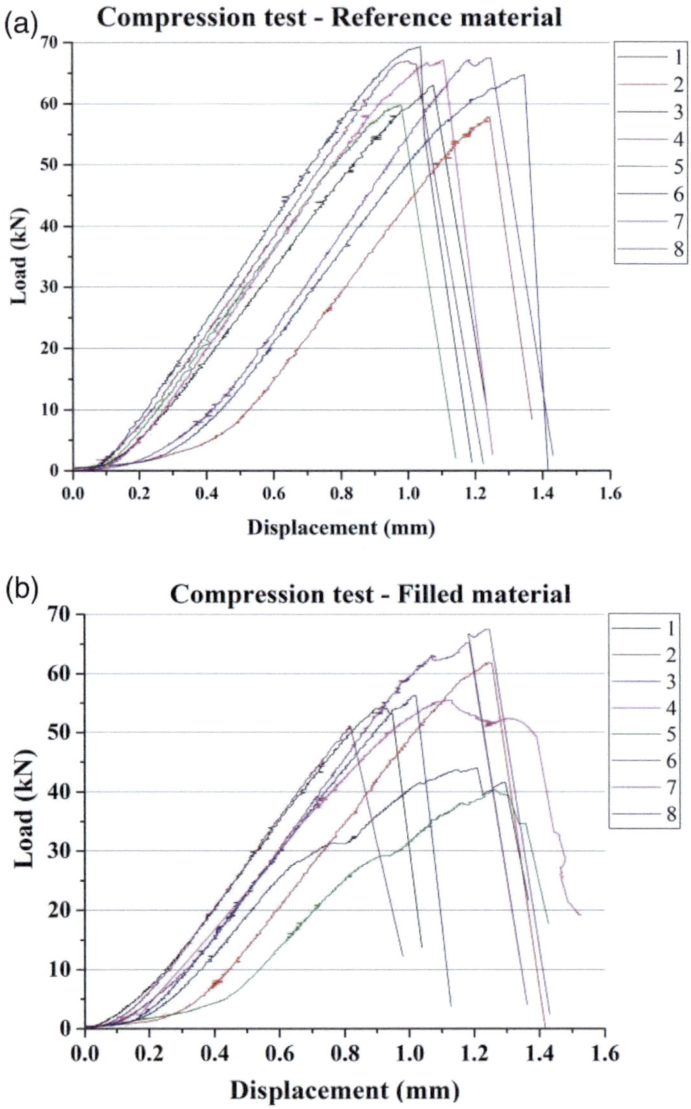

Fig. 8.21 Load-displacement curves for compression test of reference (**a**) and enriched (**b**) material

ethylene maleic anhydride (EMA) copolymer. For the preparation of the composite material, the resin was infused into a carbon fiber dry preform of 24 plies of carbon fiber cloths (SIGMATEX (UK) LDT 193GSM/PW/HTA40 E13 3K) using a modified bulk film infusion process in order to overcome the infiltration issues.

The ILSS tests were performed in accordance to the ASTM D2344 standard (ASTM D2344 2013). Four experimental series were set up; an overview of the

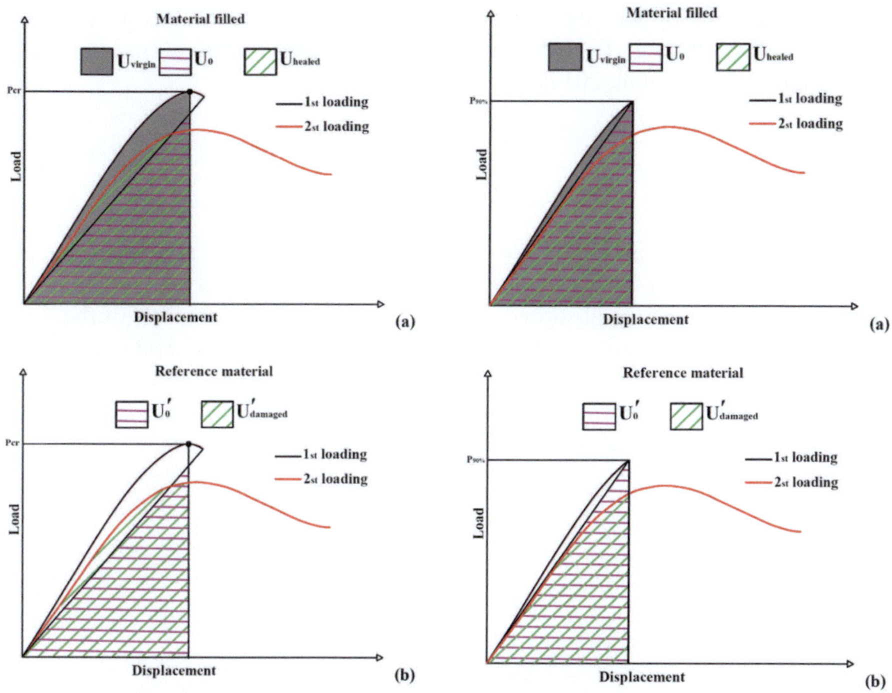

Fig. 8.22 Definition of self-healing efficiency according to [38]

performed tests may be seen in Table 8.2. The first experimental series consisting of eight unfilled and eight enhanced with microcapsules was used to obtain the reference ILSS properties of both materials. By displacement control, the specimens were loaded up to a 30% load drop (ASTM D2344 2013).

To assess the potential self-healing efficiency of the enhanced material, three more experimental series were carried out using 12 unfilled specimens and 12 specimens enhanced with microcapsules at different preloading levels to introduce different extents of damage. Preloadings of 75%, 90%, and 100% as compared to the mean ILSS strength obtained for the reference, unfilled, and enhanced materials were used. After preloading, the specimens were fully unloaded and allowed to a healing rest period of 24 h. Then, by displacement control, the specimens were loaded up to a 30% load drop.

5.1.1 Self-Healing Efficiency

For polymers, the self-healing efficiency has been defined by several authors, e.g., [16] as the ratio of the maximum load after the healing period to the maximum load before the healing period. For CFRPs, irreversible damage (e.g., fiber breakages) has to be excluded by the self-healing definition. Therefore, the self-healing efficiency is defined through the assessment of the absorbed energy during the specimen loading (Sanada et al. 2017):

$$h = \frac{U_{\text{healed}} - U_{\text{damaged}}}{U_{\text{virgin}} - U_{\text{damaged}}} \tag{8.1}$$

where U_{virgin} is the absorbed energy during the first loading of the filled material obtained by measuring the area below the load-displacement curves with regard to the P_{critical} and U_{healed} is the absorbed energy during the second loading of the filled material obtained by measuring the area below the load-displacement curves with regard to the P_{critical} as shown in Fig. 8.2a. The value for U_{damaged} is obtained by assuming the same damage severity a as for the case of reference material, as shown in Fig. 8.22. The a value is given by Eq. (8.2) and U_{damaged} by Eq. (8.3):

$$a = \frac{U'_{\text{damaged}}}{U'_0} \tag{8.2}$$

$$a = \frac{U_{\text{damaged}}}{U_0} \tag{8.3}$$

Combining Eqs. (8.2) and (8.3), U_{damaged} is given by:

$$U_{\text{damaged}} = \frac{U'_{\text{damaged}}}{U'_0} U_0 \tag{8.4}$$

For preloadings of 75% as well as 90% of the mean ILSS value, the self-healing efficiency is also defined according to Eq. (8.1) taking into account the P_{critical} as the maximum load during the first loading of the filled material, as shown in Fig. 8.22c.

5.2 *ILSS Behavior*

Representative load-displacement curves for the R.1-3 as well as for the SH.1-3 series (Table 8.3) are shown in Fig. 8.23. Preloadings at 75%, 90%, and 100% of the mean ILSS strength as well as the loading after the rest period for both materials are presented, respectively. Preloading at 75% has no effect on the ILSS behavior of the specimen, as it can be seen in Fig. 8.23a. The load-displacement curves after the rest period fall together with the load-displacement curve obtained during preloading to 75%. In this case, no extensive damage has occurred, and hence, no healing effect can be expected. However, a preloading increase to 90% of the mean ILSS strength induces a limited damage to the material leading to the degradation of the ILSS

Table 8.3 Matrix of ILSS tests

Material			Type of tests		
Unfilled	Filled	No. of tests	Preloading	Rest period	Loading
R	SH	8	No preloading	No rest period	30% load drop
R.1	SH.1	3	75% preloading	24 h	30% load drop
R.2	SH.2	3	90% preloading	24 h	30% load drop
R.3	SH.3	6[a]	100% preloading	24 h	30% load drop

[a]It includes three experiments from the first batch (SH)

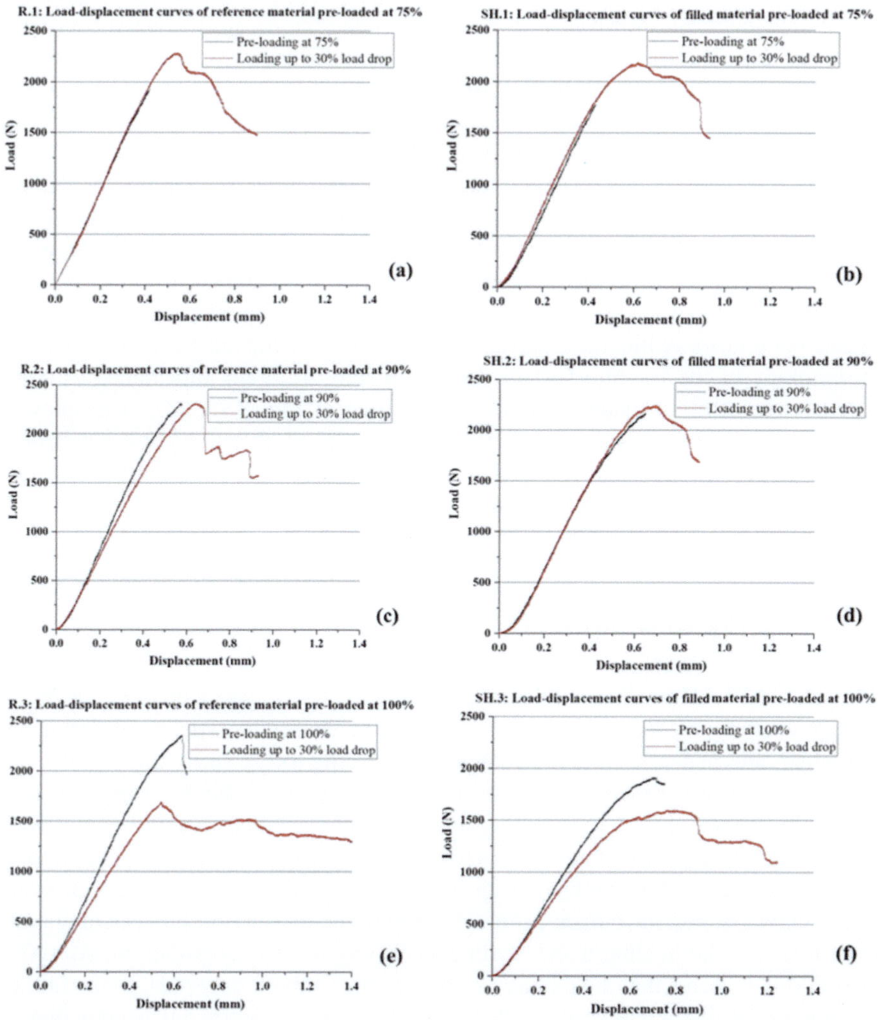

Fig. 8.23 Load-displacement curves of representative samples before and after the rest period. First loading at 75% of the mean short-beam strength (**a**), (**b**) first loading at 90% of the mean short-beam strength (**c**), (**d**) and first loading at 100% of the mean short-beam strength (**e**), (**f**) of reference and filled material, respectively

behavior of the unfilled material, as shown in Fig. 8.23c. On the other hand, in the case of the filled material, it seems that the damage occurred by the preloading at 90% has been eliminated due to the self-healing mechanism activation. Extensive damage has occurred leading to a significantly degraded behavior of both materials during preloading at 100%, as shown in Fig. 8.23. Nevertheless, a part of this extensive damage seems to have been recovered in the case of the filled material as it can be seen in Fig. 8.23f. Polymerized self-healing agent has been revealed by the SEM analysis results confirming the self-healing activation. The SEM analysis revealed the existence of the microcapsules into the polymer matrix before the experimental procedure (Fig. 8.23a) as well as polymerized microcapsules at the fracture surface of the specimens as shown in Fig. 8.23. Debonded microcapsules have been also observed (Fig. 8.23b). The polymerization of the self-healing agent proves that the self-healing mechanism has been activated triggered by matrix cracks during the loading of the specimen. However, a nonuniform dispersion of the microcapsules into the polymer matrix is also observed in Fig. 8.23.

The SEM analysis has revealed the surface morphology and the microstructure of the material which is enhanced with microcapsules before rupture, as shown in Fig. 8.24. The appearance of the neat resin is almost smooth without rough areas (Fig. 8.24a). On the other hand, when microcapsules are embedded into the resin, the surface morphology becomes appreciably rough, and the microcapsules are dispersed nonuniformly into the material (Fig. 8.24b). After the mechanical testing, the morphology of the fracture surface has completely different appearance as compared to the unfilled material, as it may be seen in Fig. 8.25. Microcapsules of 1.5 μm mean diameter appear to have ruptured, and their healing agent seems to have been polymerized and diffused close to them (Fig. 8.25b) forming rough surfaces completely different as compared to the unfilled material (Fig. 8.25a).

5.3 *Healing Efficiency*

The calculated self-healing efficiency according to Eq. (8.1) is given in Fig. 8.26. At 75% preloading, no discrepancies from the linear behavior have occurred during the reloading, and hence, the curves coincide. According to Eq. (8.1), this result in an undefined self-healing efficiency, and therefore, it is excluded from the graph of Fig. 8.26. Almost 115% self-healing efficiency of the filled material at the 90% preloading has been calculated. It seems that only micro-cracks have occurred at this preloading level which have been filled by the healing agent. Moreover, some micro-cracks which have been developed during the production process may have been filled by the healing agent leading to an improved microstructure. At 100% preloading, it seems that the damage is irreversible for both materials, as the failure extends to the fibers. However, a 26.5% self-healing efficiency has been achieved for the filled material which may represent the healing of some matrix micro-cracks.

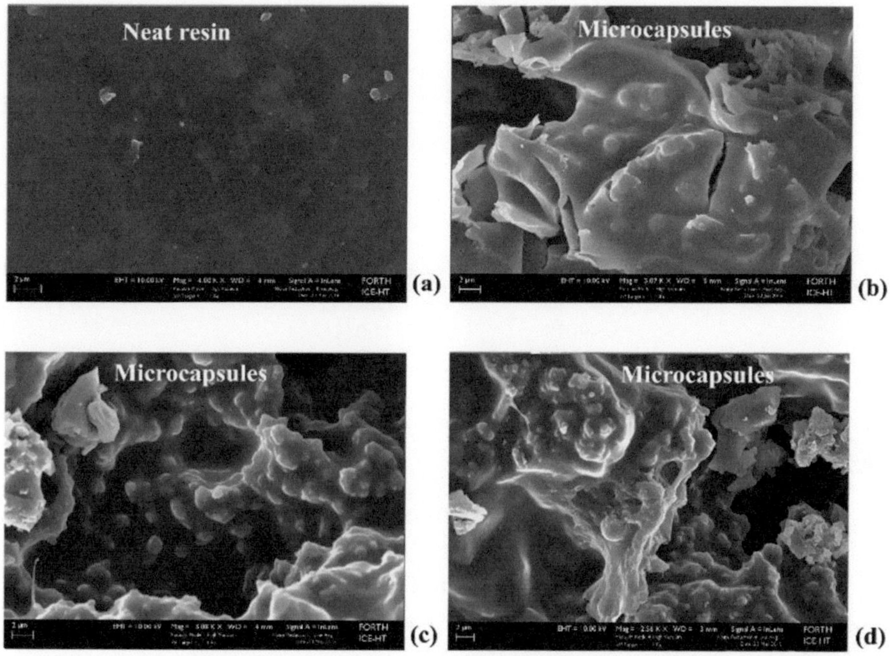

Fig. 8.24 Surface of (**a**) reference material and (**b–d**) material filled with microcapsules

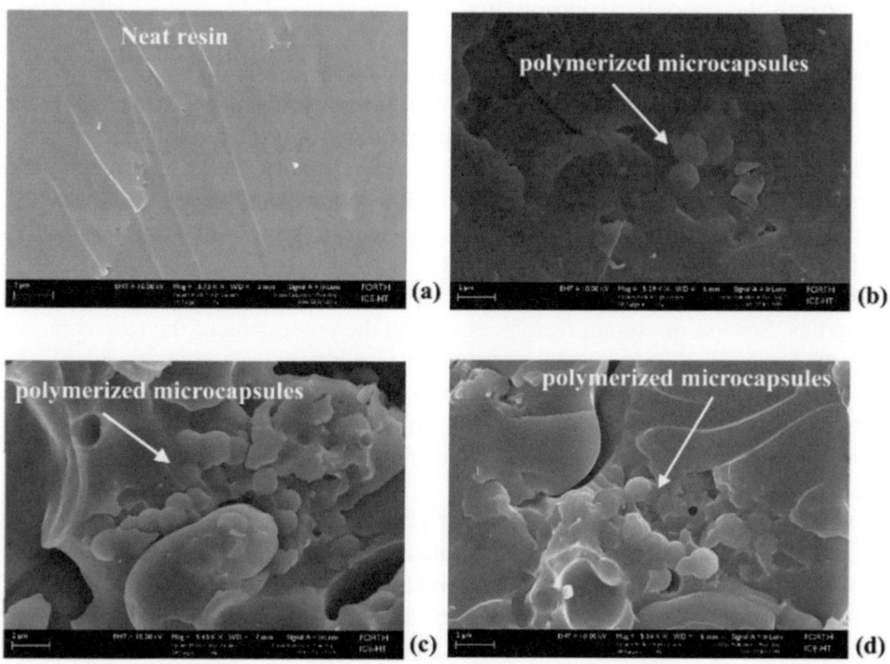

Fig. 8.25 Fracture surface of (**a**) reference material and (**b–d**) material filled with microcapsules

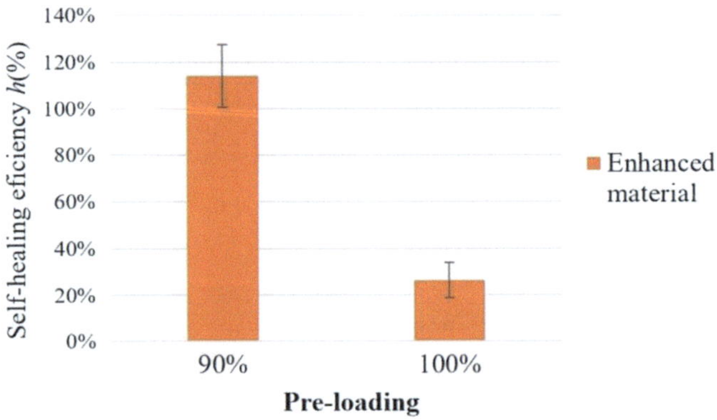

Fig. 8.26 Self-healing efficiency of the filled material as compared to the unfilled material during ILSS test

6 Multi-scale Modeling of Nanocomposites

6.1 Towards a Simulation-Driven Design of Nano-enabled MM

The ability to tailor the properties of MM depends on several factors. The first is the property being tailored. Both mechanical and electrical properties seem to be more sensitive to the nanofiller concentrations and geometrical parameters when compared with thermal properties. On the contrary, the thermal conductivity of cured nanofilled-epoxy composites shows a minimal and near insignificant increase with nanofiller content. A second aspect to consider when tailoring the properties is the choice of nanofiller and polymeric medium to which it is dispersed. The enhancement in thermal conductivities appears to be greater for some nanofillers. Furthermore, the mechanical properties of nanocomposites have been shown to increase significantly when the nanofillers are chemically modified to form reactive bridges with the surrounding polymer chains, a process known as functionalization. The addition of nanofiller in polymeric materials does not always result in improved properties. Several important factors relating to the processing of the nanocomposite also play a significant role. One of the most important aspects to consider is the homogeneous dispersion of the nanofillers in the polymeric matrix. For instance, carbon nanotubes (CNTs) tend to exhibit an enormous surface area being several orders of magnitude larger than the surface of conventional fillers. The large surface area of CNTs leads to two counteracting effects: one desirable offering increased stress transfer and the other undesirable leading to strong attractive intermolecular and van der Waals forces between the nanofillers resulting in excessive agglomeration.

It is obvious that there are several important factors to consider in the design and fabrication of multifunctional nanocomposite materials as it is highly possible in some cases the choices made to tailor one property to dislocate another property. At present, the development of multifunctional materials is being done through a trial-and-error process comprising repeated manufacturing trials and tests. Such a process is highly cost- and time-consuming. Upon this background, it is very important to develop simulation-based design tools whose implementation will lead to a significant reduction of cost and time required for the development of multifunctional nanocomposites.

A very good framework for simulation-driven design tools of multi-scale materials such as the nano-enabled MM is provided by Olson's three-link chain concept [39] proposed for simulation-based design of hierarchically structured materials (Fig. 8.27). Olson's concept (Olson 1997) integrates four critical elements of materials science and engineering: processing, structure, properties, and performance. The deductive approach of understanding material properties and performance from the structure and processing is explained by means of the flowchart of Fig. 8.28. The deductive approach is implemented through multi-scale modeling. The main goal of the approach is to determine the input and output functions that describe the two hubs shown in Fig. 8.28. The inductive approach is built on the knowledge gained from deductive approach on the interaction of different processing and structural factors and their effects on the properties and performance of the multifunctional

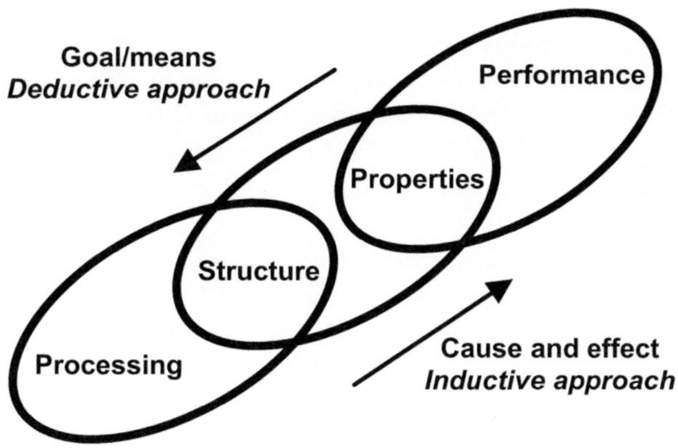

Fig. 8.27 Olson's concept of "materials by design" [39]

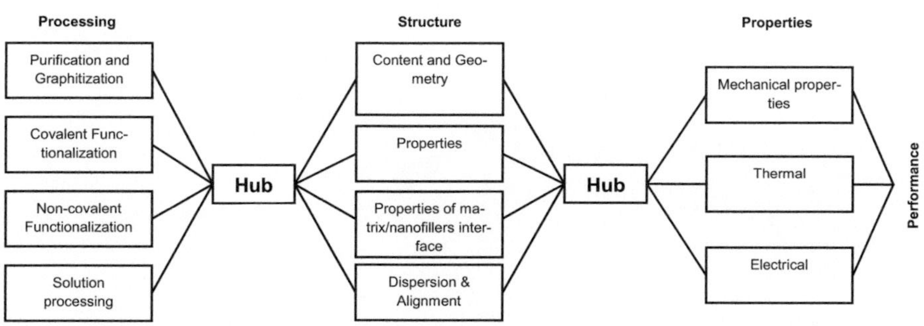

Fig. 8.28 Flowchart describing the deductive approach for multifunctional nanocomposites

material. The goal of inductive approach is the development of a methodology that will take as input the performance requirements and material specifications to provide a path towards manufacturing of multifunctional materials with desired performance and properties.

Although extensive research was carried out on multi-scale modeling of nano-enabled MM, a complete model implementing the Olson's concept has not been reported. Most of the models have been concentrated on the development of links from processing towards performance (inductive approach); not all links have been covered into a single model. On the other hand, no modeling work has been reported on the deductive direction.

In the following sections, the work that has been done by the authors on parametric evaluation of mechanical properties of MWCNT/CFRP parts and the parametric evaluation of the mechanical, electrical, and thermal properties of MWCNT/polymers is presented.

6.2 Prediction of Mechanical Properties of MWCNT/CFRP Parts

In a work performed within the SARISTU project, a continuum-based multi-scale model has been developed to predict the mechanical properties of MWCNT-doped CFRP parts. The reinforced structural part is a CFRP panel with a CFRP omega stringer reinforced by CNT-doped interlayers for enhanced delamination resistance. The flowchart of the model containing the different scales and materials is illustrated in Fig. 8.29.

In the nanoscale, the mechanical properties of the isolated CNTs have been computed using novel finite element (FE) models (Tserpes and Papanikos 2005). Young's modulus has been found in the range of 1.0–1.1 TPa (Tserpes and Papanikos 2005), shear modulus in the range of 0.2–0.5 TPa, and tensile strength in the range

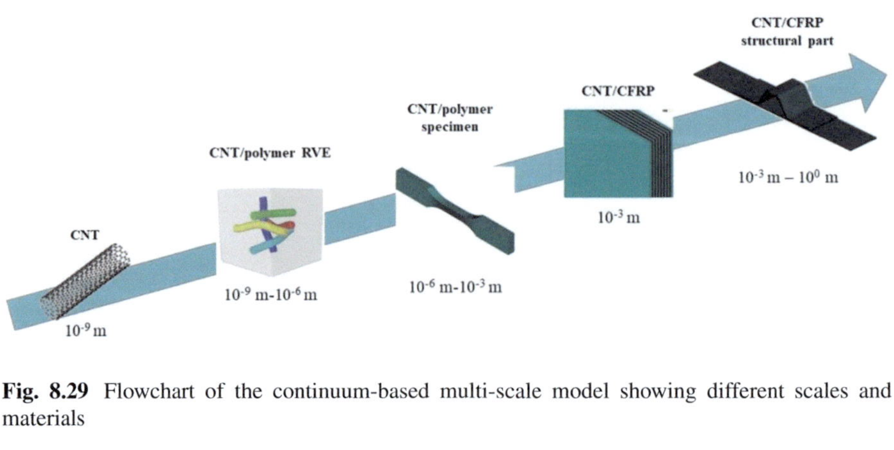

Fig. 8.29 Flowchart of the continuum-based multi-scale model showing different scales and materials

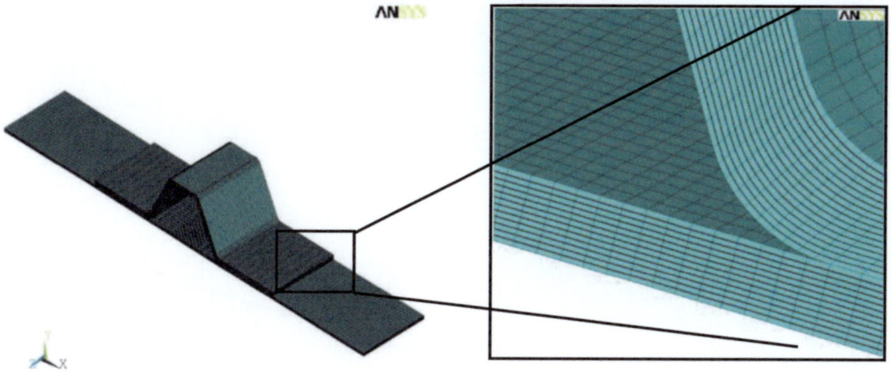

Fig. 8.30 FE mesh of the stiffened panel and a close view showing the modeled CNT/polymer interlayers (red color)

of 100–120 GPa (Tserpes et al. 2006). The tensile strength of CNTs has been found to highly depend on the presence of defects such as vacancies (Tserpes et al. 2006) and Stone-Wales defect (Tserpes and Papanikos 2007).

In the micro-scale, the mechanical properties of the MWCNT/polymer have been predicted using representative volume elements (RVEs) of single CNTs (Tserpes and Chanteli 2013) and CNT agglomerates (Chanteli and Tserpes 2015; Tserpes et al. 2017a, b) developed using SEM and AFM presented in Sect. 4 of the chapter. The properties of the nanocomposites have been correlated with material and geometrical parameters such as the CNT content, the CNT aspect ratio, the CNT orientation, the size of the agglomerate, and the thickness and properties of the CNT/polymer interphase. In summary, the results have shown that the elastic properties are mainly determined by the CNT content, that the agglomeration of the CNTs leads to a significant decrease of mechanical properties of the nanocomposite, and that the strength of the nanocomposite is mainly determined by the properties of the CNT/polymer interphase. All predictions of the micro-scale models were validated against experimental results presented in Sect. 4 of the chapter.

Using the modeling experience and implementing some of the findings from the micro-scale, RVEs have been developed at the mesoscale to predict the mechanical properties of the CNT/polymer interlayer which was placed between the CFRP layers of the stiffened panel. Figure 8.30 shows a cross section of the RVE with the single CNT oriented at 45° to account for an average alignment of CNTs. Figure 8.30 shows the FE mesh of the panel and the modeled CNT/polymer interlayers. Aiming to evaluate the reinforcement efficiency, the bending behavior of the panel was simulated using the progressive damage modeling method, and the numerical results were compared with results from tests. Figure 8.31 compares the evolution of surface strain as predicted by the model and measured by strain gauges during the bending test. Both the model and the test show a decrease of strain for the reinforced panel which is an indication of reinforcement of the bending stiffness due to the presence of the CNT/polymer interlayers. Nevertheless, the complete set of failure data reveal the need for improvements on the reinforcement efficiency of the CFRP material by CNTs.

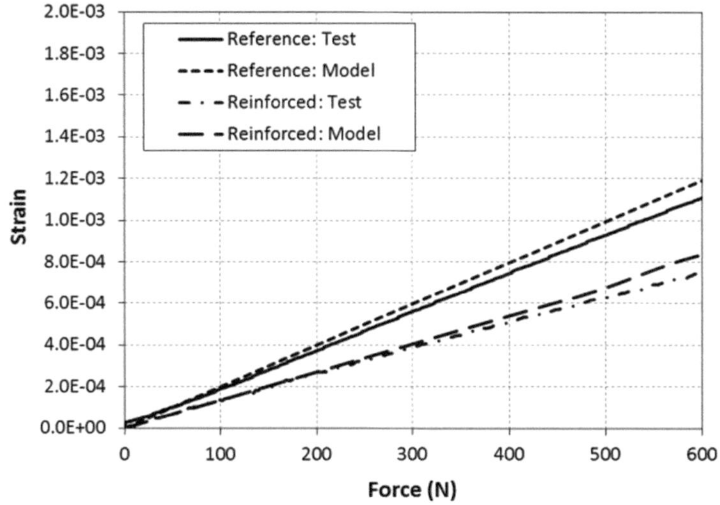

Fig. 8.31 Evolution of strain in the CFRP stiffened panel

6.3 Parametric Evaluation of Elastic, Thermal, and Electrical Properties of CNT/Polymers

After having evaluated the effect of material and geometrical parameters on the mechanical properties of the CNT/polymer nanocomposites, the effect of the same parameters on the elastic, thermal, and electrical properties have been investigated (Tzatzadakis and Tserpes 2017). This investigation is one step further towards the completion of the inductive path (see Fig. 8.28) for nanocomposites. The parameters and properties considered in the investigation are shown in Fig. 8.32. Investigation has been carried out by means of RVEs created using the DIGIMAT software. Figure 8.33 illustrates a typical RVE and its FE mesh.

The results of the investigation are summarized in Table 8.4. As expected, the CNT content has a negative effect on all properties although it is well-known that the higher the CNTs content, the larger the CNT agglomerations formed. The same stands for the CNT's aspect ratio. Given the diameter of MWCNTs is almost fixed, the larger the length of the CNTs, the more effective the reinforcement will be. CNT waviness hinders the mechanical properties of the nanocomposite. CNT orientation and properties (isotropic vs. orthotropic) do not affect the mechanical properties of the nanocomposite. Strength of the nanocomposite is governed by the properties of the CNT/polymer interphase. Perhaps the most interesting finding is the contradictory finding on the effect of the size of the agglomerates (number of CNTs) on the mechanical, thermal, and electrical properties. This is because the larger the number of CNTs, the higher the possibility for the formation of thermal and electrical networks.

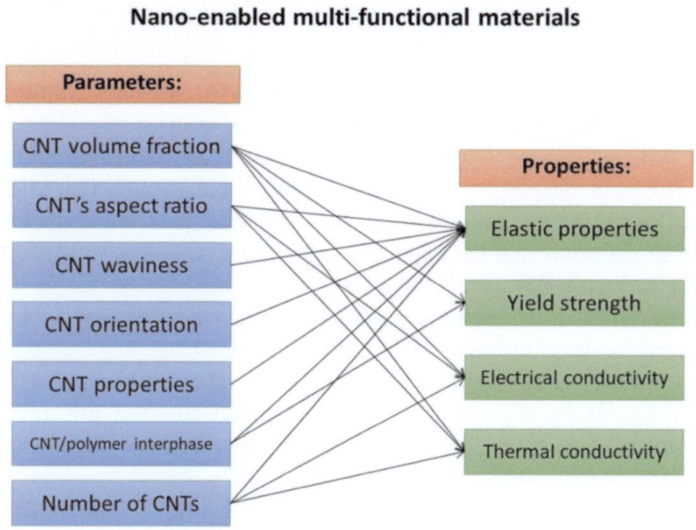

Fig. 8.32 Parameters and properties investigated

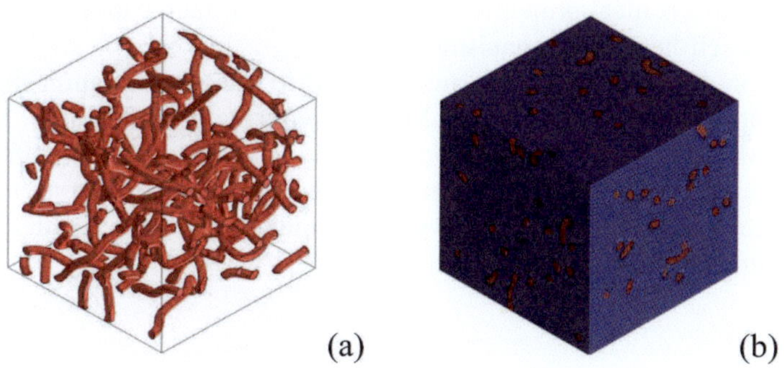

Fig. 8.33 (**a**) RVE with curved CNTs and (**b**) FE mesh of the RVE

Table 8.4 Effects of different parameters on the properties of the nanocomposites

	Elastic properties	Yield strength	Electrical conductivity	Thermal conductivity
CNT content	↑	N/A	↑	↑
CNT's aspect ratio	↑	N/A	↑	↑
CNT waviness	↓	N/A	N/A	N/A
CNT orientation	–	N/A	N/A	N/A
CNT properties	–	N/A	N/A	N/A
Interphase	–	↑	N/A	N/A
Number of CNTs	↓	↓	↑	↑

7 Conclusions

From the first section of the chapter, it can be concluded that the technology of MM is evolving very fast. The technological readiness level of the existing MM is increasing, whereas new MM with additional functionalities are continuously added in the list. However, despite the progress achieved, there are very few MM that have managed to reach the production level. Nevertheless, MM are considered as the future aircraft materials and are expected to be in the core of aeronautical research activities in the next years.

Nano-enabled MM and especially CNT/polymer nanocomposites have shown a great potential as they possess significantly enhanced mechanical, thermal, and electrical properties compared to the neat polymer. The results from the extensive characterization efforts of these materials have shown that the uniform dispersion and the chemical bond between the nanofiller and the matrix are very critical for the efficiency of reinforcement.

There have been reported many types of SHM, but none of them have reached up to the level of proof of concept. Microencapsulated SHM are more technologically matured; however, much progress remains to be done in order for them to reach the production level. Two major drawbacks that must be resolved are the reduction of the reference mechanical properties due to addition of microcapsules and the efficiency of the self-healing mechanism.

Simulation-driven design tools are expected to contribute significantly to the development and implementation of MM by reducing cost and time. For nano-enabled MM, multi-scale models have managed to go through the inductive path which lies in the relation between processing parameters and performance. What remains is to prove their efficiency by going through the deductive path to describe the production process for a given performance.

References

Andersson H, Keller M, Moore J, Sottos N, White S (2007) Self healing polymers and composites. Springer Series in Materials Science, pp 19–44

Chanteli A, Tserpes K (2014) Tensile behaviour of carbon nanotube/polypropylene composite material. Plast Rubber Compos 43(10):330–336

Chanteli A, Tserpes K (2015) Finite element modeling of carbon nanotube agglomerates in polymers. Compos Struct 132:1141–1148

Chen Y et al (2006) Development and verification of real-time controllers for F/A-18 vertical fin buffet load alleviation. In: SPIE, smart structures and materials 2006: smart structures and integrated systems, San Diego

Erturk A, Inman D (2011) Piezoelectric energy harvesting

Ferreira A, Nóvoa P, Marques A (2016) Multifunctional material systems: a state-of-the-art review. Compos Struct 151:3–35

Ghosh S (2009) Self-healing materials. Wiley-VCH, Weinheim, Germany

Guadagno L, Raimondo M, Naddeo C, Longo P, Mariconda A, Binder W (2014a) Healing efficiency and dynamic mechanical properties of self-healing epoxy systems. Smart Mater Struct 23(4):045001

Guadagno L, Raimondo M, Vittoria V, Vertuccio L, Naddeo C, Russo S, De Vivo B, Lamberti P, Spinelli G, Tucci V (2014b) Development of epoxy mixtures for application in aeronautics and aerospace. RSC Adv 4(30):15474–15488

Guadagno L, Raimondo M, Vietri U, Vertuccio L, Barra G, De Vivo B, Lamberti P, Spinelli G, Tucci V, Volponi R, Cosentino G, De Nicola F (2015) Effective formulation and processing of nanofilled carbon fiber reinforced composites. RSC Adv 5(8):6033–6042

Guadagno L, Raimondo M, Vietri U, Naddeo C, Stojanovic A, Sorrentino A, Binder W (2016) Evaluation of the mechanical properties of microcapsule-based self-healing composites. Int J Aerosp Eng 2016:1–10

Guadagno L, Naddeo C, Raimondo M, Barra G, Vertuccio L, Sorrentino A, Binder W, Kadlec M (2019) Development of self-healing multifunctional materials

Hager M, Greil P, Leyens C, van der Zwaag S, Schubert U (2010) Self-healing materials. Adv Mater 22(47):5424–5430

Hayes S, Zhang W, Branthwaite M, Jones F (2007) Self-healing of damage in fibre-reinforced polymer-matrix composites. J R Soc Interface 4(13):381–387

Inman DJ, Johnson K (2013) Genomics of multifunctional structures and materials for flight. In: 54th AIAA/ASME/ASCE/AHS/ASC structures, structural dynamics, and materials conference 8–11 April 2013, Boston, Massachusetts

Jha D, Kant T, Singh R (2013) A critical review of recent research on functionally graded plates. Compos Struct 96:833–849

Kousourakis A, Mouritz A (2010) The effect of self-healing hollow fibres on the mechanical properties of polymer composites. Smart Mater Struct 19(8):085021

Nemat-Nasser S, Amirkhizi A, Nemat-Nasser S, Plaisted T, Starr A (2005) Multifunctional materials (Chapter 12). In: BIOMIMETICS: biologically inspired technologies, pp 309–338

Olson G (1997) Computational design of hierarchically structured materials. Science 277(5330):1237–1242

Pantelakis S, Katsiropoulos C, Polydoropoulou P (2016) Assessing the compression after impact behaviour of innovative multifunctional composites. Nanomater Nanotechnol 6:184798041667962

Pereira T, Scaffaro R, Nieh S, Arias J, Guo Z, Hahn H (2006) The performance of thin-film Li-ion batteries under flexural deflection. J Micromech Microeng 16(12):2714–2721

Polydoropoulou P, Katsiropoulos C, Pantelakis S (2016) The synergistic effect of CNTs and flame retardants on the mechanical behavior of aerospace epoxy resin. Polym Eng Sci 57(5):528–536

Polydoropoulou P, Katsiropoulos C, Loukopoulos A, Pantelakis S (2018) Mechanical behavior of aeronautical composites containing self-healing microcapsules. Int J Struct Integr 9(6):753–767

Raimondo M, Russo S, Guadagno L, Longo P, Chirico S, Mariconda A, Bonnaud L, Murariu O, Dubois P (2015) Effect of incorporation of POSS compounds and phosphorous hardeners on thermal and fire resistance of nanofilled aeronautic resins. RSC Adv 5(15):10974–10986

Sanada K, Suyama T, Nassho Y (2017) Interlaminar shear strength and self-healing of spread carbon fiber/epoxy laminates containing microcapsules. J Soc Mater Sci Jpn 66(4):299–305

Sofla A, Meguid S, Tan K, Yeo W (2010) Shape morphing of aircraft wing: status and challenges. Mater Des 31(3):1284–1292

Sun L, Huang W, Ding Z, Zhao Y, Wang C, Purnawali H, Tang C (2012) Stimulus-responsive shape memory materials: a review. Mater Des 33:577–640

Thomas J, Qidwai M, Matic P, Everett R, Gozdz A, Keennon M (2002) Multifunctional approaches for structure-plus-power concepts. In: 43rd AIAA/ASME/ASCE/AHS/ASC structures, structural dynamics, and materials conference

Toohey K, Sottos N, Lewis J, Moore J, White S (2007) Self-healing materials with microvascular networks. Nat Mater 6(8):581–585

Trask R, Williams G, Bond I (2007) Bioinspired self-healing of advanced composite structures using hollow glass fibres. J R Soc Interface 4(13):363–371

Tserpes K (2011) Strength of graphenes containing randomly dispersed vacancies. Acta Mech 223(4):669–678

Tserpes K, Chanteli A (2013) Parametric numerical evaluation of the effective elastic properties of carbon nanotube-reinforced polymers. Compos Struct 99:366–374

Tserpes K, Papanikos P (2005) Finite element modeling of single-walled carbon nanotubes. Compos Part B 36(5):468–477

Tserpes K, Papanikos P (2007) The effect of Stone–Wales defect on the tensile behavior and fracture of single-walled carbon nanotubes. Compos Struct 79(4):581–589

Tserpes K, Papanikos P, Tsirkas S (2006) A progressive fracture model for carbon nanotubes. Compos Part B 37(7–8):662–669

Tserpes K, Chanteli A, Floros I (2017a) Prediction of yield strength of MWCNT/PP nanocomposite considering the interphase and agglomeration. Compos Struct 168:657–662

Tserpes K, Moutsompegka E, Murariu O, Bonnaud L, Chanteli A (2017b) Experimental investigation of the effect of hygrothermal aging on the mechanical performance of carbon nanotube/PA6 nanocomposite. Plast Rubber Compos 46(6):239–244

Tzatzadakis V, Tserpes K (2017) Numerical and analytical evaluation of mechanical, thermal and electrical properties of CNT/polymer multifunctional nanocomposites using representative unit cells. In: Proceedings of the ICCS20 conference, 4–7 September, Paris, France

Varley R, Craze D, Mouritz A, Wang C (2013) Thermoplastic healing in epoxy networks: exploring performance and mechanism of alternative healing agents. Macromol Mater Eng 298(11):1232–1242

Wang Y, Inman D (2013) Simultaneous energy harvesting and gust alleviation for a multifunctional composite wing spar using reduced energy control via piezoceramics. J Compos Mater 47(1):125–146

Wu D, Meure S, Solomon D (2008) Self-healing polymeric materials: a review of recent developments. Prog Polym Sci 33(5):479–522

Zhu D, Rong M, Zhang M (2015) Self-healing polymeric materials based on microencapsulated healing agents: from design to preparation. Prog Polym Sci 49–50:175–220

Chapter 9
Biopolymers and Biocomposites

Xiaosu Yi, Jianfeng Tong, Xvfeng Zhang, Jin Zhu, Xiaoqing Liu, Guijun Xian, Yan Li, Fangbo Ding, Chris Rudd, Xiaoling Liu, and Pooria Khalili

1 Introduction

With the present state of composite technological development, biocomposites are understood as composites that consist of biopolymer matrices, i.e., bio-sourced resins and/or natural fiber reinforcements, e.g., plant fibers (PFs) (Fig. 9.1). Initially, biocomposites, and plant-fiber-reinforced composites (PFRCs) in particular, were developed in automotive industry as nonwoven given their low density, low cost, and environmentally friendly interior materials (Powers 2000; Drzal et al. 2001).

X. Yi (✉)
The University of Nottingham Ningbo China, Ningbo, China

AVIC Composite Corporation Ltd.,/ACC Tech, Beijing, China
e-mail: Xiaosu.yi@nottingham.edu.cn

J. Tong · X. Zhang
AVIC Composite Corporation Ltd.,/ACC Tech, Beijing, China

J. Zhu · X. Liu
Ningbo Institute of Materials Technology and Engineering, Ningbo, China
e-mail: jzhu@nimte.ac.cn; liuxq@nimte.ac.cn

G. Xian
Harbin Institute of Technology, Harbin, China
e-mail: gjxian@hit.edu.cn

Y. Li
Tongji University, Shanghai, China
e-mail: liyan@tongji.edu.cn

F. Ding
AVIC XAC Commercial Aircraft Co. Ltd., Xi'an, China

C. Rudd · X. Liu · P. Khalili
The University of Nottingham Ningbo China, Ningbo, China
e-mail: Chris.rudd@nottingham.edu.cn; Pooria.khalili@nottingham.edu.cn

S. Pantelakis, K. Tserpes (eds.), *Revolutionizing Aircraft Materials and Processes*, https://doi.org/10.1007/978-3-030-35346-9_9

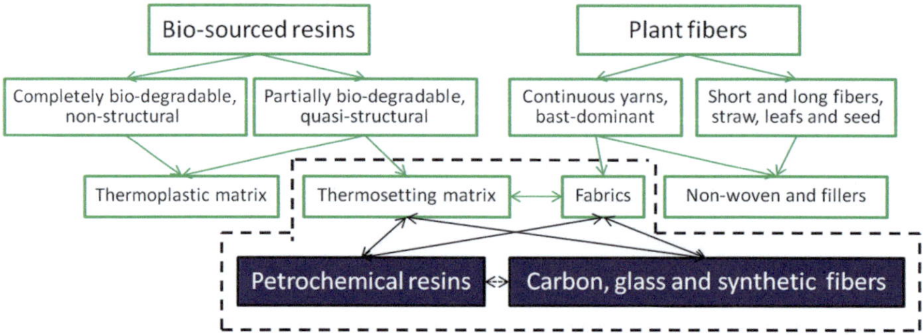

Fig. 9.1 Biocomposite system and the research scope of the present study

Manufacturers, however, have since learned that these materials offer both structural and damping benefits. This may render them an economical and environmentally friendly alternative to glass fiber-reinforced composites for quasi-structural applications.

On the other hand, in the air and ground transportation sectors, new environmental regulations and societal concerns have triggered a search for new products and processes that complement resources and the environment. To address this issue, the present work gives an overview on biocomposite in general but, in particular, on the development progress of biocomposite materials by an internationally joint project, ECO-COMPASS (Eco-compass.eu 2016; Bachmann et al. 2018) (Ecological and Multifunctional Composites for Application in Aircraft Interior and Secondary Structures, 2016–2019), which is co-funded by the Chinese Ministry of Industry and Information Technology (MIIT) and the European Union. The focus of the book chapter is shown in the dashed frame of figure.

2 Bio-based Polymers

2.1 Bio-based Thermoplastics

2.1.1 Polylactic Acid (PLA)

Polylactic acid (PLA) is one of the most promising biodegradable eco-friendly polymers, and PLA has the largest volume in production and lowest price. PLA is gained from corn starch, and it will not produce extra CO_2 during the processing of production and consumption. This is because the water and CO_2 produced during degradation will be reused by plant's photosynthesis. Therefore, PLA offers low carbon footprints throughout the cycle of production to consumption (Natureworksllc.com 2019).

PLA is mostly used in packaging applications, and modified PLA can be used as engineering plastics in electronics, automobiles, and buildings according to literature reports with improved properties (Huang et al. 2009). In the latter case scenario, PLA can be considered as quasi-structural plastic which is hardly biodegradable, referred to Fig. 9.1.

Although PLA has many promising applications, there are drawbacks limiting its wide use such as low crystallization rate, high mold shrinkage, low processing and dimensional stabilities, brittleness, and low heat distortion temperature (HDT) (Ikada and Tsuji 2000; Tsuji and Miyauchi 2001). In particular, its low HDT (less than 55 °C) is the biggest problem since the molded PLA parts would be deformed during transportation due to high temperature (more than 60 °C) in cabins. For this defect, heat-resistant polylactic acid has been developed (Tang et al. 2011), which is obtained through direct extrusion blending without addition of any inorganic filler.

The crystallization rate of heat-resistant PLA was much faster than neat PLA. Isothermal differential scanning calorimeter (DSC) was used to get crystallization half-time. Crystallization half-time of the modified PLA at 105 °C was significantly reduced from 16 to 2.5 min (see Fig. 9.2). Although PLA with a nucleating agent demonstrated faster crystallization rate than neat PLA, it is still not as good as heat-resistant PLA. Experiments from polarizing microscope showed that the crystal size of the modified PLA was significantly smaller than neat PLA at crystallization temperature of 105 °C (see Fig. 9.3).

The heat-resistant PLA was molded at mold temperature of 105 °C for 40 s. The properties of the injection molded samples were measured according to ISO standard test. The properties are shown in Table 9.1. The density of the heat-resistant PLA was 1.24 g/cm^2. The HDT was more than 120 °C, which meets the user requirements of conventional heat-resistant plastics. Heat-resistant PLA remained good mechanical strength and modulus of PLA. It is noteworthy that the impact strength of heat-resistant PLA increased. The heat-resistant experiment of heat-resistant

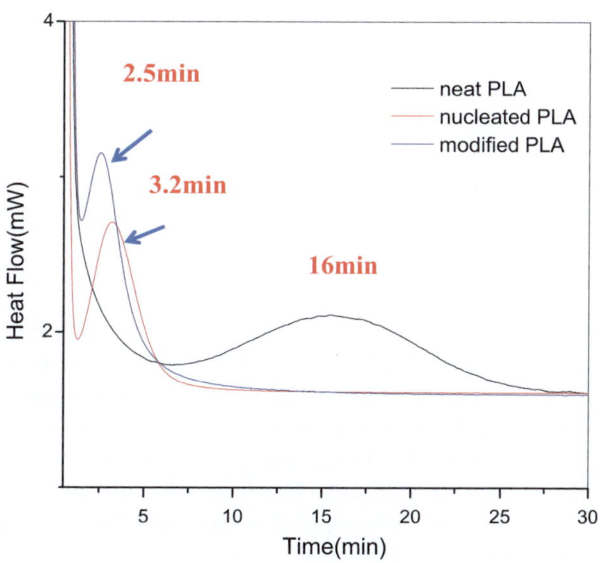

Fig. 9.2 Isothermal DSC curves of heat-resistant PLA

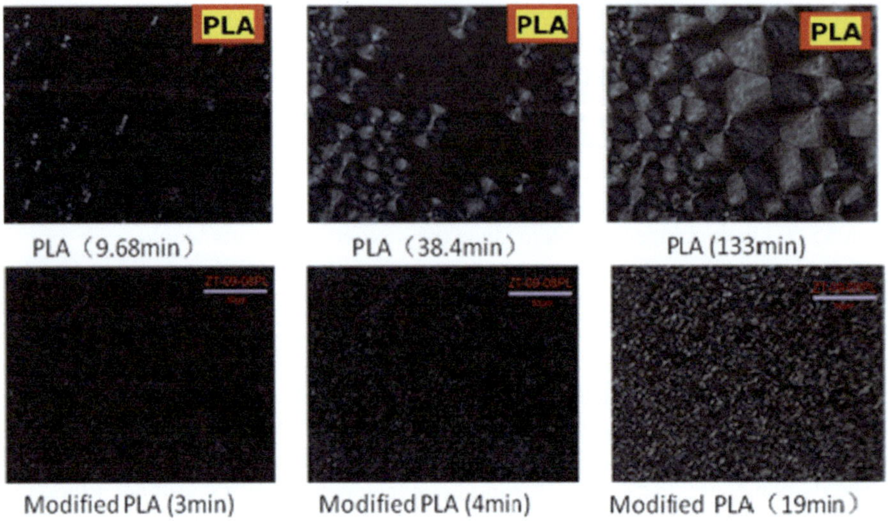

Fig. 9.3 Crystal size of modified PLA with time at 105 °C

Table 9.1 The physical and mechanical properties of the modified PLA

	PLA	Modified PLA	Test methods
Physical properties			
Density (g/cm^3)	1.24	1.24	ISO 1183
Melt flow index (g/10 min, 190 °C/2.16 kg)	5–15	1–5	ISO 1133
Melting point (°C)	155–170	155–170	DSC
HDT (°C) (0.45 MPa)	55	120–140	ISO 75
Mechanical properties			
Tensile strength at break (MPa)	60	50–65	ISO 527
Tensile modulus (MPa)	3400	3500–4000	ISO 527
Flexural strength (MPa)	110	95–105	ISO 527
Flexural modulus(MPa)	3400	3500–4000	ISO 175
Elongation at break (%)	3	3–6	ISO 175
Notched impact strength (KJ/m^2)	3	4–6	ISO 179

PLA was done at 120 °C. The result was shown in Fig. 9.4. The specimen of heat-resistant PLA held well while that of neat PLA was distorted upon heating at 120 °C.

Meanwhile, the heat-resistant PLA was able to maintain good clarity. As shown in Fig. 9.5, the transparency of the heat-resistant PLA was much better than neat PLA and nucleated PLA. In Fig. 9.5, all the samples were annealed to achieve their highest crystallinity. The light transmittance of heat-resistant PLA showed 82.6%. The good clarity of heat-resistant PLA was due to the small size of the spherulites.

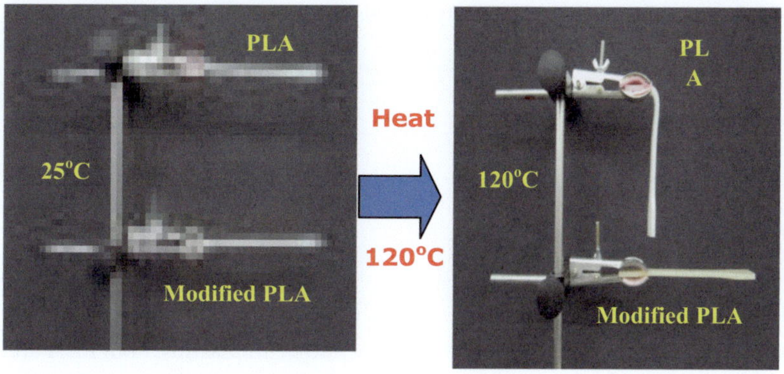

Fig. 9.4 Heat test of heat-resistant PLA

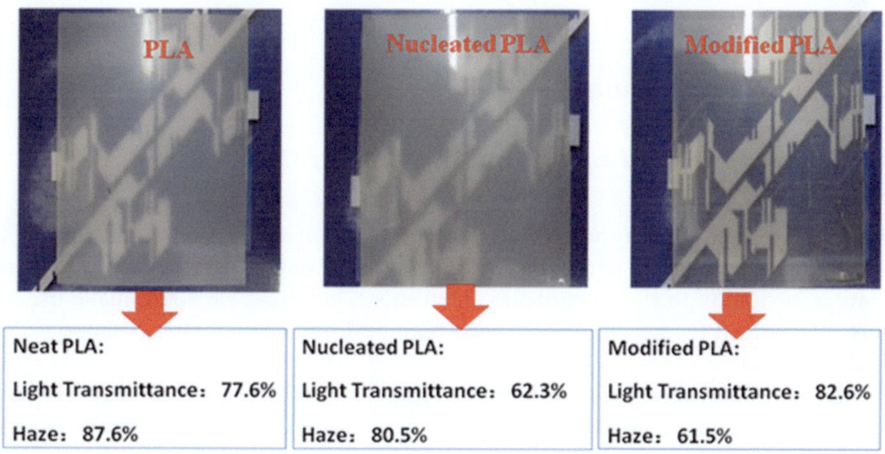

Fig. 9.5 Clarity of heat-resistant PLA

2.1.2 Polyhydroxybutyrate (PHB) and the Others

Bodros et al. (2007) studied the glass transition temperature (T_g), melting temperature (T_m), and density of commonly used bio-thermoplastics as well as polypropylene (PP) for comparison purpose (Table 9.2). The bio-based thermoplastic polymers investigated were PLA, poly (butylene succinate) (PBS), poly(butylene adipate-co-terephthalate) (PBAT), poly(3-hydroxybutyrate) (PHB), poly-L-lactic acid (PLLA), and starch thermoplastic (Mater-Bi® Z). The synthetic PP was blended with 5% of PP-g-maleic anhydride (MA).

Differential scanning calorimeter (DSC) was used to measure T_g and T_m of polymers, and a pycnometer was employed for density measurement. It was observed that T_g of PLA thermoplastics showed greater values than that of other bio-based thermoplastics and PP. Among bio-based thermoplastics, PLAs had the highest T_m,

Table 9.2 T_g, T_m, and density of commonly used bio-thermoplastics vs. PP

Bio-thermoplastic	Tg (°C)	T_m (°C)	Density (g/cm^3)
PLA 4042D	56	175	1.27
Bionolle 1020 (PBS)	−40	115	1.26
Ecoflex (PBAT)	n.a.	115	1.27
PHB P226	−10	161	1.25
PLLA L9000	65	169	1.25
Mater-Bi® ZF03U/A	−50	55	1.25
PP-g-MA (5%)	−10	170	0.91

followed by PHB biopolymer. For the density of thermoplastics, both PLA and PHB showed approximately the same value, and this was almost the same for other bio-based thermoplastics. However, synthetic PP demonstrated the lowest density of 0.91 g/cm^3 relative to that of bio-thermoplastics.

3 Bio-based Thermosets

3.1 Rosin-Sourced Epoxy and Curing Agents

Rosin is an important natural product. It is composed of *ca.* 90% acidic and *ca.* 10% neutral compounds. The hydrogenated phenanthrene ring structure in rosin acids is similar in rigidity to rigid petroleum chemicals. Therefore, we could make use of this bio-sourced feedstock to synthesize high-performance polymeric materials Yi and Li (2017).

Two rosin derivatives with different functionality, maleopimaric acid (MPA) and methyl maleopimarate (MMP) (Fig. 9.6), were synthesized and studied as curing agents for the curing of a commercial epoxy resin (Liu et al. 2018). For comparison, two commercial epoxy curing agents, 1,2,4-benzenetricarboxylic anhydride (BTCA) and 1,2-cyclohexanedicarboxylic anhydride (CHDB) which resemble MPA and MMP in structure and functionality, respectively, were also studied for the curing of the same epoxy (Liu et al. 2009). In addition, a rosin-sourced epoxy with three functional groups was synthesized and cured with a rosin-sourced curing agent (MPA). Its mechanical and thermal properties were investigated in detail (Liu et al. 2012a, b).

Figure 9.6 shows the chemical structures for the rosin-based curing agents (MMP and MPA) and their counterparts. Obviously, they possess similar rigid ring structures and functionalities. In order to investigate and compare their behaviors in curing epoxy, the epoxy DER332 was cured with them, and the dynamical thermal properties of the resulting resins were studied.

Figure 9.7 displays the dynamic mechanical analysis (DMA) curves for the resins cured with different curing agents. The resin cured with MPA had the highest T_g, which was followed by the resin cured with BTCA, MMP, and CHDB. The

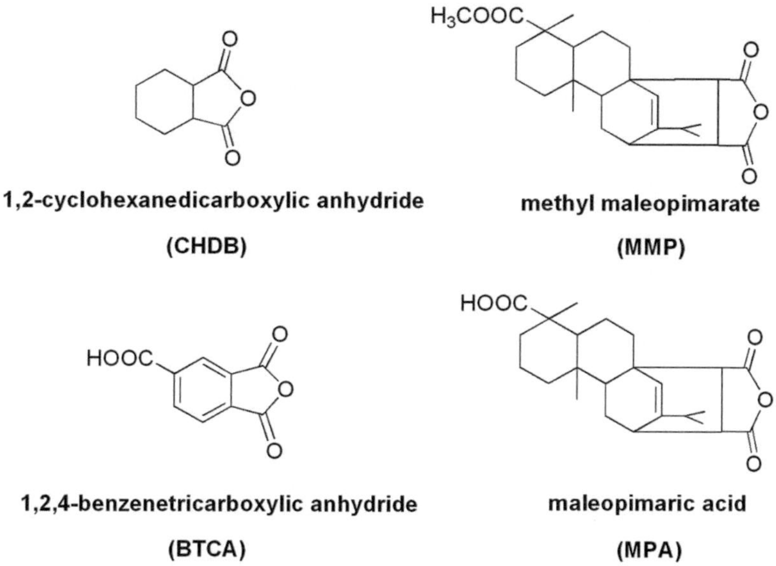

Fig. 9.6 Chemical structures for the rosin-based curing agents (MMP and MPA) and their counterparts (CHDB and BTCA)

difference in T_g values of the cured resins corresponded to the differences in chemical structure and functionality of the curing agents. Because the functionality of BTCA or MPA ($f = 3$) was higher than that of CHDB or MMP ($f = 2$), the epoxy resins cured with BTCA or MPA possessed higher cross-link density than the resins cured with the CHDB or MMP, displaying high T_g. It is also noted that the resins cured with rosin-based curing agents exhibited higher T_g (*ca.* 10 °C) than the resins cured with their corresponding commercial counterparts.

Figure 9.8 shows the dynamic mechanical properties of triglycidyl ester of maleopimaric acid cured with maleopimaric acid. This material had high glass transition temperature (T_g) and high storage modulus of 164 °C and 3.2 GPa at room temperature, respectively. To the best of our knowledge, this is the first fully bio-based thermosetting resin that has demonstrated outstanding dynamic mechanical properties and thermal properties.

3.2 *Epoxy Resin Based on Itaconic Acid*

Itaconic acid, which is also referred to as methylenesuccinic acid, is typically produced through the fermentation of carbohydrates such as glucose or starch using *Aspergillus terreus*. Given its strong capacity to replace petrochemicals in the chemical industry, it has been selected as one of the top 12 potential bio-based platform chemicals by U.S. Department of Energy (Werpy and Petersen 2004). To the best of

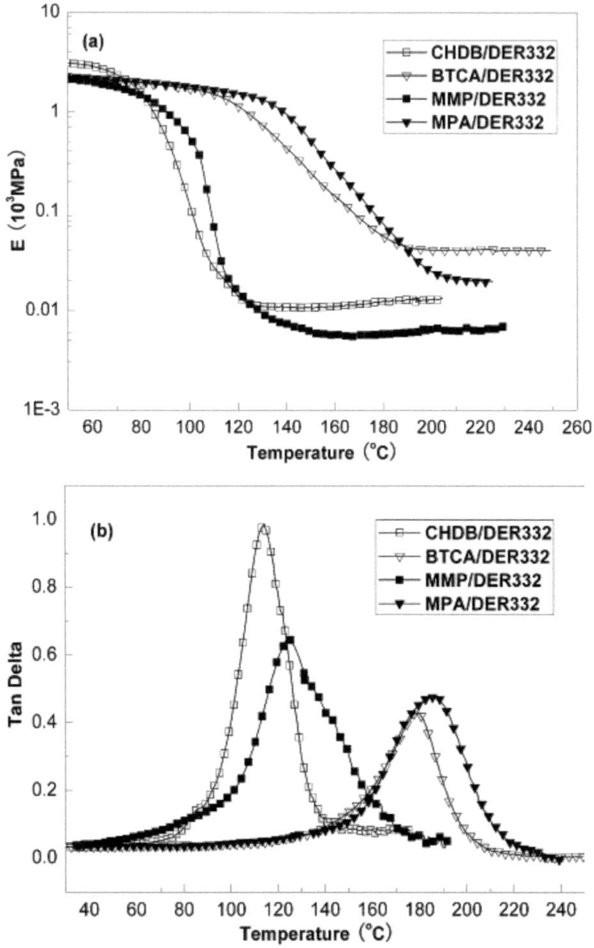

Fig. 9.7 Storage modulus vs. temperature (**a**) and tan δ vs. temperature (**b**) for DER332 cured with different curing agents

our knowledge, it has been widely used in the production of styrene-butadiene-acrylonitrile and acrylate latex in the paper and coating sectors.

EP resin (EIA) derived from itaconic acid can be synthesized following the synthetic route shown in Fig. 9.9. To evaluate its properties, EIA and commercial bisphenol A diglycidyl ether (DGEBA) (epoxide equivalent weight of 182–192 g/eq.) were cured with methylhexahydrophthalic anhydride (MHHPA), respectively. The results show that EIA presents higher epoxide (0.625) and higher reactivity values than DGEBA. Relative to DGEBA, the cured EIA shows comparable or higher tensile strength (87.5 MPa), elongation at break (7.1%), flexural strength (152.4 MPa) and modulus (3430.8 MPa), and glass transition temperature (T_g = 130 °C). In addition, after comonomers such as divinylbenzene (DVB) and acrylated epoxidized soybean oil (AESO) were introduced into the EIA/MHHPA system, dual-curing systems were formed, and rigidity and toughness could be

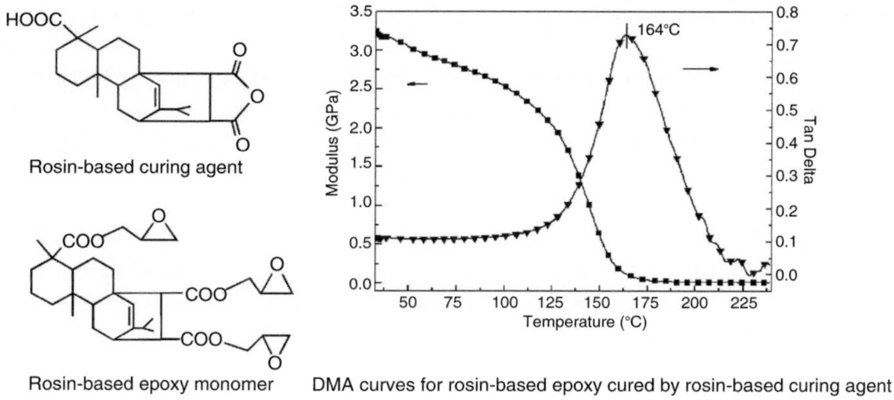

Fig. 9.8 Rosin-based epoxy and DMA curve for the cured resin

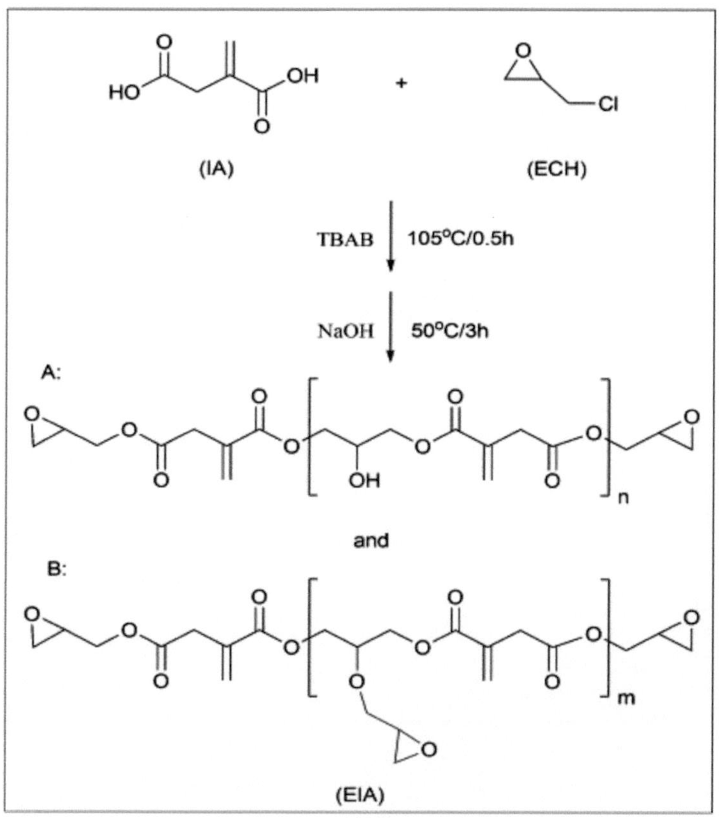

Fig. 9.9 Synthetic route and chemical structures of itaconic acid-based epoxy resin (EIA)

manipulated further via various contents of rigid DVB or soft AESO, as shown in Fig. 9.10.

As shown in Fig. 9.11, direct reactions between itaconic acid and epichlorohydrin generate resin EIA, which is a mixture of different oligomers of varying molecular weights. To make the best use of the carboxyl groups and itaconic acid double bond, a trifunctional EP monomer trifunctional epoxy resin of itaconic acid (TEIA) was designed and synthesized (Fig. 9.11), and it generated an extremely high epoxide value of 1.16 and a low viscosity of 0.92 Pa s at 25 °C. It is well-known that low resin viscosities are beneficial for manufacturing. Lower viscosities in TEIA render it easier to process than DGEBA. In Table 9.3, flexural properties of TEIA and DGEBA are listed. When flexible poly(propylene glycol) bis(2-aminopropyl ether)

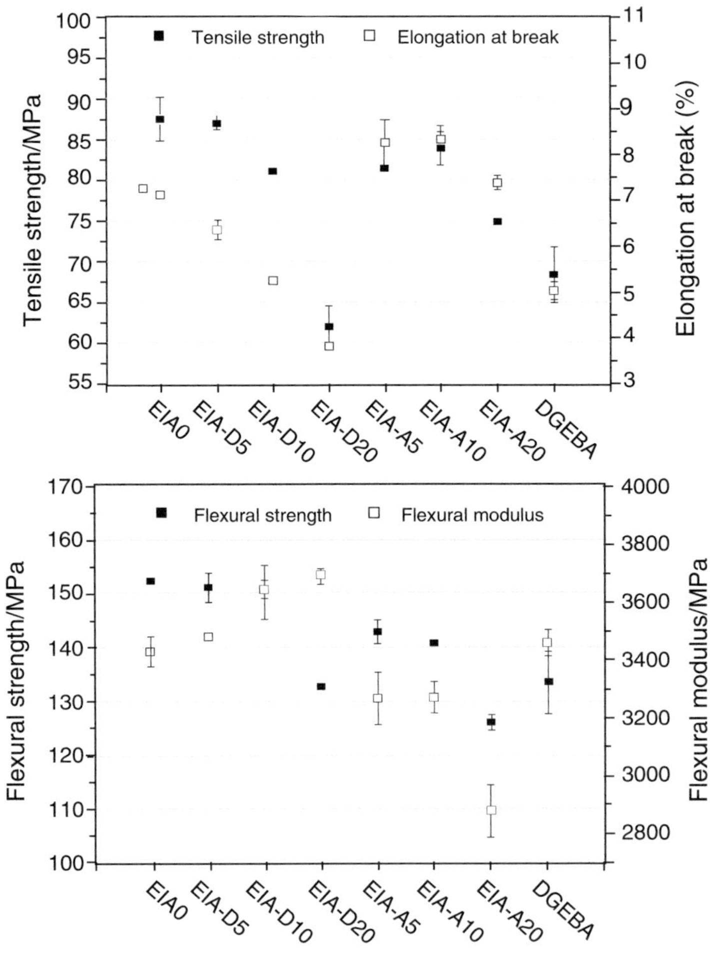

Fig. 9.10 Mechanical properties of the cured EP resins. EIA0 and DGEBA refer to the cured samples without comonomers; D and A denote the DVB and AESO comonomers, respectively

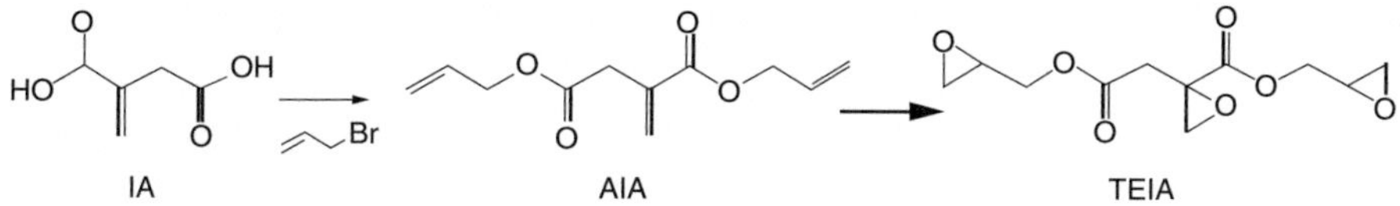

Fig. 9.11 Synthesis of the trifunctional epoxy resin of itaconic acid (TEIA)

Table 9.3 Flexural properties of cured EP resins with D230 curing agents

Samples	Flexural strength (MPa)	Flexural modulus (MPa)	Elongation at break (%)
DGEBA/D230	121 ± 1	2952 ± 18	13 ± 2
TEIA/D230	117 ± 2	3603 ± 77	21 ± 1

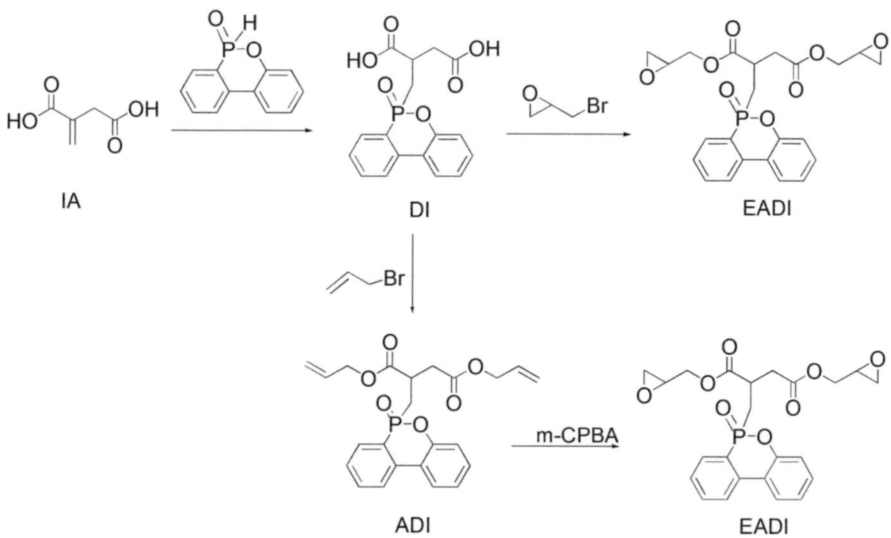

Fig. 9.12 Synthetic route of phosphorus-containing itaconic acid-based epoxy resin

(D230) is used as a curing agent, the TEIA/D230 system shows higher flexural modulus, higher strain at break, and comparable flexural strength than DGEBA/D230. When a rigid MHHPA was employed as a curing agent, flexural strength, modulus, and elongation at break of the TEIA/MHHPA system exceeded those of the DGEBA/MHHPA system. These results denote that TEIA may be used as a high-performance epoxy resin (Ma et al. 2013a, b, c).

Given the presence of double bonds in itaconic acid and the low flame retardancy of EP resins, a flame-resistant DOPO was chemically incorporated into the itaconic acid-based EP resin, and a phosphorus-containing EP resin (EADI) was obtained (Fig. 9.12). The study shows that the cured EADI network presents comparable glass transition temperature and mechanical properties with those of the DGEBA system. In addition, excellent flame retardant capacities with UL94 V-0 grade obtained from vertical burning tests were observed for the EADI system. EADI may be used as a bio-based candidate for flame-retarded EP (Ma et al. 2013a, b, c).

4 Biocomposites with Rosin-Sourced Epoxy as Matrix Resin

In this study (Zhang et al. 2017), an anhydride-type epoxy curing agent, the maleopimaric, was synthesized as the hardener from rosin acid. This was supplied by Ningbo Institute of Materials Technology and Engineering. A resin mixed with an E51-type epoxy and a solid phenolic epoxy was then prepared as main component, together with an accelerator mixed of two amino imidazole salts as thermally latent curing agent and also as catalyst for the anhydride. Figure 9.13 exhibits the glass transition behavior of the formulated resin system. As shown, the curing degree increases with the curing temperatures for a constant curing time of 3 h, as well as the glass transition temperature. The formulated matrix resin is finally designated as AGMP3600, with a bio-content of about 30%.

Figure 9.14 shows the typical viscosity behavior of the trial product AGMP3600. It works well in the film manufacturing and subsequent prepreg production (Yi et al. 2018). The process condition for the prepreg using AGMP3600 as matrix in autoclave is shown in Fig. 9.15.

The mechanical properties of AGMP 3600 laminates reinforced with different kinds of fibers and weaves were determined, and they are listed in Tables 9.4, 9.5, 9.6 (Yi et al. 2018), in some cases compared with the state-of-the-art counterparts as reference.

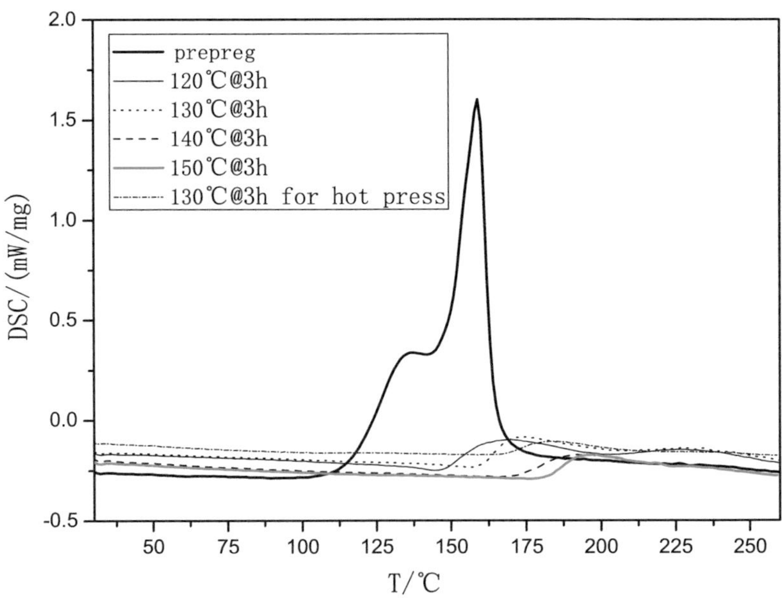

Fig. 9.13 DSC curves of the rosin-sourced epoxy resin system designated as AGMP3600 at different temperature conditions (Yi et al. 2018)

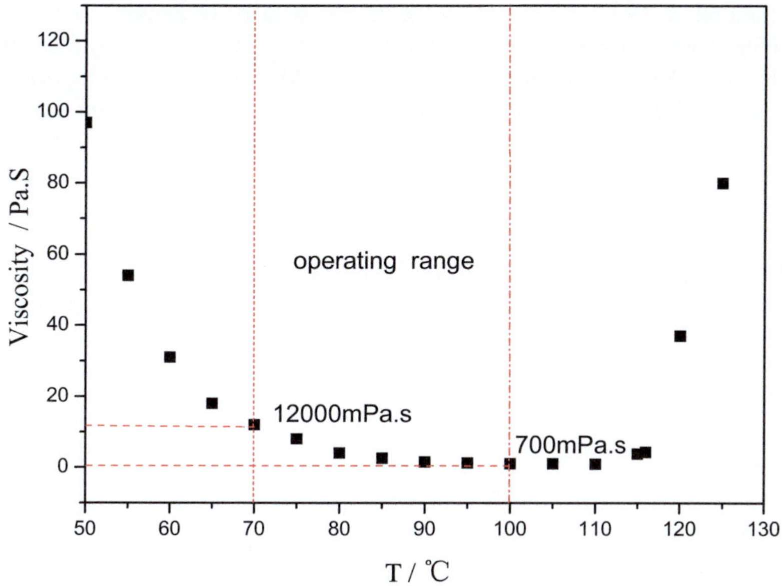

Fig. 9.14 Viscosity vs. temperature of AGMP3600, a rosin-sourced epoxy resin system (Yi et al. 2018)

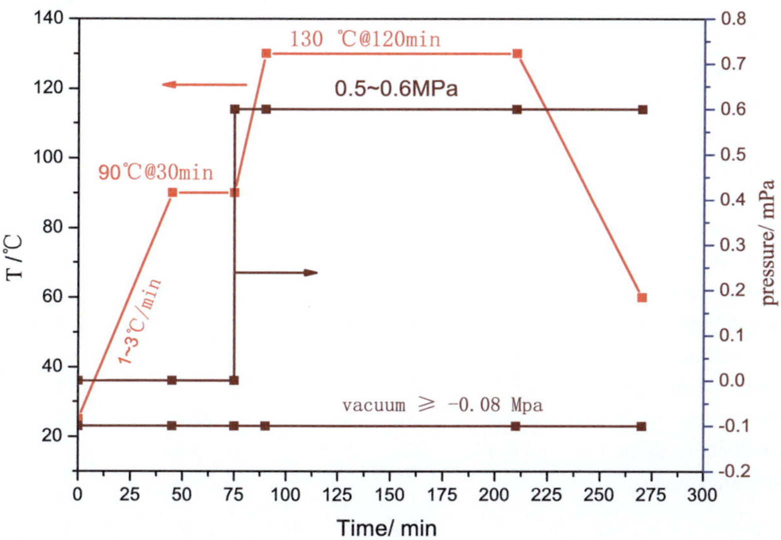

Fig. 9.15 Process parameters for curing of AGMP3600 prepreg in autoclave (Yi et al. 2018)

In parallel, a 180 °C/2 h cure rosin epoxy is also developed, designated as AGMP5600, with a higher bio-content of about 40%. It is particularly interesting to note that the glass transition temperature of AGMP5600 matrix composite reinforced with EW250F glass weave is about 220 °C. Table 9.7 lists the mechanical properties of the laminated composite. The temperature-dependent mechanical properties were also studied, as well as those after the hot/wet exposure (1000 h/70 °C/85% RH).

Table 9.4 Mechanical properties of AGMP3600/EW250F (glass fabric, 8 satin, 47 vol.%), a rosin-sourced epoxy laminate, and of 3233B/EW250F, a standard intermediate-temperature epoxy laminate as reference in comparison

Mechanical properties	Unit	3233B/EW250F	AGMP3600/EW250F
Ply thickness	mm	0.241	0.237
Bending strength	MPa	696	674
Bending modulus	GPa	19.9	21.1
Short-beam shear strength	MPa	45	60.9
Tensile strength	MPa	470	540
Tensile modulus	GPa	22.5	25.3
Compression strength	MPa	474	483
Compression modulus	GPa	23.9	27

Table 9.5 Mechanical properties of AGMP3600/CW3011 (carbon fiber plain weave, 200 g/m^2, 57 vol.%), a rosin-epoxy laminate and a corresponding reference laminate for comparison

Property and test condition		Unit	Reference[a]	Test result	Standard
Tensile strength warp	RT/dry	MPa	≥500	707	ASTM D3039
Tensile modulus warp	RT/dry	GPa	65 ± 8	62.3	
Tensile strength weft	RT/dry	MPa	≥500	557	
Tensile modulus weft	RT/dry	GPa	65 ± 8	60.9	
Compression strength warp	RT/dry	MPa	≥300	509	ASTM D6641
Compression modulus warp	RT/dry	GPa	58 ± 8	61.2	
Compression strength weft	RT/dry	MPa	≥280	362	
Compression modulus weft	RT/dry	GPa	57 ± 8	57.7	
Bending strength warp	RT/dry	MPa	≥650	883	ASTM D 790
Bending modulus warp	RT/dry	GPa	58 ± 8	56.8	
Short-beam shear strength	RT/dry	MPa	≥50	55.7	ASTM D2344
In-plane shear strength	RT/dry	MPa	≥45	72.6	ASTM D3518
In-plane shear modulus	RT/dry	GPa	3.5 ± 1	3.84	

[a]A commercial product

Table 9.6 Mechanical properties of AGMP3600/A38 (carbon fiber, twill fabric, 200 g/m^2, 42 vol.%), a rosin-epoxy laminate, at a hydrothermal condition of 1000 h/70 °C/85% RH

	Unit	Humidity	Temperature (°C)	Reference[a]	AGMP 3600/A38	Standard
Ply thickness	mm		–	0.26	0.265	
Flexural strength warp	MPa	Dry	−55	–	949	ASTMD 790-03
			RT	–	947	
			70	–	903	
		70 °C/wet 85%	70	–	826	
Flexural modulus warp	GPa	Dry	−55	–	50.9	
			RT	–	51.4	
			70	–	58.7	
		70 °C/wet 85%	70	–	51.8	
Interlaminar shear strength warp	MPa	Dry	−55	65	62.4	ASTMD 2344/D2344M-00(2006)
			RT	60	74.8	
			70	40	65.3	
		70 °C/wet 85%	70	19	46.4	
Tensile strength warp	MPa	Dry	−55	500	618	ASTMD 3039/D3039M-e1
			RT	625	667	
			70	590	714	
		70 °C/wet 85%	70	560	674	
Tensile modulus warp	GPa	Dry	−55	52 ± 6	55.25	
			RT	52 ± 6	54.6	
			70	52 ± 6	53	
		70 °C/wet 85%	70	52 ± 6	49	
Poisson ratio	–	Dry	RT	0.05 ± 0.005	0.052	

(continued)

Table 9.6 (continued)

	Unit	Humidity	Temperature (°C)	Reference[a]	AGMP 3600/A38	Standard
Compression strength warp	MPa	Dry	−55	600	758	ASTMD 6641/D6641M-14
			RT	535	651	
			70	430	630	
		70 °C/wet 85%	70	310	605	
Compression modulus warp	GPa	Dry	−55	46 ± 6	52.5	
			RT	46 ± 6	53	
			70	46 ± 6	52.5	
		70 °C/wet 85%	70	46 ± 6	55	
In-plane shear strength	MPa	Dry	−55	100	136	ASTMD 3518/D3518M-94
			RT	95	114	
			70	80	96.8	
		70 °C/wet 85%	70	60	89	
In-plane shear modulus	GPa	Dry	−55	4.5 ± 0.35	4.74	
			RT	3.65 ± 0.35	4.21	
			70	3.5 ± 0.35	2.63	
		70 °C/wet 85%	70	1.25 ± 0.35	2.75	
CAI (layup (+/0/−/90)2 s, energy 25 J)	MPa	Dry	RT	180	185	ASTMD 7136/7137
Filled hole tension strength	MPa	Dry	RT	180	317	ASTMD3518
Filled hole compression strength	MPa	Dry	RT	250	No destroy, displacement 4 mm	ASTMD6742

[a]A commercial product

Table 9.7 Mechanical properties of AGMP5600/EW250F (glass fabric, 8 satin, 47 vol.%), a 180 °C cure rosin-sourced epoxy laminate, at different hydrothermal conditions

Mechanical properties	Unit	Test condition	Reference[a]	AGMP5600/EW250F
Ply thickness	mm	–	0.26	0.265
Flexural strength warp	MPa	−55	_	864
		RT	_	687
		70	_	650
		70 °C/wet 85%	_	431
Flexural modulus warp	GPa	−55	_	21.4
		RT	_	23.4
		70	_	22
		70 °C/wet 85%	_	22.2
Interlaminar shear strength warp	MPa	−55	88	74.6
		RT	68	59.1
		70	59	49
		70 °C/wet 85%	39.6	48.3
Tensile strength warp	MPa	−55	500	573
		RT	410	510
		70	330	460
		70 °C/wet 85%	315	–
Tensile modulus warp	GPa	−55	27 ± 6	23.9
		RT	24 ± 6	24.4
		70	24 ± 6	22.3
		70 °C/wet 85%	24 ± 6	–
Compression strength warp	MPa	−55	800	589
		RT	660	456
		70	550	397
		70 °C/wet 85%	470	384
Compression modulus warp	GPa	−55	26 ± 3	26.7
		RT	25 ± 3	25.8
		70	25 ± 3	24
		70 °C/wet 85%	25 ± 3	24.7
In-plane shear strength	MPa	−55	110	113
		RT	85	90
		70	88	71
		70 °C/wet 85%	77	56.4
In-plane shear modulus	GPa	−55	5.7 ± 1	4.8
		RT	4.8 ± 1	3.26
		70	3.9 ± 1	3.1
		70 °C/wet 85%	3.5 ± 1	3.13

[a]A commercial product

5 Plant-Fiber-Reinforced Biocomposites

5.1 Plant Fibers as Reinforcing Component of Biocomposites

Plant fibers (PFs), when used as reinforcing components of composites, are inexpensive, lightweight, and biodegradable and present unique mechanical, acoustic, and damping properties owing to their chemical and structural characteristics.

In relation to man-made fibers, PFs are nonuniform with irregular cross sections; ramie, which is characterized by oval-shaped cross sections, serves as a typical case (Fig. 9.16). Ramie fiber (RF), commonly known as China grass, is one of the oldest natural textile fibers grown primarily in China. Ramie surfaces are naturally coarse and bumpy. The average diameters are estimated between 100 and 300 μm, with the presence of dozens of microfibers of varying diameters.

Figure 9.17 compares ramie cross sections with those of the other three fiber types examined (Li et al. 2008). Ramie has a unique single-cell structure, i.e., a single-cell fiber with a lumen interior, whereas jute, kenaf, and sisal fibers are composed of several to several tens of cell fiber bundles.

5.1.1 Surface Treatment for Plant Fibers

It is widely acknowledged that the inherent polar and hydrophilic nature of lignocellulose fibers and nonpolar hydrophobic characteristics of most polymeric resins cause poor interfacial bonding between fibers and matrix resins, impairing mechanical properties of the composite. Silane, permanganate, and alkali are standard chemicals used to pre-treat PFs to improve interfacial properties. It is well studied that the ramie extracted from a woven fabric has been rendered much coarser as a

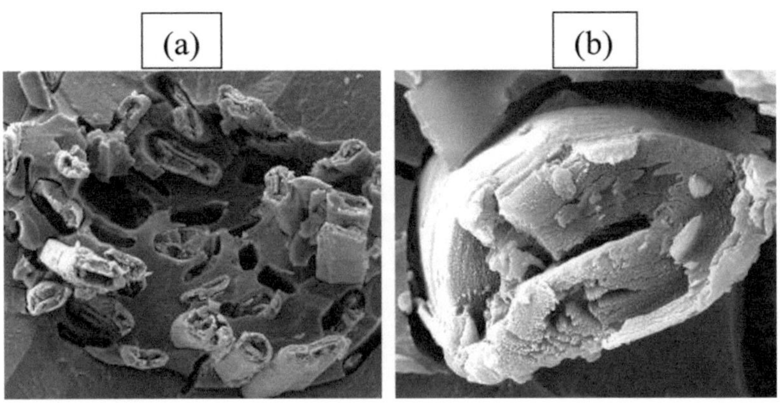

Fig. 9.16 SEM micrographs of fractured surfaces of (**a**) ramie and (**b**) a single ramie fiber embedded in epoxy

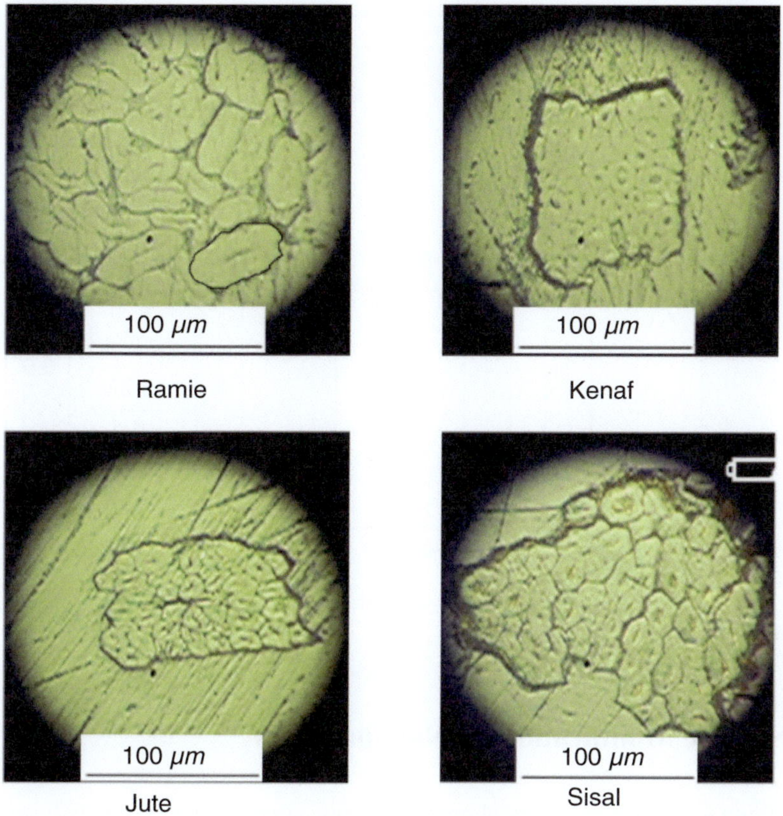

Fig. 9.17 Cross-sectional views of (**a**) ramie, (**b**) kenaf, (**c**) jute, and (**d**) sisal fibers

result of the weaving process. Chemical treatments with $KMnO_4$, NaOH, flame retardant (FR), and silane gradually rendered the ramie surface scaly, rough, and bumpy.

After applying the pre-treatments to different plant fibers, resulting interfacial shear strengths of the PFRCs were used to quantify effectiveness levels (Li et al. 2005), as shown in Fig. 9.18. Overall, interfacial shear strength levels were improved considerably by the corresponding treatments, among which the silane treatment generated the highest values for the fiber types tested. The silane (3-aminopropyltriethoxy silane and glycidoxypropyltrimethoxysilane) and alkali treatments also improved ramie-fabric-reinforced phenolic resin composite (RFRC) tensile strength, flexural strength, and impact toughness levels (Yu et al. 2010). The Fourier transform infrared spectroscopy results showed that via the silane treatment, siloxane groups were grafted on RFs, while the alkali treatment improved fiber surface adhesion levels by removing natural and artificial impurities.

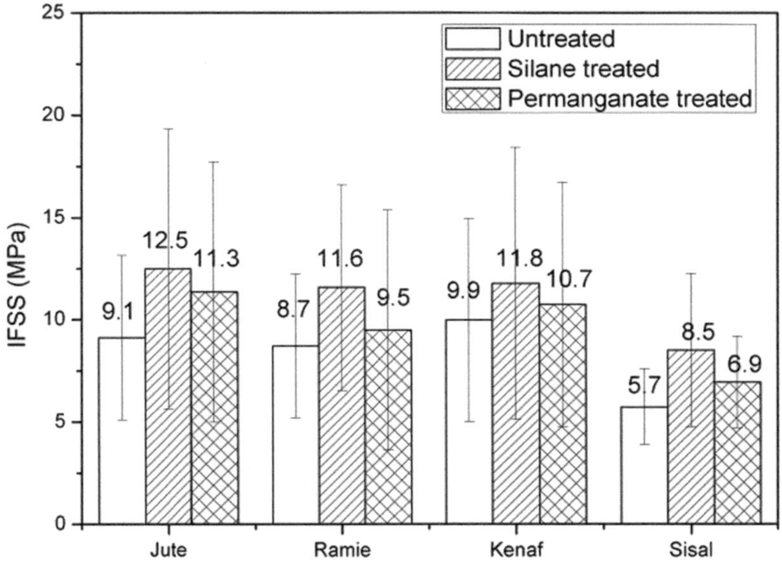

Fig. 9.18 Interfacial shear strengths of different PF-reinforced phenol laminates via the two surface treatments

6 Surface Modification by Nanoparticles

In addition to the use of chemical pre-treatments, attempts have recently been made to explore the use of nanoparticles for the modification of fibers in order to rectify limitations of PFs while enhancing PFRC properties. In this study, zirconia and titanium dioxide nanoparticles were designed to graft onto flax fibers (FF) (Wang et al. 2015). Zirconia (ZrO_2) is of great significance due to its high melting point, superior mechanical properties, low thermal conductivity levels, and high ionic conductivity levels; it can also withstand higher temperatures without emitting toxins (Zhu et al. 2008). ZrO_2 also serves as an excellent catalyst, as it is chemically stable, presents oxidation-reduction properties, and demonstrates surface acidity and alkalescence properties, contributing to its antimicrobial capabilities (Tian et al. 2011; Jangra et al. 2012).

Hydrous zirconia nanoparticles were synthesized via hydrolysis of $ZrOCl_2$ solution and were grafted to FF surfaces through hydrogen bonds between abundant hydrogen groups on FF surfaces and $ZrO_2 \cdot nH_2O$. HCl and ammonium were added to adjust pH values and in turn alter the size of synthesized nanoparticles (Matsui and Ohgai 2005). Flax yarns were surface-modified by hydrous zirconia nanoparticles through a simple chemical process involving the hydrolysis of zirconium oxychloride solution under mechanical stirring and ultra-sonication conditions.

Figure 9.19a–d presents SEM images of the treated and untreated FFs. As shown in Fig. 9.19b, nanoparticles are shown on the FF surface but in a remarkably

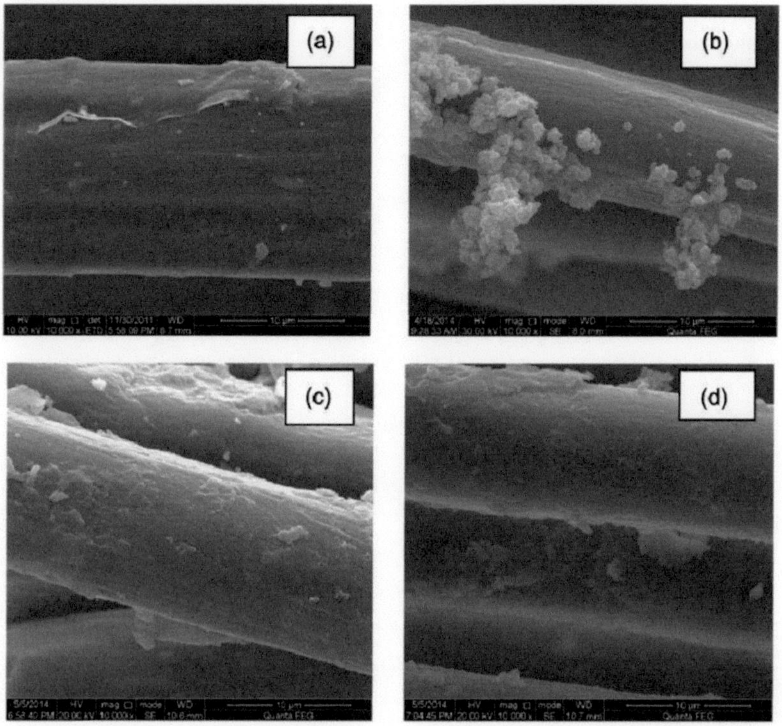

Fig. 9.19 SEM images of surfaces of (**a**) control flax fibers, (**b**) fibers grafted under non-stirring conditions, (**c**) fibers grafted with mechanically stirring, and (**d**) fibers grafted via ultrasound sonication

agglomerated form. In comparison with results shown in Fig. 9.19b, grafting assisted with stirring (Fig. 9.19c) or sonification (Fig. 9.19d) resulted in more homogenous nanoparticle formation patterns on the fiber surfaces. This can be attributed to the fact that stirring and sonification mixing promote the deposition of $ZrO_2 \cdot nH_2O$ particles on fiber surfaces, resulting in reduced particle agglomeration.

Figure 9.20 shows TEM photographs of a cross section of a flax fiber grafted with $ZrO_2 \cdot nH_2O$ particles via ultrasonic processes (the same sample as that shown in Fig. 9.19d). The $ZrO_2 \cdot nH_2O$ particles were grafted onto the FF surfaces in aggregation, and the diameter of the fine particles can be measured in nanometers.

To explore the effect of ZrO_2 surface grafting on mechanical properties of the FF, single-fiber tensile tests were performed. Table 9.8 compares tensile strength levels, Young's modulus values, and elongation patterns at breakage points of the control and grafted fibers. It is clear that $ZrO_2 \cdot nH_2O$ grafting effectively increased tensile strength levels. FF tensile strength improvements were found for the fiber surface treatment due to the removal of fiber surface defects (Xia et al. 2014). In the present study, fibers only treated via sonification (without $ZrOCl_2$ in acid and following with alkaline solution) also showed an improvement in tensile strength levels that

Fig. 9.20 TEM photograph of the ultrasonically treated flax fiber (the same sample as Fig. 9.19d). Note that the left-hand image presents a cross section of the FF and of cracks formed during slicing. The black particles are nanoparticle aggregations

Table 9.8 Single-fiber tensile test results for the control and grafted flax fibers

Samples	Tensile strength (MPa)	Young's modulus (MPa)	Elongation at break (%)
Control fiber	399.7	30.4	1.56
Fiber grafted with non-stirring	512.9	32.8	1.82
Fiber grafted with mechanically stirring	450.3	29.2	1.43
Fiber grafted with sonication	723.4	33.6	1.69
Fiber treated with sonication	428.8	30.2	1.57

can be attributed to sonification effects on fiber individualization properties and dissolving some small molecules on surfaces (e.g., lignin and hemi-cellulose), which reduced the number of defects found on fiber surfaces (Kalia et al. 2009).

The tensile modulus of the FFs, presented in Table 9.8, was found to range from 29 to 33.6 GPa, and fiber treatments do not influence the modulus. This is a reasonable result, as zirconia nanoparticle grafting was mainly found on the fiber surfaces, while fiber stiffness levels were found to be mainly dependent on cellulose microfibrils in the cell wall (Summerscales et al. 2010), which were not affected by the present treatment as expected.

Mechanical properties of polymer composites can be modified by adding nano-size additives into the multi-scale structure. The schematics of experiment methodology is showed in Fig. 9.21. The results of such an experiment are presented in Table 9.8 for PFRC. In this experiment, multiwalled carbon nanotubes (MWCNTs) were added to a ramie/EP laminate. The addition of 0.6 wt% of MWCNTs was found to enhance interlaminar shear strength (ISS), flexural strength, and flexural modulus levels by roughly 38%, 34%, and 37%, respectively (Shen et al. 2014).

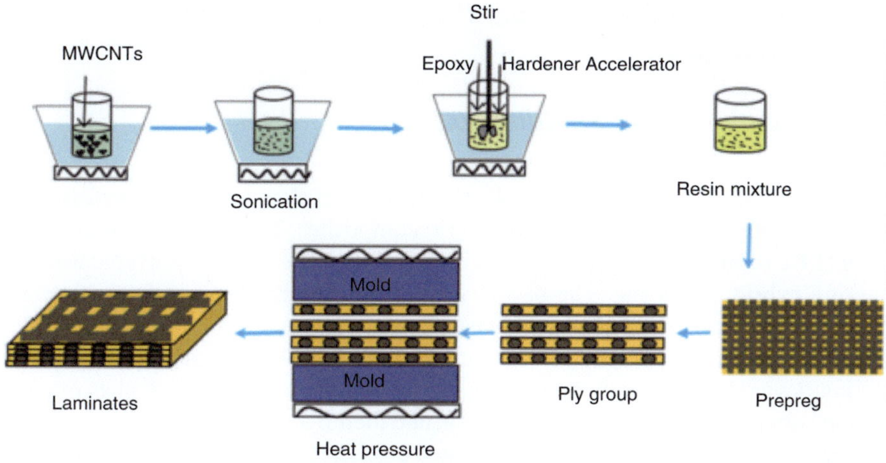

Fig. 9.21 Flowchart for creating multiwalled carbon nanotube-modified ramie/EP composites (Shen et al. 2014)

Fig. 9.22 Results of both control and grafted flax-fiber-reinforced epoxies in the nutrient agar medium after incubation for 3 weeks

More recently, the interfacial microstructure between the plant fibers and the polymer matrix has been explained by hierarchical multi-scaled fracture modes (Li et al. 2018).

Figure 9.22 presents antimicrobial activity patterns of ZrO_2 in the composites. In the control samples, a significant expansion of fungal colonies was observed throughout the composite, whereas in the modified samples, almost no colony growth was found. This reveals the antifungal activity of ZrO_2 nanoparticles (Khan et al. 2014). Thus, composites with significant antimicrobial properties are to be synthesized and are expected to exhibit long-term stability features.

7 Structural and Functional Properties of Laminated Biocomposites

7.1 Interply Hybrid Modification

Combining different fibers into a polymer matrix can alter composite performance levels, leading to the development of hybrid composites. The symmetrical layer-by-layer (LBL) configuration of PF plies with glass fiber plies serves as a typical example of a hybrid composite. The basic hybridization mechanism involves using and balancing characteristic features of plies at a global scale.

In this study (Zhang et al. 2013a, b, c), a key facet of hybridization was found in the interlayer between plant fiber and glass or carbon fiber layers. Ten plies of plain woven laminates of neat flax, glass, carbon, and their hybrid laminates were prepared (Fig. 9.23). The hybrid composites generated possessed varying hybrid ratios but the same total fiber volume fraction. As shown (Fig. 9.23 left, A and B), interlaminar

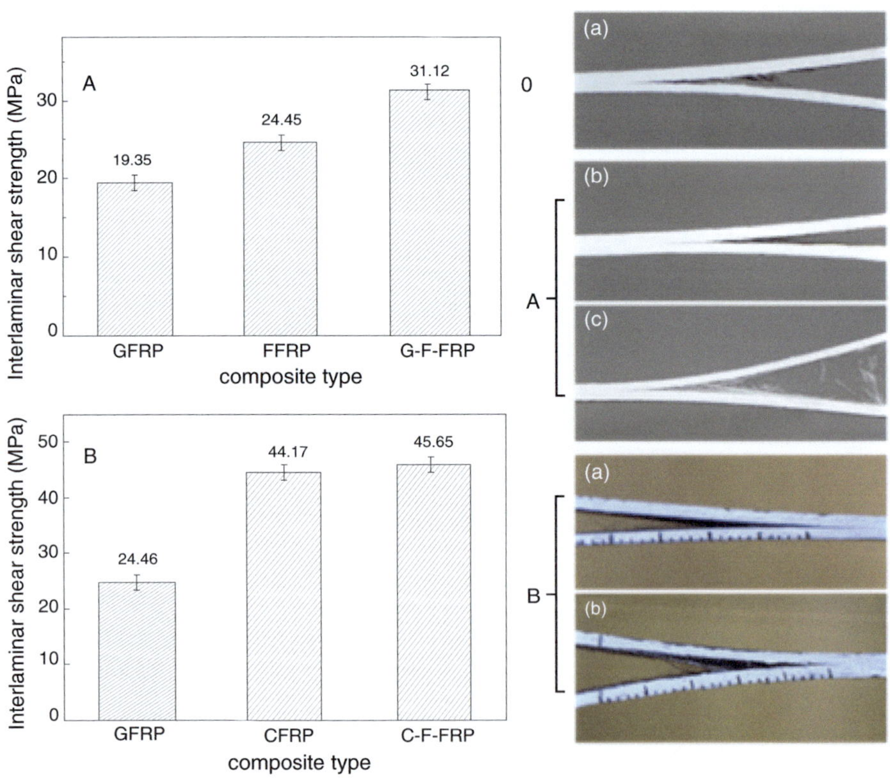

Fig. 9.23 Comparisons of interlaminar shear strengths of composite laminates of neat glass (GFRP), neat flax (FFRP), neat carbon (CFRP), and their hybrids (G-F-FRP and C-F-FRP) composites (left A and B) and crack opening graphs (right, O, A, B) showing various fiber pull-push, breakage, and bridging phenomena

shear strength (ISS) values vary by fiber type. The neat flax laminate generated much higher ISS values than the neat glass laminate (Fig. 9.23 left, A), and the highest value was found in the neat carbon laminate (Fig. 9.23 left, B). The hybrid laminates composed of flax with glass, and of flax with carbon in particular, always generated higher ISS values than their neat-fiber counterparts (compare Fig. 9.23 left, A and B). This exciting result may be attributed to the fact that thicker flax yarns and rougher surfaces of flax plies may interlock adjacent plies, regardless of fiber types involved in hybridization processes.

Optical micrographs confirm this assumption (Fig. 9.23 right). Significant fiber pull-push, breakage, and bridging processes were identified through the crack opening test. This effect is more significant for the flax/glass hybrid laminate than for the flax/carbon laminate (see Fig. 9.23 right, compare A(a) and B(b)). Though the cause of this result is currently unknown, it may be attributed to the fact that the basic ISS of the neat carbon laminate was higher than that of the glass, thus generating the highest ISS value of roughly 46 MPa among the samples tested.

7.2 Structural Damping and Acoustic Properties

Energy absorption by mechanical damping is highly relevant to vibrations that can be dampened in mechanical systems. High-damping materials are used to reduce vibration levels in aircrafts and other machinery, thus extending the service life of components and reducing weight and noise levels. Structural materials generally exhibit low-damping and high modulus properties. Rubbery materials do not exhibit structural properties, and they are thus mainly used for damping purposes. Materials that present both high-damping and stiffness properties are uncommon. Fiber composites are nonhomogeneous materials, and thus major potential sources of composite damping would include the following: the viscoelastic nature of matrix and/or fiber materials, damping as a result of interphase properties, frictional damping as a result of delamination or slippage in unbounded regions between fiber and matrix interfaces, etc. In comparison with man-made fibers, plant fibers exhibit multi-scale and porous structures that may generate a larger number of damping sources. PF laminates were thus studied for the development of structural damping composites.

Figure 9.24 presents representative damping behaviors of a ramie- and jute-fabric-reinforced epoxy laminate placed under free and forced vibration conditions, respectively. The matrix resin used was a 120 °C curing EP for liquid molding. The test samples were manufactured by vacuum infusion. We obtained the ramie laminate loss factor (η) of 0.0129 for the free vibration condition, which is over 30% higher than that of the jute laminate of 0.0099, whereas in the forced vibration condition, η values of the first and second order are 0.143 (η_1) and 0.032 (η_2), respectively, for the ramie and 0.118 (η_1) and 0.025 (η_2), respectively, for the jute. These results suggest that ramie may exhibit higher levels of damping efficiency than jute, though both maintain their inherent structural properties (Fig. 9.25).

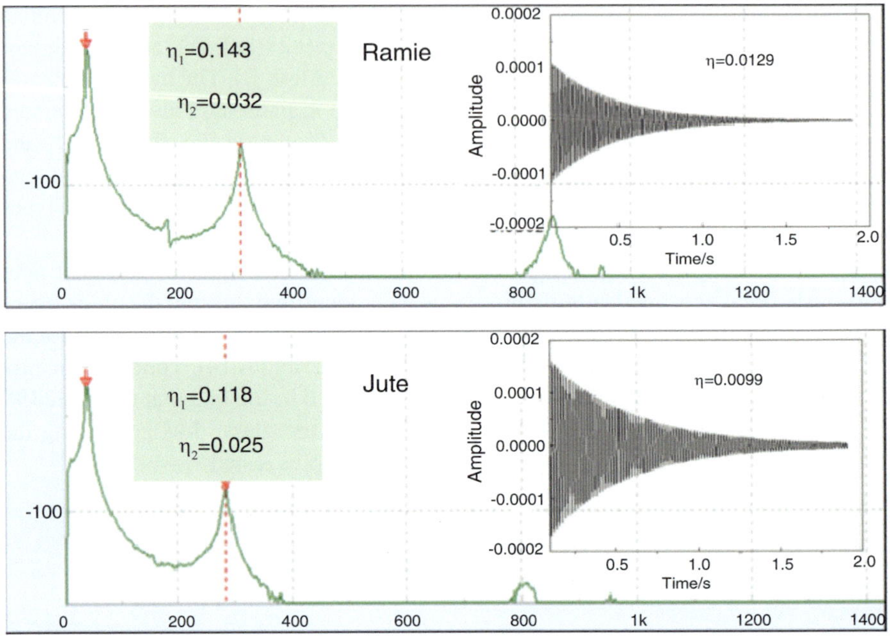

Fig. 9.24 Representative damping behaviors of ramie and jute fabric laminate under free (left) and forced vibration (right) conditions

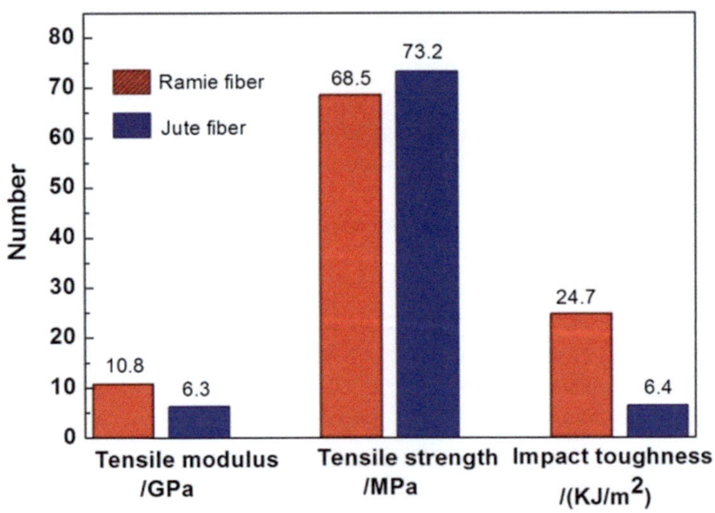

Fig. 9.25 Comparisons between typical mechanical properties of ramie and jute fabric samples

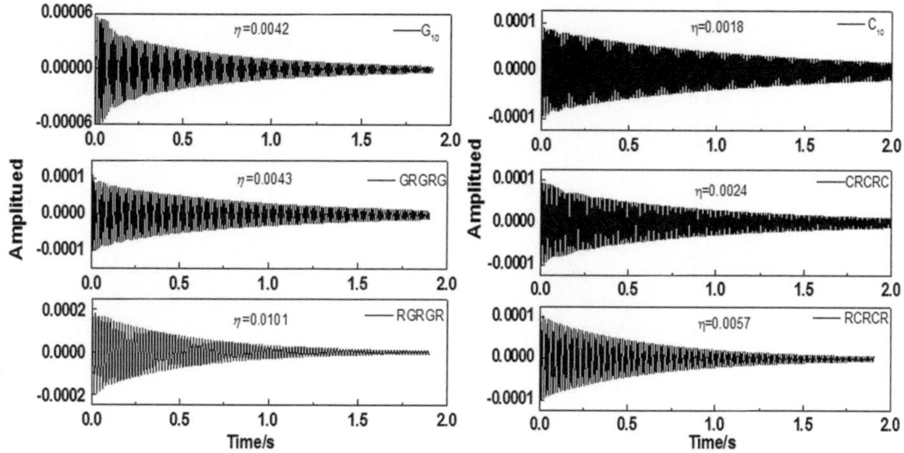

Fig. 9.26 Typical vibration responses of (left) neat glass (G10, 10-ply glass laminate), glass/ramie hybrids (GRGRG, two-ply glass/two-ply ramie/two-ply glass/two-ply ramie/two-ply glass laminate, and RGRGR, with the same configuration), and (right) neat carbon (C10) and carbon/ramie hybrids

Vibration behaviors of a multilayer composite laminate can be controlled via hybridization through the use of alternating symmetrical or asymmetrical plies (Ni et al. 2015). In this study, free and forced vibration responses of neat-fiber laminates composed of ramie, glass, carbon, and their hybrids were examined. Figure 9.26 shows typical vibration responses of neat glass and glass/ramie hybrids and neat carbon and carbon/ramie hybrids. G, R, and C denote two-ply glass, ramie, and carbon layers arranged in a symmetrically alternating manner, respectively. Ten layers were used for each laminate sample.

As shown, vibration levels of the neat glass (G10) (η of 0.0042) and neat carbon (C10) (η of 0.0018) samples were clearly lower than the level found for the neat ramie sample (η of 0.0129) (comparison with Fig. 9.24). Given this difference, which is basically associated with the intrinsic stiffness of the fibers, we predicted their hybrids to behave in a controlled manner: the more ramie plies, the greater the damping response of hybrid laminates. This was found to be true, as the results demonstrated this behavior (Fig. 9.26). Moreover, the hybrid ratio of the specific fiber types and the symmetrical layer stacking sequence significantly affected the resulting vibration behaviors. Generally speaking, arranging ramie layers symmetrically on both laminate surfaces was found to be more effective in terms of damping than covering both surfaces with glass or carbon. The quantitative difference in η appears to be remarkable for the surface configurations. η_1 and η_2 measurements for the forced vibration samples further confirm these results.

Porous material and fiber composite sound absorption occurs due to viscous effects on interfaces and thermal losses resulting from heat transfer processes between different fibers. Sound absorption coefficients of three types of PFs, i.e., ramie, flax, and jute fabrics, were measured using the two-microphone transfer function of the impedance tube. The results were compared with those found for the

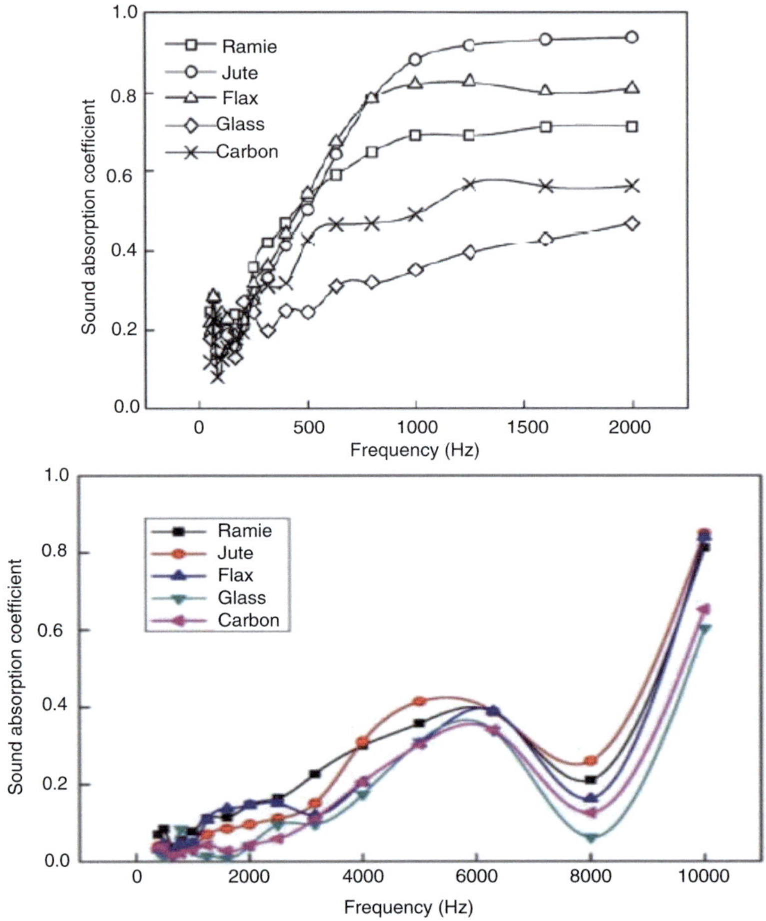

Fig. 9.27 Sound absorption coefficients of the ramie, flax, jute, glass, and carbon fabric samples

glass and carbon fibers, as shown in Fig. 9.27. The dry PF samples showed clearly superior noise-reduction capabilities (Yang and Li 2012). Sound absorption properties were then modeled using the *Delany-Bazley* and *Garai-Pompoli* equations, and good agreement with the experimental results was found.

8 Flame-Retarded Biopolymers and Biocomposites

8.1 Flammability Study of Natural Fiber Composite

Biocomposites are inherently hydrophilic and highly flammable. Previous studies have identified barriers to the use of PFRC, and especially in areas that present flame hazards and risks (e.g., interior spaces). Various strategies can be applied to improve the flame retardancy of biocomposites, e.g., the chemical modification of polymer matrices, the addition of surface treatments to PFs, the use of inherently flame-resistant polymer matrices, and the direct incorporation of flame retardants (FRs) into materials.

The main substance found in PFs is cellulose, which undergoes thermal decomposition when exposed to flames or high-intensity heat sources. The natural cellulosic fiber combustion process is highly complex (Price et al. 1997): At low temperatures (300–400 °C), cellulose decomposition proceeds through two competing pathways, aliphatic carbon forms after the dehydration and carbonization of cellulose, while the other rapidly depolymerizes into active levoglucosan; into flammable gases (such as CH_4, C_2H_2); and finally into water, carbon dioxide (CO), and carbon monoxide (CO_2) in the presence of oxygen. Aliphatic carbon dehydrates into a more stable form of aromatic carbon and eventually into water, CO, and CO_2 at high temperatures (400–600 °C). Char substances formed through the first pyrolysis process act as a thermal barrier, limiting heat transfer to the sample interior and thus reducing the extent of pyrolysis. The burning process involves heating, decomposition, ignition, combustion, and propagation. Flame retardancy can be achieved by interfering with any stage that can terminate the process before ignition occurs (Sain et al. 2004).

Traditional dipping treatments are conducted by soaking fabrics in a solution of flame retardants, catalysts, resins, surfactants, and other additives. After soaking and subsequent drying processes, fabrics should exhibit flame retardancy properties. The above processes can be repeated until the required flame-retardant level is achieved. Novel halogen-free flame retardants containing phosphorus, nitrogen, and silicon were synthesized and applied to cotton fabrics via dipping, improving flame retardancy levels (Yang et al. 2011; Liu et al. 2012a). Dipping methods are uncomplicated, cost-effective, and widely applicable but compromise washing durability levels.

Grafting copolymerization serves as an effective technique through which small functional groups and macromolecular compounds can be directly grafted to diversified underlying substrates. Numerous grafting methods are now applied in the textile flame-retardant field (e.g., chemical grafting, high-energy radiation grafting, and plasma-induced grafting polymerization (PIGP)). Vinylphosphonic monomer (Opwis et al. 2011) and synthetic phosphorus-containing flame retardants (Yuan et al. 2012) were grafted to the surface of cotton fabric via UV-induced photografting and UV-cured techniques. The photochemically modified textiles showed high levels of flame-retardant performance and passed a vertical flammability test for protective clothing. All novel phosphorous monomers and macromolecules have been synthesized using the PIGP technique (Jimenez et al. 2010; Lam et al. 2010; Edwards et al. 2012), which appears to be the most promising and versatile method available for industrial exploitation.

The sol-gel technique, a versatile synthetic method, is based on a two-step reaction (viz., hydrolysis and condensation) occurring at or near room temperature that begins with (semi)metal alkoxides and ends in the formation of fully inorganic or hybrid organic-inorganic coatings (Sakka 2003). In the textiles sector, sol-gel processes have been exploited to confer multifunctional properties, though their use for textile flame retardancy purposes has been very recent. It is worth noting that fabrics have soft and porous structures with limited thickness, and thus-deposited protective coatings have a limited shielding effect on the underlying polymers, while this issue can be solved by combining the sol-gel oxidic phases (typically silica) with other active flame-retardant species (e.g., phosphorus- and/or nitrogen-containing compounds). Synergistic effects of sol-gel-derived inorganic architectures doped with aluminum phosphinate; of a mixture of aluminum phosphinate, melamine poly(phosphate), and zinc and boron oxide; and of ZrP nano-platelets have been systematically investigated. As an example, the presence of at least 5 wt % phosphorus-based compounds in a sol-gel precursor improves the flame retardancy of cotton remarkably (TTI levels increase from 14 (untreated cotton) to 40 s) (Alongi et al. 2011). In turn, more P- and N-based chemicals have been added to silica for the preparation of hybrid organic-inorganic coatings via one-step or multistep sol-gel processes. During combustion processes, these coatings behave like char promoters and thermal shields, improving the flame retardancy of cellulose fabrics significantly (Alongi et al. 2012a, b, 2013; Brancatelli et al. 2011).

The layer-by-layer (LBL) assembly method is another novel approach that is suitable for application with various substrates such as fabrics, ceramics, plastics, and metals. It simply involves the alternating immersion of substrates into oppositely charged polyelectrolyte solutions or suspensions, resulting in the construction of multiple layers on substrate surfaces. To further improve FR performance levels, LBL architectures contain fully organic coatings. Namely, intumescent LBL coatings (Guin et al. 2014; Alongi et al. 2012a; Zhang et al. 2013a, b, c), such as PAA/PSP, chitosan/PSP, chitosan/APP, and PEI/APP, have been developed for plant fabrics. Additionally, intumescent multilayer nano-coatings composed entirely of renewable biomass polyelectrolytes have recently been applied to cotton and have remarkably decreased peak heat release rate (PHRR) and total heat release (THR) levels (Laufer et al. 2012; Pan et al. 2014). When referring to cotton fabrics, intumescent systems have been identified as the most promising flame-retardant strategies.

Intumescent multilayer nano-coatings not only improve fabric flame retardancy levels but also enhance interfacial adhesion between fabrics and certain polymer matrices. As shown in Fig. 9.28, fiber pullout effects disappeared after surface treatments. Thus, both mechanical properties and flame-retardant performance of PFRCs can be improved using intumescent coatings (Zhang et al. 2013a, b, c) (Fig. 9.29).

8.2 *Flammability Study by Using Microcrystalline Cellulose*

Recently, a study was conducted to investigate the simultaneous effect of fire retardants, PFs, and microcrystalline cellulose (MCC) on thermal, flammability, and mechanical properties of epoxy-based composites. In this study (Khalili et al. 2017a), MCC (average particle diameter and bulk density were about 2–20 μm and 0.4 g/cm^3, respectively), ammonium polyphosphate (APP; the degree of polymerization of more than 1000, density of 1.9 g/m^3, decomposition temperature of more than 275 °C, and particle size of below 50 μm), alumina trihydrate

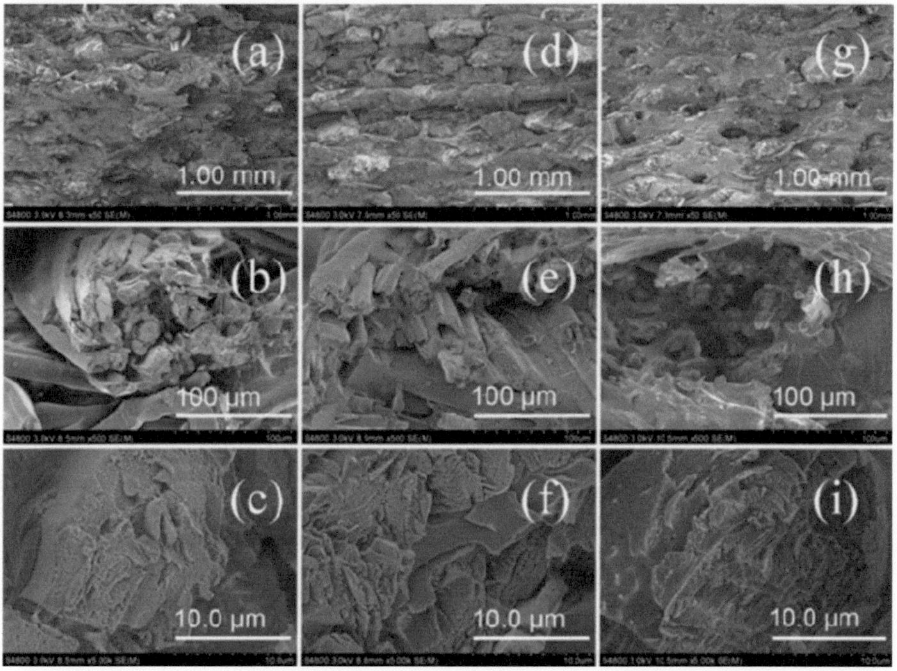

Fig. 9.28 SEM images of the fracture surfaces of benzoxazine resin (BZ)/ramie (**a**–**c**), BZ/ramie/MWNT/APP (**d**–**f**), and BZ/ramie/PEI/APP (**g**–**i**) laminates

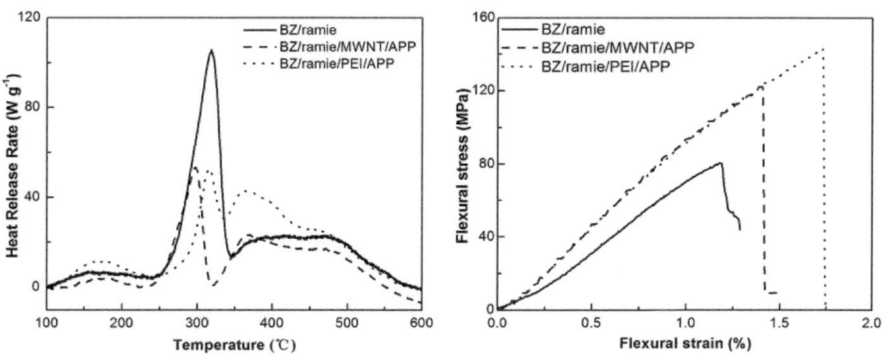

Fig. 9.29 Heat release rate curves (l) and flexural stress levels vs. strain curves (r) for benzoxazine resin/ramie fabric laminates

(ATH; specific gravity of 2.42 g/m^3, decomposition temperature of 220 °C, and average particle size of 1 µm), and an ambient cure two-part epoxide resin system Ultimeg 2020 were used. The specific gravity and the viscosity of mixture of epoxy and hardener were measured 1.1 g/m^3 and 0.2–0.6 Pa s, respectively. Pulverized palm empty fruit bunch (EFB) fiber was used as reinforcement. The EFB fiber had

Table 9.9 Sample designation (values represent wt%)

Formulations	Natural fiber	MCC	APP	ATH	Epoxy
Control	20	–	10	5	65
3MCC	20	3	10	5	62
5MCC	20	5	10	5	60
7MCC	20	7	10	5	58

Table 9.10 The results of vertical Bunsen burner test for the composites

Formulations	Burn length (mm)	Total flame time (s)	Drip flame time (s)	FAR requirement of the test
Control	2.5	0	0 (no drip)	Passed
3MCC	~2	0	0 (no drip)	Passed
5MCC	~2	0	0 (no drip)	Passed
7MCC	~2	0	0 (no drip)	Passed

the average density of 1.07 g/m^3, diameter of 100–400 μm, and length of 1–3 mm. Sodium hydroxide (NaOH) possessing 99% alkalimetric titration was used for the chemical treatment of natural fibers. The composites prepared based on the weight percentage of constituent in this study are shown in Table 9.9.

The outcome of vertical Bunsen burner test for control, 3MCC, 5MCC, and 7MCC composites is shown in Table 9.10. This shows that MCC additives possess the capability of enhancing the flame resistivity of natural fiber epoxy composites, which is attributed to the formation of more charred barrier in the burning area. This could be further explained through the results obtained from thermogravimetric analysis (TGA) study.

Total flame time and drip flame time were recorded as 0 s for all four formulations (Table 9.10). For the burn length, control obtained 2.5 mm, whereas 3MCC, 5MCC, and 7MCC specimens demonstrated 2 mm. All composites were observed to meet FAR requirements. This is due to the presence of flame retardants, i.e., APP and ATH. The combination of APP and ATH at this specific weight percentage as shown in the previous work (Khalili et al. 2017a, b) led to a synergy upon combustion, formed an intumescent char, and then self-extinguished the test specimens.

Upon the incorporation of MCC particles at any weight concentrations, the burn length reduced to approximately 2 mm for all composites. This shows that MCC additives possess the capability of enhancing the flame resistivity of natural fiber epoxy composites, which is attributed to the formation of more charred barrier in the burning area. This could be further explained through the results obtained from TGA study.

The changes in decomposition patterns and thermal stability of the control, 3MCC, 5MCC, and 7MCC were evaluated by TGA. Figure 9.30 shows the thermal degradation of all composites, and Fig. 9.31 displays their derivative thermogravimetric analysis (DTGA) curves.

The TGA curves demonstrated one thermal degradation step, as displayed in Fig. 9.30. All of the samples showed similar decomposition profiles and degradation temperature up to about 55 wt% mass loss, suggesting the slight effect of MCC fillers on the thermal stability as compared to the control. The addition of MCC particles shifted the TGA curves to higher mass residue, in particular from approximately 400 to 700 °C. The gradual enhancement with increasing addition of MCC additives was due to the formation of more char than that of the control, which inhibited the heat and mass transfer by acting as a barrier between the burning surface and the bulk of composite. This is attributed to the bond formation between epoxy resin and MCCs (Xiao et al. 2014). As shown in Fig. 9.31, the maximum

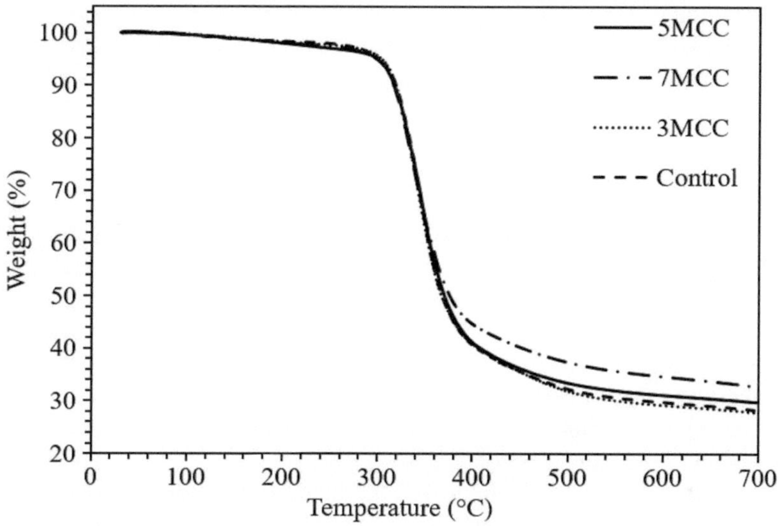

Fig. 9.30 TGA results of the control and MCC-filled composites

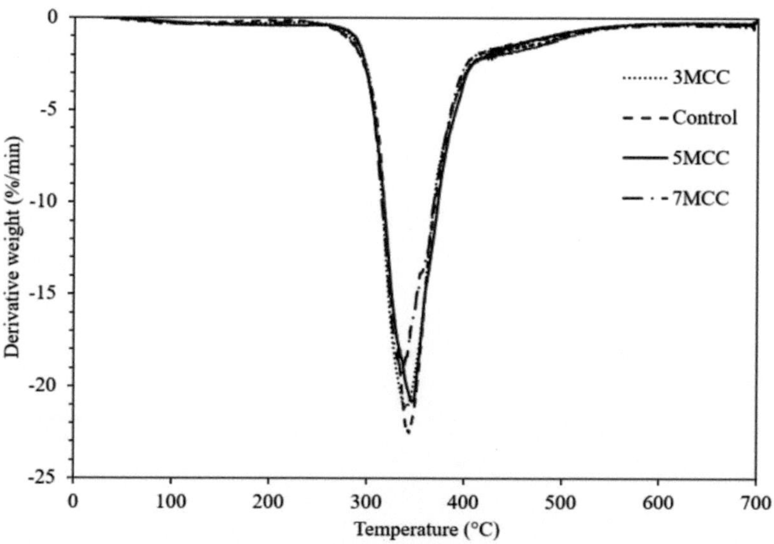

Fig. 9.31 DTGA results of the control and MCC-filled composites

degradation temperature (T_{max}) had no significant changes upon the inclusion of MCC additives, and T_{max} remained almost at the same value of 340 °C. However, increasing incorporation of MCC particles reduced the weight loss rate as compared to the control, indicating the improvement of thermal stability with the addition of MCC fillers.

The storage modulus (E') of the control, 3MCC, 5MCC, and 7MCC composites as a function of temperature is displayed in Fig. 9.32. The study was conducted to measure the impact of MCC particles on the viscoelastic properties of the control.

The addition of MCC particles was observed to decrease E' at glassy-state region as compared to the control. With the rise in temperature, a slight reduction in E' of composites was obtained, followed by a sudden decrease in the transition region, which is due to the transition of composite samples from glassy state to rubbery region.

The loss factor (Tan δ) is shown in Fig. 9.33 and is defined as the ratio of the loss modulus to E'. The peak of damping coefficient (loss factor) curves displays the glass transition temperature (Tg). The control was found to show T_g of 77.1 °C, and the inclusion of MCCs increased T_g of composites. Increasing addition of MCC was observed to shift the Tg to higher temperatures. This is attributed to the mobility level of matrix segments (Khalili et al. 2017a, b; Jana and Zhong 2009), which means the enhancement in T_g is correlated with the decrease in the mobility of polymeric chains owing to the presence of MCC particles that constrained the mobility of micromolecules of matrix. The cross-linking network between MCCs and epoxy matrix hindered the movement of chains.

It was observed that MCCs improved the flame retardancy and thermal performances of composites. This could be attributed to the good interfacial network between the epoxy-based matrix and MCC fillers.

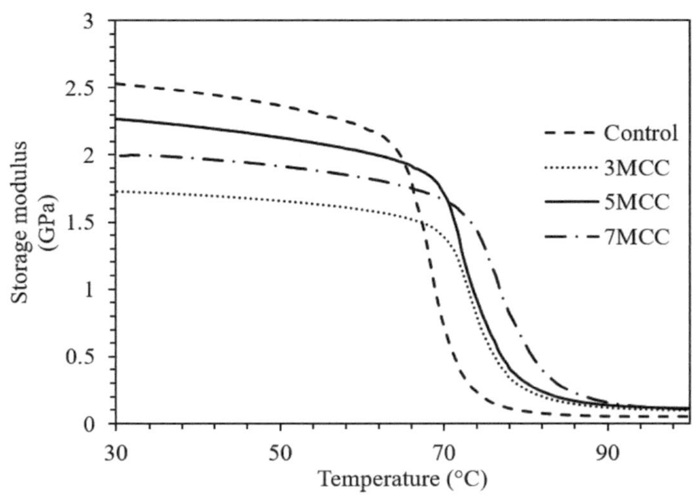

Fig. 9.32 Storage modulus vs. temperature for the control and composites containing MCC

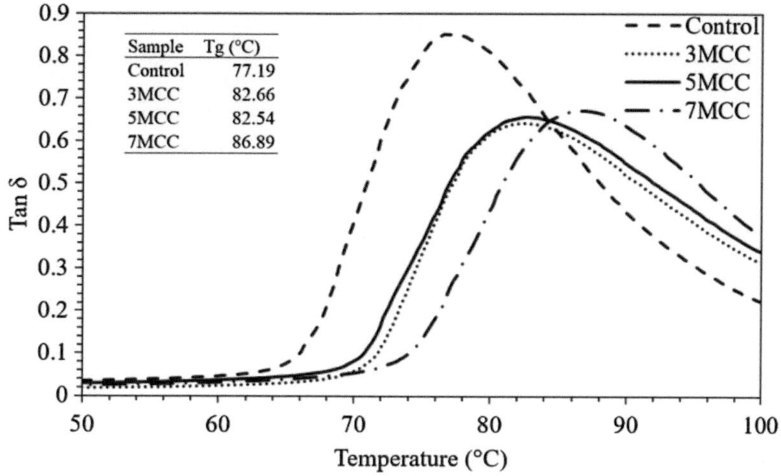

Fig. 9.33 Tan δ as a function of temperature for all composites

9 Hygrothermal Aging of PFRCs

The principal components of PFs are cellulose, hemicellulose, lignin, and pectin, which are composed of large numbers of hydroxyl groups (Liu et al. 2007; Summerscales et al. 2010). Hydroxyl groups tend to combine with water molecules through hydrogen bonds and are responsible for strong hydrophilic features of the PFs (Li et al. 2009). Pectin in particular serves as a polyanion that is responsible for the majority of absorbed water in the cell wall. A main concern regarding the use of PFRCs pertains to their susceptibility to moisture absorption. Hygrothermal aging may lead to PF swelling and to the formation of voids and micro-cracks within the fiber-matrix interface, leading to the deterioration of mechanical properties and dimensional stability levels (Alamri and Low 2012; Dhakal et al. 2007).

To examine hygrothermal aging processes of a PFRC, a model sample with ramie-fabric-reinforced phenolic resin composite (RFRC) was prepared through hot compression and was examined in terms of water uptake and mechanical property under hygrothermal conditions. The plate was 2.77 mm thick, and the ramie fabric volume was roughly 40.4%. Three exposure environments were considered, i.e., humidity, immersion, and alternating humidity/temperature levels (Wang et al. 2014; Yang et al. 2014; Xian et al. 2014). The test results are shown in Fig. 9.34.

Figure 9.34a presents the water uptake curves of the RFRC plates exposed to 50, 80, and 98% relative humidity levels at room temperature. The coefficient of moisture diffusion in RFRC was determined using a curve fitting method based on the simplified Fick's law equation (Shen and Springer 1976). The determined saturated moisture content ($M\infty$) and coefficient of diffusion (D) levels are summarized in Table 9.11. Saturated moisture content variations as a function of humidity (ϕ) can be described by the following equation (Loos and Springer 1979):

$$M\infty = a\phi b \tag{9.1}$$

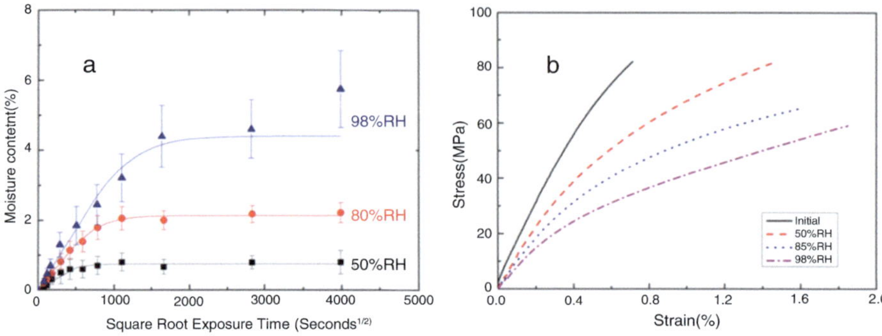

Fig. 9.34 (**a**) Moisture uptake curves of ramie/phenol samples under three relative humidity (RH) conditions at room temperature. Note that the solid lines denote curve fitting results of the classic Fick's equation. (**b**) Typical tensile stress–strain curves of the control and aged RFRCs in the humidity environments at room temperature for 6 months

Table 9.11 Maximum moisture absorption (M_m) and diffusion coefficient (D) levels of ramie/phenol samples subjected to three relative humidity (RH) conditions over 90 days

RH (%)	M_m (wt%)	D (mm²/s)
50	0.73	5.50×10^{-06}
85	2.09	2.08×10^{-06}
98	4.40	0.83×10^{-06}

where a and b are constants depending on the material type. For CFRCs, the value of a is roughly 0.01~0.02 and that of b is roughly 1. Much higher values of $a = 0.12$ and $b = 2.48$ found for the RFRC samples denoting that RFRC possesses superior hydrophilic properties relative to the synthetic fiber-based FRPs.

Variations in tensile strength, tensile modulus, elongation at break, and short-beam shear strength (SBS) levels of the RFRC samples as a function of a 180-day exposure period under the three hydrothermal conditions were studied. The results are summarized against moisture content levels in Fig. 9.35. It is interesting to highlight the presence of a strong linearity relationship between moisture content and tensile strength levels and between modulus and SBS strength levels when the last high moisture points (moisture content of 5.76% under 98% RH for 6 months) were negligible. The retention of each mechanical property as a function of water content (M) can be obtained by linear fit. Tensile strength (σr) retention is reported as σr = 1.033–0.065 M, tensile modulus (Er) retention is reported as Er = 0.81–0.10 M, and SBS strength (σsr) retention is reported as σsr = 1.02–0.08 M. The modulus is clearly more susceptible to moisture content levels, while moisture content effects on tensile and SBS strength levels are similar and less significant.

It is worth noting that mechanical property degradation as a result of humidity exposure can be mitigated to some extent, as shown in Table 9.12. The remarkable recovery of mechanical RFRC properties due to drying denotes that permanent degradation of fiber and resin and their bonding is limited. Despite this, at 98% RH, permanent tensile strength and modulus degradation levels are much more serious compared to those found for the other conditions.

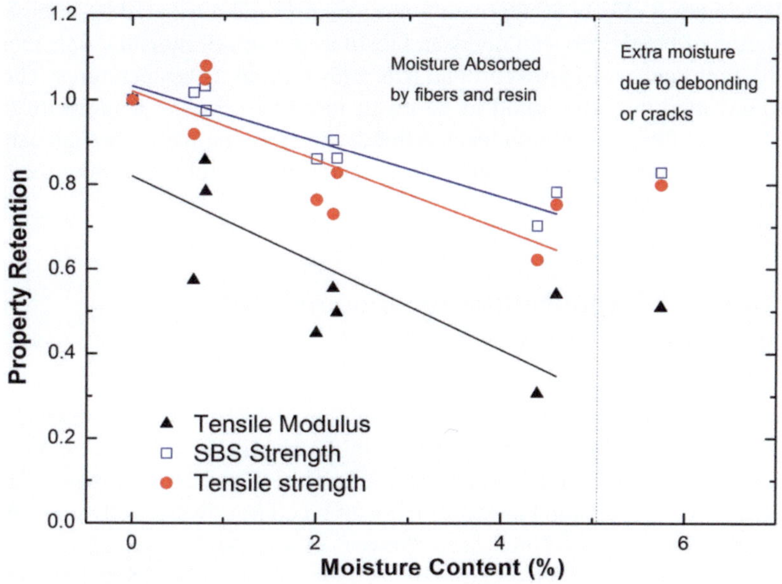

Fig. 9.35 Variations in RFRC tensile modulus and short-beam shear strength (SBS) as a function of moisture content

Table 9.12 Variations in mechanical properties of aged RFRCs exposed to humidity conditions for 3 months and of dried RFRCs exposed to humidity conditions for 3 months at 60 °C compared to those of the un-aged samples

	Tensile strength		Tensile modulus		SBS strength	
Exposure humidity (% RH)	Variation due to exposure (%)	Variation after drying (%)	Variation due to exposure (%)	Variation after drying (%)	Variation due to exposure (%)	Variation after drying (%)
50	4.8	6.2	−14.2	−5.7	3.3	2.7
85	−27.0	4.2	−44.5	−4.7	−9.5	−9.0
98	−24.7	−0.3	−45.9	−10.8	−21.8	−7.6

Compared to humidity exposure processes, water immersion processes result in more water uptake and in RFRC mechanical property deterioration (Wang et al. 2014). At room temperature, saturated water uptake levels reached roughly 4% in the RFRC samples exposed to 98% RH and roughly 8% in the samples immersed in water. Mechanical property variations as a function of water uptake content occur over three stages: rapid decline, levelling off, and then rapid decline. This may be due to the sequent occurrence of PF degradation, fiber debonding, and fiber and resin hydrolysis during immersion (Yang et al. 2014).

Effects of alternating temperatures (from 25 to 55 °C in 24 h for a cycle) and humidity levels (98% RH) on RFRC mechanical properties were also examined as a function of moisture absorption (Xian et al. 2016). Compared to constant exposure temperatures, alternating temperatures brought higher moisture uptake levels and more serious flexural strength, flexural modulus, and short-beam shear strength degradation processes in the RFRCs exposed to the same humidity conditions.

The deteriorating effects of alternating temperatures are attributed to a heightened degradation of bonds between fibers and resin as a result of moisture uptake and as a result of internal cyclic stress around RFs with alternating temperatures. The flexural RFRC modulus was found to be much more susceptible to moisture uptake levels than the flexural strength level. After drying was complete, mechanical properties of the samples were recovered to an extent but not fully, denoting the occurrence of permanent damage.

10 Industrial Applications for Biopolymers and Biocomposites

A trial manufacturing case (Yi et al. 2019) that involved the use of sandwich panels was (1) plain-woven ramie impregnated with (2) phenol resin of Cycom 6070 and Cytec, with (3) a honeycomb core of Nomex, DuPont (Fig. 9.36). Strong conforming capacities of the panel along an extremely curved contour were well demonstrated. The ramie/phenol resin prepreg used showed an acceptable level of stiffness and strength in relation to its GFRC counterparts; more specifically, ISS and tensile modulus levels were slightly higher than those of the glass sample. A manufacturing benefit of the RFRC parts cured in an autoclave pertains to their full compatibility with state-of-the-art industrial production processes. Our flammability study generated acceptable results; therefore, the fiber reinforcement structure is being examined as a potential airplane interior material.

Figure 9.37 shows an electric race car developed in cooperation with Tsinghua University, China. In the case scenario, rosin-sourced epoxy composite was used to produce the carbon composite body with honeycomb core.

Most prior R&D studies on biocomposites and associated trial applications have focused on mechanical and functional issues without offering insight into the beneficial properties of natural fibers. Despite their moderate mechanical and functional

Fig. 9.36 Sandwich panels with plain-woven ramie as face sheets impregnated with phenol and with Nomex honeycomb as the core

Fig. 9.37 Electric race car; its body is manufactured by using rosin-sourced epoxy prepregs with honeycomb core

properties, ramie fibers are one of the premium plant fibers used in the clothing industry. They are highly durable and white in color and have a silky finish, enabling their use for decoration purposes. It is a favored textile material for the construction of apparel and home fashions. Such features are promising, as ramie can be used to produce white or colorful composites, unlike known carbon composites that are black in color or aramid fiber composites, which are characteristically yellow.

In a proof-of-concept study (Yi and Liu 2015), color ramie fabrics were applied as a decorative face layer on composite laminates to produce a decorative composite. An interior side panel was then manufactured via vacuum infusion with unsaturated polyester used for a rail car (Fig. 9.38), whereby dry decorative ramie fabrics were accordingly positioned on the surface of ramie/glass hybrid laminates for the facilitation of balanced mechanical properties. For decorative purposes, the more transparent the matrix resin, the better the look of the final product. The structure-decoration-integrated panel was easily fabricated to adapt to product specifications. The decorative face layer simultaneously served as a structural component that could bear loads unlike the traditional decorative surface, which functions only decoratively. In addition to these advantages, ramie fabrics used in the face layer serve other functions, e.g., noise absorption and vibration damping.

The addition of decorations to the structure is clearly beneficial. It not only adds aesthetic appeal to the composite structure but also simplifies the manufacturing process in terms of materials, labor, and time and thus cuts costs. Another trial application was applied to the interior cabin structure of the world's largest seaplane AG600 made in China, clearly highlighting the material's aesthetic appeal (Fig. 9.39).

It is noteworthy that the use of multifunctional decorative composites for interior and quasi-structural application is by no means restricted to ramie textiles. Such composites can also be applied to other PF textiles with similar features such as flax and hemp. Figure 9.40 is an illustration of an interior side panel made of rosin-sourced epoxy prepreg with honeycomb sandwich core for MA600 airplane. The composite panels are strong, lightweight, fire-resistant, decorative, and impervious to water, mold, and insects. Structure-decoration integration methods can clearly support the production of identical or even more complex composite parts while simultaneously presenting mechanical and structural damping advantages in various applications.

Fig. 9.38 Proof-of-concept demonstration of a decorative, quasi-structural composite side panel produced with ramie fabric as the surface layer of a railcar

Fig. 9.39 Application of decorative, function-integrated composite panels to the interior of a seaplane, AG600

Fig. 9.40 MA600 airplane and the composite side panel made of rosin-sourced epoxy honeycomb sandwich composites

Acknowledgments The study was jointly supported by the National Basic Research Program of China (973 Program) under grant no. 2010CB631100; the National Natural Science Foundation of China (NSFC) under grants no. 51178147, 11172212, and 11625210; the AVIC Innovation Foundation and the joint project ECO-COMPASS co-funded by Chinese MIIT Special Research Program under grant no. MJ-2015-H-G-103; and European Union's Horizon 2020 Research and Innovation Program under grant agreement no. 690638. They are greatly acknowledged.

References

Alamri H, Low I (2012) Effect of water absorption on the mechanical properties of n-SiC filled recycled cellulose fibre reinforced epoxy eco-nanocomposites. Polym Test 31(6):810–818

Alongi J, Ciobanu M, Malucelli G (2011) Novel flame retardant finishing systems for cotton fabrics based on phosphorus-containing compounds and silica derived from sol–gel processes. Carbohydr Polym 85(3):599–608

Alongi J, Carosio F, Malucelli G (2012a) Layer by layer complex architectures based on ammonium polyphosphate, chitosan and silica on polyester-cotton blends: flammability and combustion behaviour. Cellulose 19(3):1041–1050

Alongi J, Colleoni C, Malucelli G, Rosace G (2012b) Hybrid phosphorus-doped silica architectures derived from a multistep sol–gel process for improving thermal stability and flame retardancy of cotton fabrics. Polym Degrad Stab 97(8):1334–1344
Alongi J, Colleoni C, Rosace G, Malucelli G (2013) The role of pre-hydrolysis on multi step sol–gel processes for enhancing the flame retardancy of cotton. Cellulose 20(1):525–535
Bachmann J, Yi X, Gong H, Martinez X, Bugeda G, Oller S, Tserpes K, Ramon E, Paris C, Moreira P, Fang Z, Li Y, Liu Y, Liu X, Xian G, Tong J, Wei J, Zhang X, Zhu J, Ma S, Yu T (2018) Outlook on ecologically improved composites for aviation interior and secondary structures. CEAS Aeronaut J 9(3):533–543
Bodros E, Pillin I, Montrelay N, Baley C (2007) Could biopolymers reinforced by randomly scattered flax fibre be used in structural applications? Compos Sci Technol 67(3–4):462–470
Brancatelli G, Colleoni C, Massafra M, Rosace G (2011) Effect of hybrid phosphorus-doped silica thin films produced by sol-gel method on the thermal behavior of cotton fabrics. Polym Degrad Stab 96(4):483–490
Dhakal H, Zhang Z, Richardson M (2007) Effect of water absorption on the mechanical properties of hemp fibre reinforced unsaturated polyester composites. Compos Sci Technol 67(7–8):1674–1683
Drzal L, Amar Mohanty A, Manjusri Misra M (2001) Bio-composite materials as alternatives to petroleum-based composites for automotive applications. https://www.researchgate.net/publication/228474911_Bio-composite_Materials_as_Alternatives_to_Petroleum-based_Composites_for_Automotive_Applications. Accessed 21 Feb 2019
Eco-compass.eu (2016) Eco Compass Project. http://www.eco-compass.eu. Accessed 21 Feb 2019
Edwards B, El-Shafei A, Hauser P, Malshe P (2012) Towards flame retardant cotton fabrics by atmospheric pressure plasma-induced graft polymerization: synthesis and application of novel phosphoramidate monomers. Surf Coat Technol 209:73–79
Guin T, Krecker M, Milhorn A, Grunlan J (2014) Maintaining hand and improving fire resistance of cotton fabric through ultrasonication rinsing of multilayer nanocoating. Cellulose 21(4):3023–3030
Huang J, Chang Hung Y, Wen Y, Kang C, Yeh M (2009) Polylactide/nano- and micro-scale silica composite films. II. Melting behavior and cold crystallization. J Appl Polym Sci 112(5):3149–3156
Ikada Y, Tsuji H (2000) Biodegradable polyesters for medical and ecological applications. Macromol Rapid Commun 21(3):117–132
Jana S, Zhong W (2009) Graphite particles with a "puffed" structure and enhancement in mechanical performance of their epoxy composites. Mater Sci Eng A 525(1–2):138–146
Jangra S, Stalin K, Dilbaghi N, Kumar S, Tawale J, Singh S, Pasricha R (2012) Antimicrobial activity of zirconia (ZrO2) nanoparticles and zirconium complexes. J Nanosci Nanotechnol 12(9):7105–7112
Jimenez M, Bellayer S, Duquesne S, Bourbigot S (2010) Improvement of heat resistance of high performance fibers using a cold plasma polymerization process. Surf Coat Technol 205(3):745–758
Kalia S, Kaith B, Kaur I (2009) Pretreatments of natural fibers and their application as reinforcing material in polymer composites—a review. Polym Eng Sci 49(7):1253–1272
Khalili P, Tshai K, Kong I (2017a) Natural fiber reinforced expandable graphite filled composites: evaluation of the flame retardancy, thermal and mechanical performances. Compos A: Appl Sci Manuf 100:194–205
Khalili P, Tshai K, Hui D, Kong I (2017b) Synergistic of ammonium polyphosphate and alumina trihydrate as fire retardants for natural fiber reinforced epoxy composite. Compos Part B 114:101–110
Khan B, Warner P, Wang H (2014) Antibacterial properties of hemp and other natural fibre plants: a review. Bioresources 9(2):3642–3659
Lam Y, Kan C, Yuen C (2010) Effect of zinc oxide on flame retardant finishing of plasma pre-treated cotton fabric. Cellulose 18(1):151–165

Laufer G, Kirkland C, Morgan A, Grunlan J (2012) Intumescent multilayer nanocoating, made with renewable polyelectrolytes, for flame-retardant cotton. Biomacromolecules 13(9):2843–2848
Li Y, Mai Y, Ye L (2005) Effects of fibre surface treatment on fracture-mechanical properties of sisal-fibre composites. Compos Interf 12(1–2):141–163
Li Y, Hu Y, Hu C, Yu Y (2008) Microstructures and mechanical properties of natural fibers. Adv Mater Res 33–37:553–558
Li L, Yu Y, Wu Q, Zhan G, Li S (2009) Effect of chemical structure on the water absorption of amine-cured epoxy resins. Corros Sci 51(12):3000–3006
Li Y, Yi X, Yu T, Xian G (2018) An overview of structural-functional-integrated composites based on the hierarchical microstructures of plant fibers. Adv Compos Hybrid Mater 1(2):231–246
Liu Z, Yang Y, Zhang L, Liu Z, Xiong H (2007) Study on the cationic modification and dyeing of ramie fiber. Cellulose 14(4):337–345
Liu X, Xin W, Zhang J (2009) Rosin-based acid anhydrides as alternatives to petrochemical curing agents. Green Chem 11(7):1018–1025
Liu W, Chen L, Wang Y (2012a) A novel phosphorus-containing flame retardant for the formaldehyde-free treatment of cotton fabrics. Polym Degrad Stab 97(12):2487–2491
Liu X, Huang W, Jiang Y, Zhu J, Zhang C (2012b) Preparation of a bio-based epoxy with comparable properties to those of petroleum-based counterparts. Express Polym Lett 6(4):293–298
Liu X, Yi X, Zhu J (2018) Bio-based epoxies and composites as environmentally friendly alternative materials. Thermosets, pp 621–637
Loos A, Springer G (1979) Moisture absorption of graphite-epoxy composites immersed in liquids and in humid air. J Compos Mater 13(2):131–147
Ma S, Liu X, Fan L, Jiang Y, Cao L, Tang Z, Zhu J (2013a) Synthesis and properties of a bio-based epoxy resin with high epoxy value and low viscosity. ChemSusChem 7(2):555–562
Ma S, Liu X, Jiang Y, Fan L, Feng J, Zhu J (2013b) Synthesis and properties of phosphorus-containing bio-based epoxy resin from itaconic acid. Sci China Chem 57(3):379–388
Ma S, Liu X, Jiang Y, Tang Z, Zhang C, Zhu J (2013c) Bio-based epoxy resin from itaconic acid and its thermosets cured with anhydride and comonomers. Green Chem 15(1):245–254
Matsui K, Ohgai M (2005) Formation mechanism of hydrous-zirconia particles produced by hydrolysis of ZrOCl2 solutions. J Am Ceram Soc 80(8):1949–1956
Natureworksllc.com (2019) NatureWorks | Home. https://www.natureworksllc.com/. Accessed 21 Feb 2019
Ni N, Wen Y, He D, Yi X, Zhang T, Xu Y (2015) High damping and high stiffness CFRP composites with aramid non-woven fabric interlayers. Compos Sci Technol 117:92–99
Opwis K, Wego A, Bahners T, Schollmeyer E (2011) Permanent flame retardant finishing of textile materials by a photochemical immobilization of vinyl phosphonic acid. Polym Degrad Stab 96(3):393–395
Pan H, Song L, Ma L, Pan Y, Liew K, Hu Y (2014) Layer-by-layer assembled thin films based on fully biobased polysaccharides: chitosan and phosphorylated cellulose for flame-retardant cotton fabric. Cellulose 21(4):2995–3006
Powers W (2000) FEATURES—automotive materials in the 21st century—forecast-type article is based on a presentation made by the author at ASM international's materials XXI conference. Dr. Powers is Ford Motor Co.'s vice president of research. Adv Mater Manuf Process 157(5):38–41
Price D, Horrocks A, Akalin M, Faroq A (1997) Influence of flame retardants on the mechanism of pyrolysis of cotton (cellulose) fabrics in air. J Anal Appl Pyrolysis 40–41:511–524
Sain M, Park S, Suhara F, Law S (2004) Flame retardant and mechanical properties of natural fibre–PP composites containing magnesium hydroxide. Polym Degrad Stab 83(2):363–367
Sakka S (2003) J Sol-Gel Sci Technol 26(1/3):29–33
Shen C, Springer G (1976) Moisture absorption and desorption of composite materials. J Compos Mater 10(1):2–20
Shen X, Jia J, Chen C, Li Y, Kim J (2014) Enhancement of mechanical properties of natural fiber composites via carbon nanotube addition. J Mater Sci 49(8):3225–3233

Summerscales J, Dissanayake N, Virk A, Hall W (2010) A review of bast fibres and their composites. Part 1—fibres as reinforcements. Compos A: Appl Sci Manuf 41(10):1329–1335
Tang Z, Zhang C, Liu X, Zhu J (2011) The crystallization behavior and mechanical properties of polylactic acid in the presence of a crystal nucleating agent. J Appl Polym Sci 125(2):1108–1115
Tian X, Lv C, Zhang G (2011) Preparation and antibacterial activity of Al_2O_3-ZrO_2 carrying Ag^+. Appl Mech Mater 55–57:1464–1467
Tsuji H, Miyauchi S (2001) Poly(l-lactide): 7. Enzymatic hydrolysis of free and restricted amorphous regions in poly(l-lactide) films with different crystallinities and a fixed crystalline thickness. Polymer 42(9):4463–4467
Wang H, Xian G, Li H, Sui L (2014) Durability study of a ramie-fiber reinforced phenolic composite subjected to water immersion. Fibers Polym 15(5):1029–1034
Wang H, Xian G, Li H (2015) Grafting of nano-TiO_2 onto flax fibers and the enhancement of the mechanical properties of the flax fiber and flax fiber/epoxy composite. Compos A: Appl Sci Manuf 76:172–180
Werpy T, Petersen G (2004) Top value added chemicals from biomass: volume I—results of screening for potential candidates from sugars and synthesis gas
Xia Y, Xian G, Li H (2014) Enhancement of tensile properties of flax filaments through mercerization under sustained tension. Polym Polym Compos 22(2):203–208
Xian G, Yin P, Kafodya I, et al (2014) Durability study of ramie fiber fabric reinforced phenolic plates under humidity conditions. Sci Eng Compos Mater 23(1):45–52
Xian G, Yin P, Kafodya I, Li H, Wang W (2016) Durability study of ramie fiber fabric reinforced phenolic plates under humidity conditions. Sci Eng Compos Mater 23(1)
Xiao X, Lu S, Qi B, Zeng C, Yuan Z, Yu J (2014) Enhancing the thermal and mechanical properties of epoxy resins by addition of a hyperbranched aromatic polyamide grown on microcrystalline cellulose fibers. RSC Adv 4(29):14928
Yang W, Li Y (2012) Sound absorption performance of natural fibers and their composites. Sci China Technol Sci 55(8):2278–2283
Yang Z, Fei B, Wang X, Xin J (2011) A novel halogen-free and formaldehyde-free flame retardant for cotton fabrics. Fire Mater 36(1):31–39
Yang Z, Xian G, Li H (2014) Effects of alternating temperatures and humidity on the moisture absorption and mechanical properties of ramie fiber reinforced phenolic plates. Polym Compos 36(9):1590–1596
Yi X, Liu Y (2015) A method for fabricating structure-decorative composites with plant fiber fabrics. PCT-patent in pending
Yi X, Li Y (2017) Bio-sourced resins, plant fibers and biocomposites (in Chinese). Construction Industry Press, Beijing
Yi X, Zhang X, Ding F, Tong J (2018) Development of bio-sourced epoxies for bio-composites. Aerospace 5(2):65
Yi X, Liu Y et al (2019) US20150044924A1—composite having plant fiber textile and fabricating method thereof—Google Patents. Patents.google.com. https://patents.google.com/patent/US20150044924
Yu T, Ren J, Li S, Yuan H, Li Y (2010) Effect of fiber surface-treatments on the properties of poly(lactic acid)/ramie composites. Compos A: Appl Sci Manuf 41(4):499–505
Yuan H, Xing W, Zhang P, Song L, Hu Y (2012) Functionalization of cotton with UV-cured flame retardant coatings. Ind Eng Chem Res 51(15):5394–5401
Zhang T, Yan H, Peng M, Wang L, Ding H, Fang Z (2013a) Construction of flame retardant nanocoating on ramie fabric via layer-by-layer assembly of carbon nanotube and ammonium polyphosphate. Nanoscale 5(7):3013
Zhang T, Yan H, Wang L, Fang Z (2013b) Controlled formation of self-extinguishing intumescent coating on ramie fabric via layer-by-layer assembly. Ind Eng Chem Res 52(18):6138–6146
Zhang Y, Li Y, Ma H, Yu T (2013c) Tensile and interfacial properties of unidirectional flax/glass fiber reinforced hybrid composites. Compos Sci Technol 88:172–177

Zhang X, Wu Y, Wei J, Tong J, Yi X (2017) Curing kinetics and mechanical properties of bio-based composite using rosin-sourced anhydrides as curing agent for hot-melt prepreg. SCIENCE CHINA Technol Sci 60(9):1318–1331

Zhu Y, Shi L, Liang J, Hui D, Lau K (2008) Synthesis of zirconia nanoparticles on carbon nanotubes and their potential for enhancing the fracture toughness of alumina ceramics. Compos Part B 39(7–8):1136–1141

Chapter 10
Self-Healing Mechanisms in Multifunctional Structural Materials

L. Guadagno, C. Naddeo, L. Vertuccio, E. Calabrese, G. Barra, and M. Raimondo

1 Development of Aeronautical Composites Characterized by Autonomous Self-Repair Mechanisms

The concept of self-healing materials finds a great source of inspiration by the nature, such as biological systems in which the damage triggers an autonomic healing response. The big challenge in the design of synthetic materials lies in the fundamental difference between biological systems and synthetic materials which have no metabolic capability of the former. The study of how to impart auto-repair function to polymeric materials is an emerging and fascinating area of research that could significantly extend the working life and safety of the polymeric components for a broad range of applications. Thermoplastic and thermosetting polymers represent the intelligent invention and technological driving force of the twentieth century; hence, the possibility to add self-repair function to these materials is under investigation from researchers around the globe (Binder 2013; Jin et al. 2013; Wu et al. 2008; White et al. 2001; Amendola et al. 2015; Chowdhury et al. 2015; Guadagno et al. 2010, 2011, 2014a, b, 2016; Kessler et al. 2003; Rule et al. 2005a; Raimondo et al. 2015, 2016; Everitt et al. 2015; Raimondo and Guadagno 2013; Toohey et al. 2007, 2009a, b; Liu et al. 2011; Chung et al. 2015; Cho et al. 2009).

Several self-healing concepts for polymeric materials have been published over the last 20 years. Depending on the chemical nature of the polymeric materials, different approaches have been proposed to prepare and characterize self-healing systems. Currently, most industrial polymeric materials can be used to develop self-healing materials.

However, the hardest challenge in this field is to develop "autonomic self-healing materials" for structural application. A truly self-healing material is capable of

L. Guadagno (✉) · C. Naddeo · L. Vertuccio · E. Calabrese · G. Barra · M. Raimondo
Department of Industrial Engineering, University of Salerno, Fisciano, SA, Italy
e-mail: lguadagno@unisa.it; cnaddeo@unisa.it; lvertuccio@unisa.it; elicalabrese@unisa.it; gbarra@unisa.it; mraimondo@unisa.it

S. Pantelakis, K. Tserpes (eds.), *Revolutionizing Aircraft Materials and Processes*, https://doi.org/10.1007/978-3-030-35346-9_10

repairing itself in its intrinsic environment, and it does this without any external intervention as soon as the damage occurs.

In the case of soft materials characterized by relatively low values in the glass transition temperature, the movement of the macromolecular chains (or segment of chains) makes possible the application of different self-healing mechanisms. In the structural material, the stiffness or limited movement of the chains, or part of chains among the cross-linking points in the network (for thermosetting materials), severely limits the choice of the self-healing mechanisms and the chances of successful results. In fact, the development of smart composites capable of self-repairing in load-bearing structures is still at the planning stage owing to complex issues to overcome.

One of the first promising extrinsic self-healing system for structural materials (based on epoxy resins—EPON 828) was proposed by White et al. (2001) (see Fig. 10.1). It is worth noting that, in this regard, the attention is focused on autonomous self-healing materials, where the self-healing function is inherent in the material nature and composition; therefore, in this section on state of the art, all the self-healing materials based on auto-repair mechanisms activated by external stimuli (e.g., Diel–Alder reactions) which can be of limited interest for the aim of this chapter are excluded. The self-healing system proposed by White et al. (2001) is suitable for thermosetting materials. This system consists of incorporating a microencapsulated healing agent and a catalytic chemical trigger within an epoxy matrix.

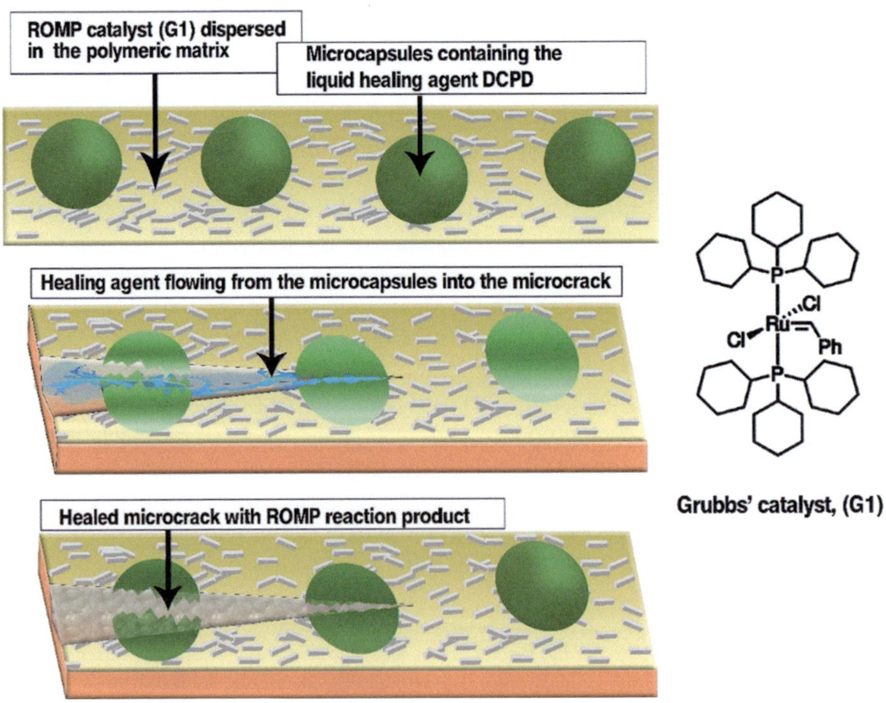

Fig. 10.1 Scheme of the first self-healing system relied on capsule-based concepts

An approaching crack ruptures embedded microcapsules releasing a polymerizer agent into the crack plane through capillary action. Polymerization of the healing agent via ring-opening metathesis polymerization (ROMP) is triggered by contact with the embedded catalyst particles, bonding the crack faces.

After the milestone work reported by White and coworkers (2001), self-healing chemistry for thermosetting polymers rapidly expanded in the past decade. A great number of self-healing materials have been demonstrated based on extrinsic or intrinsic materials. In the case of extrinsic autonomous thermosetting materials, the advancement in the knowledge and the developments of new mechanisms were much slower with respect to the self-healing systems based on soft materials and/or on the activation of the self-healing function by means of external stimuli. The employment of the healing agent has been proposed again in the subsequent literature, even if it has been stored in other types of vessels (hollow fibers, storage vessels or more complex microvascular network, etc.) and embedded into the materials in advance (Kessler et al. 2003; Rule et al. 2005a; b; Brown et al. 2004, 2005a, b; Jones et al. 2006, 2007; Kessler and White 2001; Wilson et al. 2008a). Thermosetting auto-repair polymers, which have been proposed so far, include different catalysts, for example, Grubbs' first-generation catalyst (G1); (Toohey et al. 2007; van der Zwaag 2007; Dry 2007; Motuku et al. 1999; Brown et al. 2003), Grubbs' second-generation catalyst (G2), and Hoveyda–Grubbs' first- or second-generation catalyst (HG1 or HG2) (Guadagno et al. 2011). Figure 10.2 shows the design of the self-healing system that relied on capsule-based concepts containing HG1 catalyst.

Fig. 10.2 Scheme of the first self-healing system containing HG1 catalyst particles

Wilson and coworkers evaluated the activity of G1, G2, and HG2 catalysts with various monomeric healing agents (Wilson et al. 2008b).

Guadagno et al. developed several microencapsulated self-healing systems using, as self-healing components in epoxy mixtures, microcapsules filled with 5-ethylidene-2-norbornene (ENB) and Hoveyda–Grubbs' first-generation catalyst particles (HG1) (Guadagno et al. 2014b) (see Fig. 10.3) and ENB (95%)/dicyclopentadiene (DCPD) (5%)-filled microcapsules, with the outer shell composed of poly(urea-formaldehyde) and the inner shell of ethylene maleic anhydride (EMA) copolymer (Raimondo and Guadagno 2013; Brown et al. 2003) (see Fig. 10.4) and Hoveyda–Grubbs' second-generation catalyst (HG2) (Guadagno et al. 2011) or HG1 (Guadagno et al. 2014a, b, 2016; Raimondo et al. 2015, 2016; Raimondo and Guadagno 2013).

The above-described systems represent an efficient solution for epoxy structural composites where no soft matter or thermoplastic matrix is involved: however, some drawbacks were re-evaluated in order to fully apply these systems for advanced applications where high mechanical performance is required (aerospace, aeronautics, etc.). One of these concerns is the thermal stability of the different ruthenium catalysts inside the epoxy resin during the curing cycle. The usually significantly high temperatures of the curing cycles scheduled for epoxy formulations designed for high mechanical performance, thermolytic decomposition, etc. can limit the usefulness of the ruthenium catalysts in self-healing composites based on epoxy resins. This is a crucial aspect for self-healing systems for structural applications. Generally,

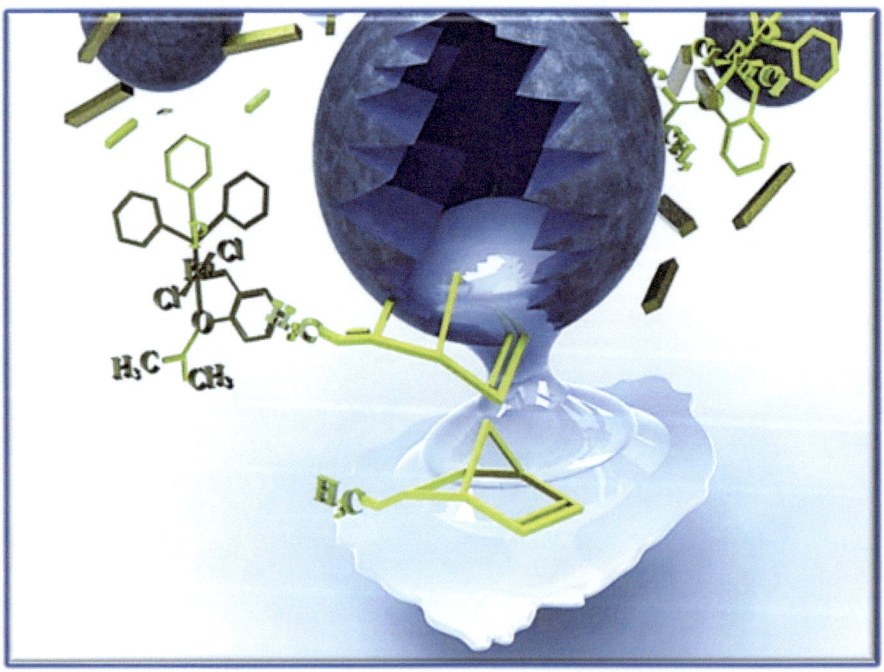

Fig. 10.3 Scheme of self-healing system based on ENB-filled microcapsules and HG1 catalyst particles dispersed in the matrix

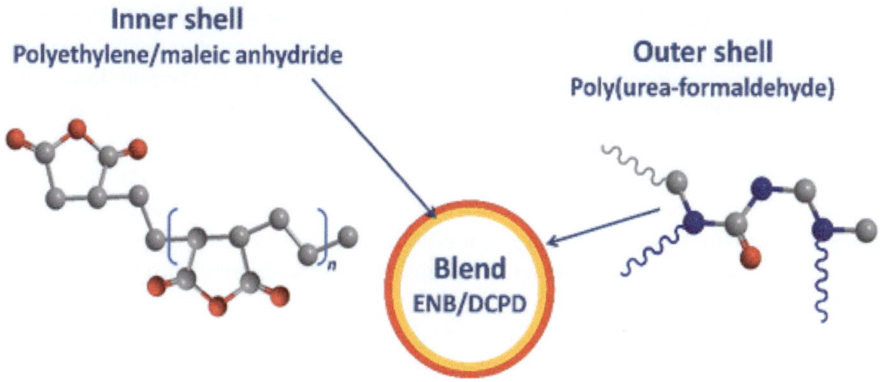

Fig. 10.4 Scheme of the microcapsules containing ENB/DCPD blend as healing agent

for the mechanical targets fixed for the materials used for structural aims, a glass transition temperature after wet aging of 170 °C/180 °C minimum is required, thus putting a curing temperature equal to or less than 100 °C as too low to this purpose. To achieve this goal, it is necessary to choose a proper chemical formulation and to make a post cure with a temperature that could be as high as 180 °C. Concerning the stability of the catalysts at different temperatures, as mentioned before, Wilson and coworkers have evaluated the reactivities of G1, G2, and HG2 catalysts with various monomeric healing agents also with a view toward improving the self-healing performances in a variety of epoxy matrices (Wilson et al. 2008b). In their work, they have evaluated the thermal stability of these three catalysts embedded into the epoxy matrices with the aim of studying their stability to processing conditions during the manufacturing conditions. To evaluate the thermal stability of the catalysts under simulated high-temperature processing conditions, the authors prepared fracture specimens with a tapered double-cantilever beam (TDCB) geometry. Different types of fracture tests were performed, among which self-activated tests in which the catalyst was embedded in the epoxy matrix and the healing agent alone was injected into the crack plane, after which the samples were left to heal for 24 h. Self-activated M1 samples (epoxy matrix EPON 828 cured with DETA) containing 1.5 wt% of an as-received catalyst were prepared and subjected to customary curing cycles followed by an additional post curing at 125 °C for 4 h before fracture testing. Samples were then fractured, and DCPD (10 μL) was injected into the crack plane. The samples were then left to heal at either room temperature (RT) or 125 °C for 24 h. Samples healed at RT showed a retention of healing performance of 77%, 84%, and 18% for G1, G2, and HG2 catalysts, respectively. Samples healed at 125 °C showed significant improvements over those healed at RT; in fact for G1, G2, and HG2 catalysts, the retention of healing performance was 109%, 216%, and 57%, respectively. Guadagno et al. (2011) have dealt with the same issue from different angles trying to understand the reason for the low retention of healing performance obtained using HG2 catalyst. In a recent paper, they have shown critical issues in the use of epoxy precursors in conjunction with Hoveyda–Grubbs' II metathesis catalyst (Guadagno et al. 2011). In recent papers, they have shown the results related to the healing efficiency and dynamic mechanical properties of self-healing epoxy formulations cured up to 170 and 180 °C using as catalyst Hoveyda–Grubbs' I (HG1) metathesis catalyst (Guadagno et al. 2014b).

It was found that the healing chemistry based on the ROMP of healing agents can be very different depending on the chemical nature of the catalyst and the healing agent. The performed experiments highlighted that the chemical nature of the epoxy matrix plays an important role in determining the mechanical properties of the resulting material.

The effect of matrix nature and curing cycle on the healing efficiency and dynamic mechanical properties of self-healing epoxy resins were also investigated for epoxy formulations solidified by choosing processing conditions that contemplate permanence at high temperatures up to 170 and 180 °C (Guadagno et al. 2014b).

As self-healing components, ENB and Hoveyda–Grubbs I catalyst particles were chosen. Microcapsules resembling snarls of tubular nanovessels (see Fig. 10.5) and catalyst particles were dispersed inside an epoxy mixture tailored to meet specific needs of structural materials (as higher storage modulus, higher glass transition temperature with respect to already published epoxy formulations).

Self-healing efficiency assessed by fracture test has been calculated for different self-healing formulations. The healing efficiency η has been evaluated by fracture tests. For quasi-static fracture conditions, crack healing efficiency, η (Eq. 10.1), is defined as the ability of a healed sample to recover fracture toughness:

$$\eta = \frac{K_{\mathrm{IChealed}}}{K_{\mathrm{ICvirgin}}} \tag{10.1}$$

where K_{ICvirgin} is the fracture toughness of the virgin specimen and K_{IChealed} is the fracture toughness of the healed specimen. Using a protocol already established in literature (Guadagno et al. 2019), η was determined for the virgin and the healed

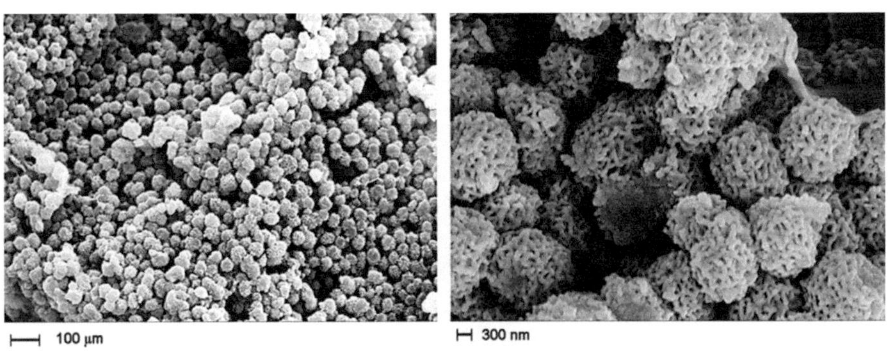

Fig. 10.5 SEM images at different magnifications of the synthesized microcapsules in the form of spherical snarls

sample by carefully controlled fracture experiments. A TDCB geometry (see Fig. 10.6) was used for these tests to ensure a controlled crack growth along the centerline of the specimen and provides a crack-length-independent measure of the fracture toughness (Brown et al. 2002).

Healing evaluation was performed first with a virgin fracture test of an undamaged TDCB sample. A precrack is introduced to sharpen the crack tip. Afterward the sample was mounted on a load frame and loaded under displacement control causing the propagation of the precrack along the centerline of the sample until failure. Subsequently, the crack was closed and allowed to heal at room temperature without external intervention. After healing, the sample is loaded again until failure. Due to the TDCB geometry, the healing efficiency measurement simply requires knowledge of the critical fracture load, P_C, of virgin and healed specimen (Eq. 10.2):

$$\eta = \frac{P_{\text{Chealed}}}{P_{\text{Cvirgin}}} \tag{10.2}$$

Equation 10.2 can be derived from Eq. 10.1 by considering that $K_{IC} = \alpha P_C$, where $\alpha = 11.2 \times 10^3\ \text{m}^{-3/2}$, for the TDCB used geometry. This procedure has been recognized to be very effective to measure the healing efficiency. In the current work, fracture specimens were tested by INSTRON mod. 5967 Dynamometer under displacement control using a load cell of 30 KN and a 250 μm/min displacement rate and measuring load and displacement values (see Fig. 10.7). Samples were loaded at first failure and then unloaded allowing the crack face to come back into contact. After 24 h, the sample was retested, and the healing efficiency was calculated using Eq. 10.2.

In all the analyzed systems, the healing efficiency measured by means of in situ tests using tapered double-cantilever beam (TDCB) specimens was found very high (also higher than 100%). In particular, it was found to be affected by microcapsule and catalyst concentrations and their morphological parameters.

Figure 10.8 shows the load–displacement curves for the sample EBA1 (curing conditions: 125 °C for 1 h and 170 °C for 2 h) with 5% of HG1 catalyst and different percentages of microcapsules (10 and 20%). EBA1 matrix was prepared by mixing

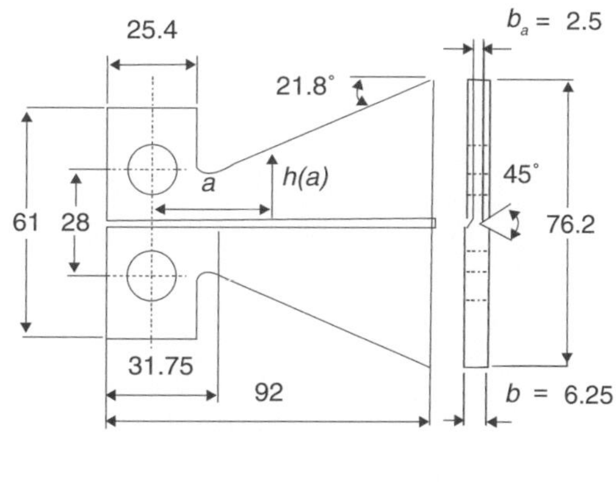

Fig. 10.6 TDCB geometry and dimensions (mm)

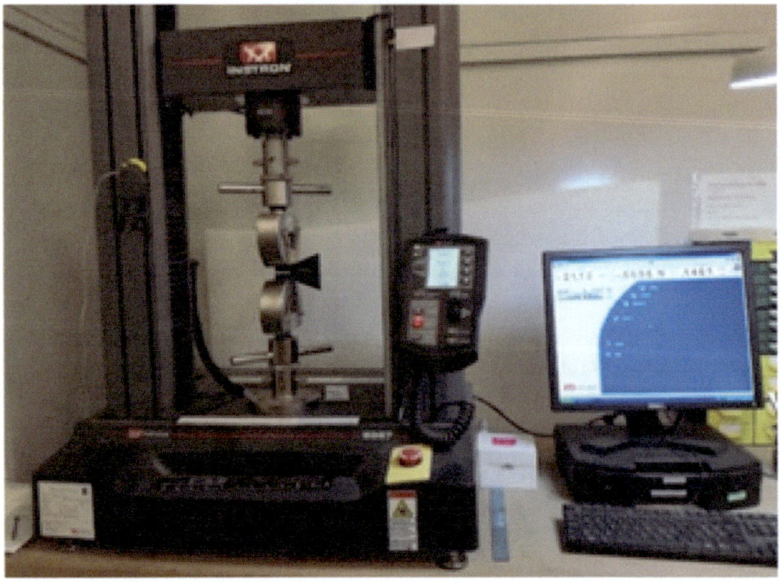

Fig. 10.7 Equipment to evaluate the healing efficiency

Fig. 10.8 Load–displacement curves detected at 25 °C for virgin and healed samples EBA1 HG1 5% Mic 20% (**a**) and EBA1 HG1 5% Mic 10% (**b**) cured up to 170 °C

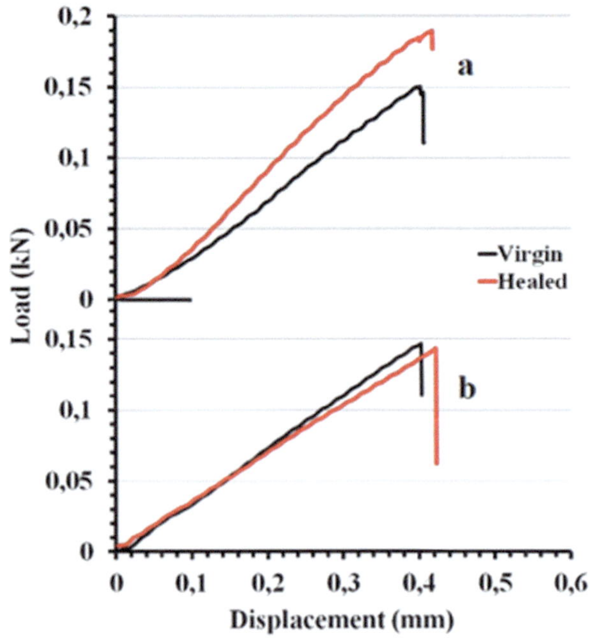

the epoxy precursor EPON 828 with an epoxy reactive monomer (1,4-butandiol diglycidyl ether) at a concentration of 90%: 10% (by wt) epoxide to flexibilizer.

The healing efficiency was found to be 97.75% for a percentage of 10% of microcapsules, 111% for a percentage of 17% of microcapsules, and 123% for a percentage of 20% of microcapsules. Considering a constant amount of catalyst (fixed at 5 wt%), an increase in the microcapsule concentration was found to improve the healing efficiency of the formulated materials. A more complete picture is shown in Fig. 10.9.

Increasing the catalyst concentration was found to improve healing efficiencies when the capsule concentration was held fixed (Brown et al. 2005b).

A very similar result was found by Brown et al. (2002) who performed experiments of healing efficiency on pure epoxy (EPON 828) cured with DETA for 24 h at room temperature, followed by 24 h at 30 °C. They used DCPD as healing agent and Grubbs' first-generation catalyst.

Figure 10.10 shows the influence of the curing cycle on the healing efficiency values for some of the analyzed formulations and for the two different curing cycles. Sample EBA1 cured up to 170 °C showed the highest values of efficiency. For sample EBA1 (10 wt% of reactive diluent) cured up to 170 °C with a microcapsule concentration higher than 10 wt%, the self-repair effect allows the healed material to have a critical load (P_{Chealed}) higher than the critical load (P_{Cvirgin}) of the virgin material, and efficiency is over 100%. The lower efficiency of the samples cured up to 180 °C, with respect to samples cured up to 170 °C, is most probably due to an initial thermolytic decomposition of the HG1 catalyst.

The authors developed the self-healing technology for use in structural materials that are not easily accessible for maintenance. These self-healing materials were designed to repair damage that begins at the micron scale, effectively stopping the

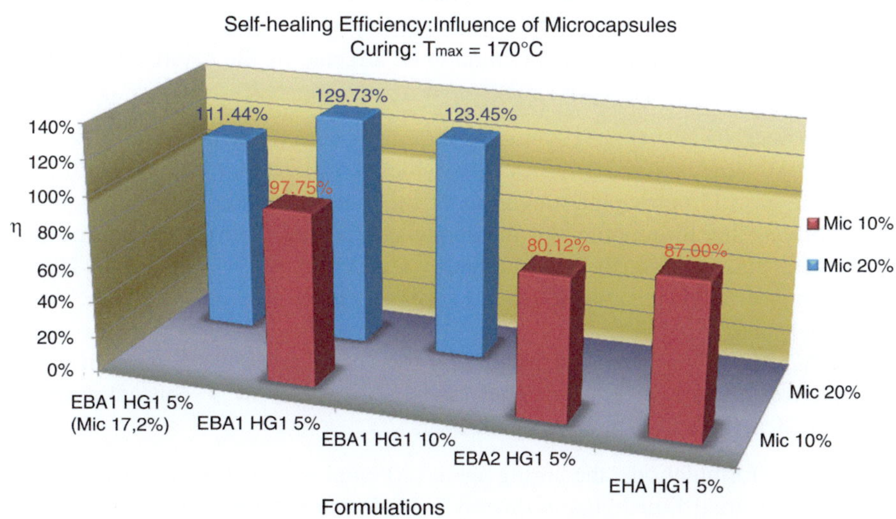

Fig. 10.9 Self-healing efficiency of samples with different amounts of microcapsules

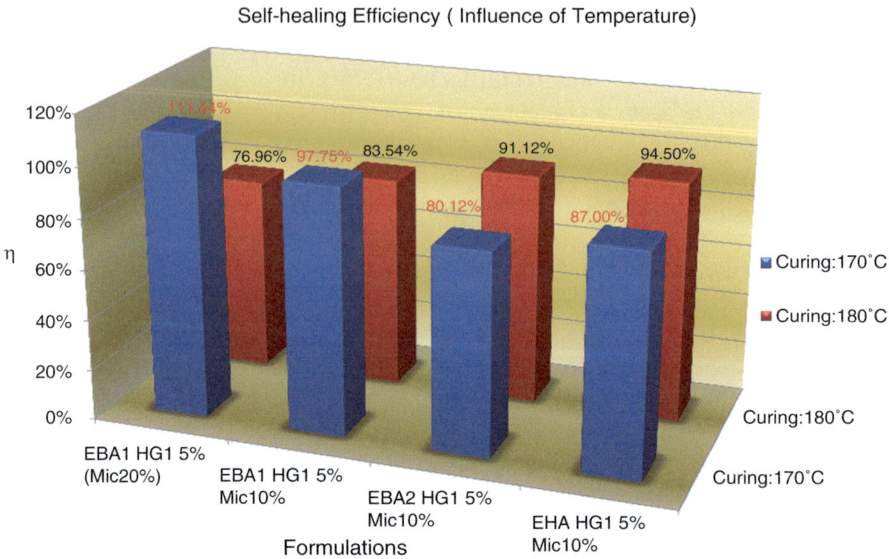

Fig. 10.10 Self-healing efficiency with different formulations and curing cycles

propagation of the damage and extending the lifetime of the material. The microcapsules were designed to be robust enough to survive the manufacturing process and the development of the material in its intended application. A range of self-healing chemistries can be microencapsulated to maximize compatibility with matrix materials. On top of this, the size of the microcapsules and the thickness of their shell walls can be customized for various applications and chemical environments. Additional functionalities can be added to the self-healing material to provide different solutions.

The healing efficiency was also found to be affected by the nature of the epoxy matrix. In particular, three epoxy matrices were investigated:

(a) EHA matrix was prepared by mixing the epoxy resin (E) with an epoxy flexibilizer (H)—trade name HELOXY 71—which was used in a percentage of 37% by weight to improve the toughness of the material and consequently growth stability. This formulation for the epoxy matrix was used for the self-healing system already described in literature (Guadagno et al. 2010, 2013 2014a; Raimondo and Guadagno 2012, 2013; Guadagno and Raimondo 2012).
(b) EBA1 matrix.
(c) EBA2 matrix was prepared by mixing the epoxy precursor E with an epoxy reactive monomer (1,4-butandiol diglycidyl ether) at a concentration of 80%/20% (by wt) epoxide to flexibilizer.

For these formulations, the curing agent (A) employed was an anionic initiator, Phenol, 2,4,6-tris[(dimethylamino)methyl] (trade name Ancamine K54), already used in literature (White et al. 2001). The catalysts used in all the epoxy matrices was Hoveyda–Grubbs' I (HG1). It was purchased from Aldrich. The catalysts were reduced in the form of thin powders before dispersion into the epoxy matrix. ENB-filled microcapsules with the outer shell composed of poly(urea-formaldehyde) and the inner shell of ethylene maleic anhydride copolymer (EMA) were prepared by in situ polymerization in an oil-in-water emulsion in accordance with previously described procedures (Brown et al. 2003).

The crack healing efficiency, η, was found to be 97.75% for a percentage of 10% of microcapsules. Slightly lower efficiency was recorded for EBA formulation cured up to 180 °C. At curing temperature up to 180 °C, high values of efficiency (94.50%) were recorded for EHA samples (see Fig. 10.11).

As found with the mechanical tests, a good adhesion between the capsules and the polymer matrix was achieved for all the formulated systems. It not only allows an effective cross-link formation along the sample but also prevents the dewetting of the microcapsules ensuring an efficient utilization during the crack propagation. The very high level of crack healing efficiency attained demonstrates the ability of these systems to regenerate the pristine properties.

Even if very good healing efficiencies were found, the developed systems required high costs for the ROMP catalyst, and most of all, they have shown to not meet some of the mechanical requirements necessary for specific scopes (structural parts of aircrafts, ships, wind turbine blades, satellite control boards, etc.). In particular, the maximum value reached in the Tg was found to be between 100 and 125 °C; the value of the storage modulus was found to range between 2500 and 2000 Mpa (in the temperature range from −50 °C to +80 °C).

These relatively poor mechanical properties (for the above listed specific applications) were found due to the impossibility to use aromatic primary amines in the developed self-healing epoxy formulations because they deactivate ROMP catalysts (Longo et al. 2017). Several issues related to transferring these self-healing systems from the laboratory scale to possible real applications in the aeronautical field, such as the possibility to choose systems with increased glass transition temperature, high

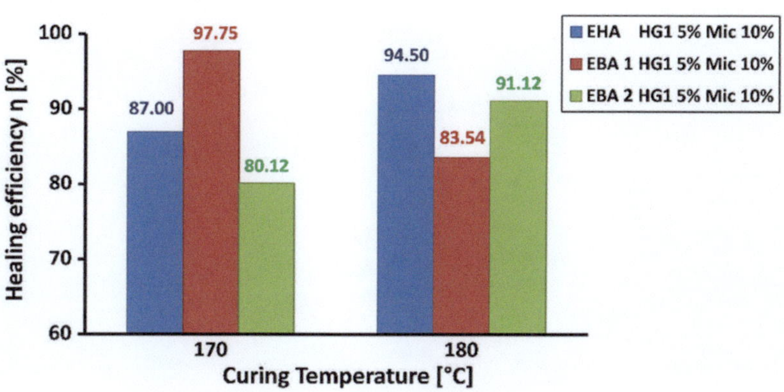

Fig. 10.11 Comparison of healing efficiency for self-healing systems with different epoxy matrices (EHA, EBA1, EBA2), at the different curing temperature

storage modulus, and high values in the healing functionality under real working conditions, have been analyzed, and alternative solutions for the formulations of multifunctional self-healing resins were proposed.

In particular, to increase the mechanical performance and to integrate other relevant functions in the material (flame resistance, electrical conductivity to enhance the lightning strike resistance), carbon nanotubes and graphene-based materials were embedded in the epoxy matrix trying to preserve the catalytic sites of the self-healing function (Mariconda et al. 2015; Guadagno et al. 2017a).

In particular, electrically conductive nanoparticles, functionalized to act as support for the catalyst and simultaneously impart self-healing and electrical conductivity properties to the polymeric matrix, have been successfully prepared. Highly dispersed copper nanoparticles, immobilized onto carbon nanomaterials, were prepared and used as a recyclable and reusable catalyst to achieve Cu (I)-catalyzed [3 + 2] cycloaddition "click" chemistry (Shaygan et al. 2015). In particular, carbon nanoparticle-supported catalysts exhibit excellent properties due to their high surface area, thermal stability, and porous surfaces. These characteristics make them particularly suitable to act as scaffold to prevent the agglomeration of the immobilized metallic particles (Shaygan et al. 2015; Sarkar et al. 2014; Gu et al. 2014; Scheuermann et al. 2009).

Among the carbon-based nanoparticles, graphene immobilized catalyst has been recently proposed for application in several technological fields (Shaygan et al. 2015; Yang et al. 2013; Eigler et al. 2013; Eigler and Hirsch 2014; Nia et al. 2015). The functionalization also favors the nanomaterial dispersibility which is improved via covalent and non-covalent functionalization (Georgakilas et al. 2012; Sahoo et al. 2010; Osim et al. 2013).

Concerning the covalent functionalization, the possibility to covalently bond ROMP catalysts to multiwalled carbon nanotubes (MWCNTs) (see Fig. 10.12) or graphene-based particles has been investigated.

Fig. 10.12 Scheme of self-healing system based on ROMP reactions activated by ruthenium catalysts covalently bonded to MWCNTs

Concerning this aspect, it was already observed that graphene-based nanomaterials can be designed as a self-assembly structure in the epoxy matrices using edge-carboxylated layers approach (Guadagno et al. 2015). The functionalization on graphene sheets facilitates the interfacial interaction between polymer and carbon layer. It also enhances the electrical percolation paths and mechanical performance of structural epoxy matrices (TBD). It has been deduced that the surface chemistry of graphene surface can be tuned at nanoscale level, to pave the way toward an effective strategy to also transfer graphene properties to polymeric matrices.

In particular, in the above-described context, graphene oxide (GO) was used to support ruthenium catalysts with the aim to activate self-healing reactions in multifunctional materials able to integrate simultaneously the healing reactions with the very interesting properties of graphene-based materials (Mariconda et al. 2015) (see Fig. 10.13).

Grubbs' first- (G1) and second-generation modified catalysts (G2o-tol), Hoveyda–Grubbs' catalysts first- (HG1) and second-generation (HG2) catalysts were covalently bonded to GO preserving the same catalytic activity of the catalysts not bonded to the graphene sheets. GO-G2o-tol and GO-G1 were found to deactivate during the process of preparation of the self-healing epoxy mixtures at 90 °C. Evidence of the self-healing activity of the various catalytic complexes was studied for both uncured and cured samples. Results showed that GO-HG1 and GO-HG2 were not deactivated, and hence they were found to be able to trigger self-healing reactions based on the ROMP of ENB. This behavior was found due to the formation of 16-electron Ru complexes that are more stable than the 14-electron complexes of GO-G1 and GO-G2 catalysts (Mariconda et al. 2015). A strong increase in the storage modulus was found (as expected), but not in the glass transition temperature. Also, in this case, it was found that no aromatic amines could be

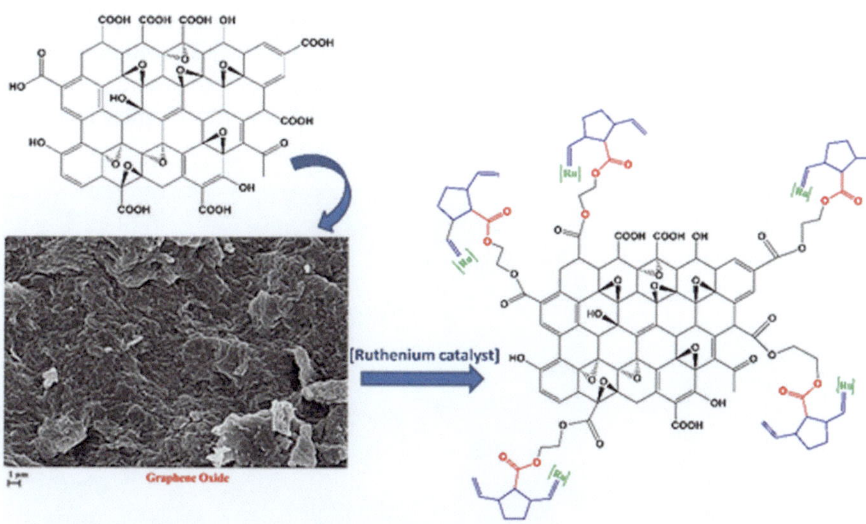

Fig. 10.13 Functionalization of graphene oxide with ruthenium catalyst

used to solidify the epoxy matrix in such a way to obtain very high mechanical performance. A strategy to overcome this drawback was experimented. The catalytic sites (containing Ru on the graphene sheets) were protected polymerizing few units of polymeric monomers around the ruthenium atoms forming a globular shell around the catalyst sites (Guadagno et al. 2017a) (see Fig. 10.14).

This system allowed high curing temperatures, but the open part of the shell is low, and this strongly influences the kinetics of the ROMP reactions inside cracks of the selected epoxy matrix.

Unfortunately, as highlighted above, many structural applications require that the function of self-healing must be active also under extreme environmental conditions and the self-healing mechanisms must be quickly activated to avoid the crack propagation (when the entity of the propagation is relevant, there is no healing of the material). Furthermore, the materials for many very interesting structural applications must be characterized by very high mechanical performances, electrical conductivity, low moisture content, long durability, etc.

Also, other problems concern the cost of the catalyst, its poisoning with aromatic primary amines, and the thermal stability which is compromised at the cure temperatures necessary to impart good mechanical performance to the epoxy matrix (cured with aromatic primary amines). Performed experiments evidenced that, in the case of the catalyst particles solubilized at molecular level (to reduce the cost of the self-healing materials), the catalyst particles which are locally in contact with the oxirane rings (during the curing reactions) deactivate, hence reducing the actual amount of active catalyst. Additional experiments were performed on this issue to better understanding the reason of the deactivation for the most promising ROMP catalysts (catalysts active in the ring-opening metathesis polymerization). An equimolecular reaction between epoxide ring and alkylidene of ruthenium compound was found to be responsible of the deactivation (Raimondo et al. 2015).

It is possible to activate self-healing reactions in epoxy matrices cured at high temperature (between 130 and 170 °C) using solid catalyst particles which retain an intact heart of catalyst which is not deactivated in contact with the oxirane rings of the epoxy matrix. This strategy was adopted by Leonardo-Finmeccanica to

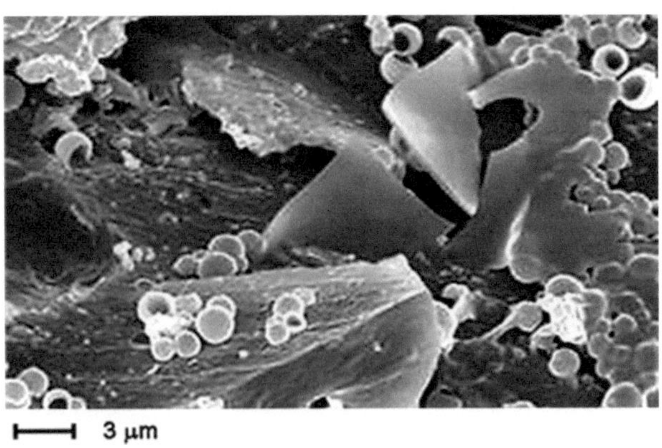

Fig. 10.14 Globular shell around the catalyst sites attached on graphene sheets

manufacture carbon fiber-reinforced composites (CFRCs) for the realization of structural aeronautical components in order to address their vulnerability to impact damage in the real service conditions (Raimondo et al. 2016). The developed self-healing panels were based on ring-opening metathesis polymerization reaction of microencapsulated 5-ethylidene-2-norbornene/dicyclopentadiene cyclic olefins using Hoveyda–Grubbs' first-generation catalyst as catalyst. The self-healing resin was infused into a carbon fiber dry preform using an unconventional bulk film infusion technique which allowed to minimize the filtration effects via a better compaction and reduced resin flow paths. Infrared spectroscopy proved to be a useful technique to identify metathesis products and therefore catalyst activity in the self-healing panel after damage. The damage resistance of the manufactured CFRPs was evaluated through hail and drop tests (see Fig. 10.15). Although aerospace and aviation technology allow aircraft to avoid a great many storms and other dangers, wildlife impacts and hail still cause millions of dollars in plane crashes and aircraft property damage—not to mention the danger to human life. Efficient self-healing

Fig. 10.15 Tool drop test setup (on the top) and depth measuring after impact (on the bottom)

mechanisms can help to avoid this type of risk. The developed panels have been tested for different values of impact energy (tool drop test), after which it has been possible to compare the characteristics of damage resistance and to measure the maximum residual deformation. The estimate of the latter (also said indentation) has been obtained through the use of a digital comparator. For the impact of 20 J (using a tip diameter of 1 in.), it has been possible to measure an indentation of 0.14 mm without drilling. Figure 10.15 shows the tool drop test setup and the depth measuring after impact.

Hail impact tests have been performed considering the requirements of hail impact for the design of a fuselage in composite material (ground hail). They are the following: (a) hail diameter 2.4 in. (Cum Prob 0.98) and (b) energy 50 J. The panels have been subjected to impact tests "gas gun" at different levels of energy and then inspected by means of nondestructive inspection (NDI).

To simulate the damage caused by the impact of hail on aeronautical structures, a plant for the launch of pneumatic darts has been used. Table 10.1 shows the experimental values concerning the performed experiments. Figure 10.16 shows the photo of one of the tested panels.

Figure 10.17 shows the hail impact test setup and test fixture.

Figure 10.18 shows the photos of the hailstone+sabot weighing.

Figure 10.19 shows the following photos: sabot in the gun (on the left at the top), photocell station for hailstone velocity (on the right at the top) and measuring and test equipment (at the bottom).

The result shows that, with the energies reported in Table 10.1, the panel was not damaged because it returns in a perfectly elastic way in its undeformed condition after the impact. It is worth noting that the mechanisms proposed for the auto-repair action are very fast. From the first NDI, no cracks are observed in the panels also for maximum energy value tested.

Table 10.1 Experimental values concerning the performed experiments

Panel	Impact	Pressure valve	Hail weigh	Speed	Energy
[N°]	[Location]	[bar]	[g]	[m/s]	[j]
1	A	1.00	163	41.6	141
2	A	0.78	108	36	70
2	B	0.90	163	38	118

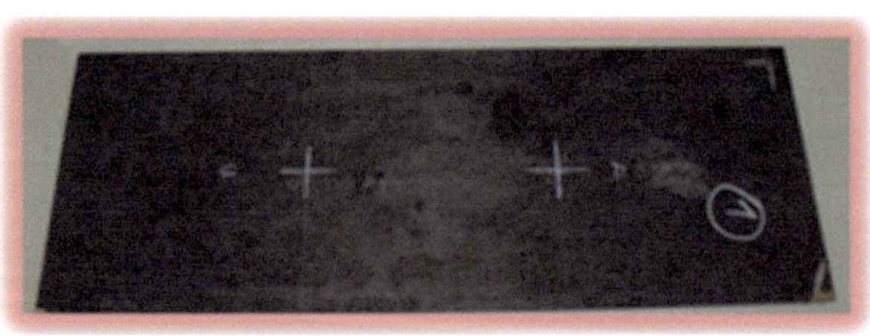

Fig. 10.16 Photo of the one of the tested panels

Fig. 10.17 Hail impact test setup (on the top) and test fixture (on the bottom)

The self-healing manufactured panels showed, after damage, catalytic activity with the metathesis product formation poly(ENB/DCPD) (see Fig. 10.20), as also evidenced by an infrared peak at 966 cm^{-1}.

The damage response of CFRPs, detected in accord with the requirements of hail impact for the design of a fuselage in composite material, was found very good, also at low working temperatures (values as low as −50 °C). These results encouraged further research to reduce the cost of the CFRCs and further enhance the stability of the catalyst during the curing cycles of the panels.

Apart from the excess of catalysts to be used, another problem addressed by Leonardo-Finmeccanica was the development of a new catalyst able to resist in presence of the common industrial hardeners (primary aromatic amines such as DDS) at high values of temperature (during the curing cycle) (Guadagno et al. 2017b).

A reduction of material costs and the possibility to apply the ROMP initiators in epoxy resins aimed at manufacturing load-bearing structures were obtained using the initiator $HG2_{Ph\text{-}Mes}$ specially synthesized for this purpose (Longo et al. 2017; Guadagno et al. 2017b). The synthetized catalyst proved to be characterized by high thermal stability and tolerance toward the epoxy groups and the aromatic primary amines. The $HG2_{Ph\text{-}Mes}$ initiator also proved to be active in the synthesis in situ of polymers (PN/NB) inside thermosetting resins characterized by improved mechanical performance. In particular, the initiator solubilized in the form of molecular complex proved to be active in structural resins hardened with aromatic primary

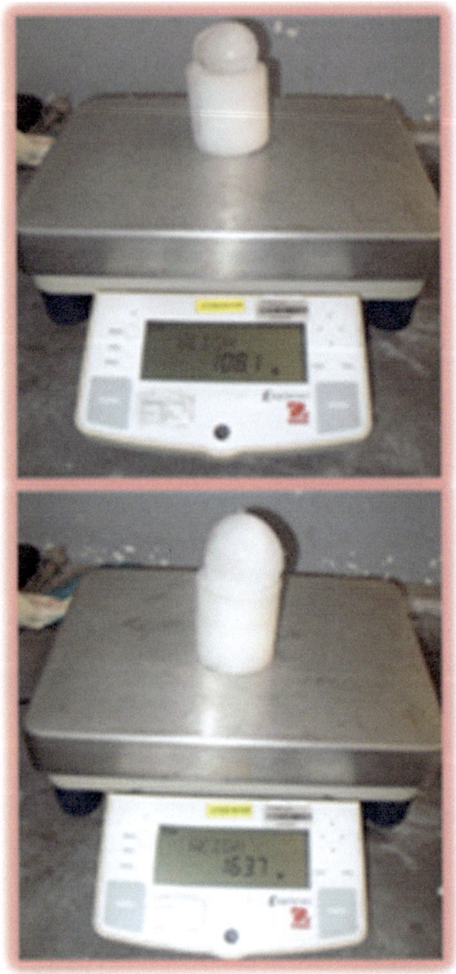

Fig. 10.18 Photos of the hailstone+sabot weighing

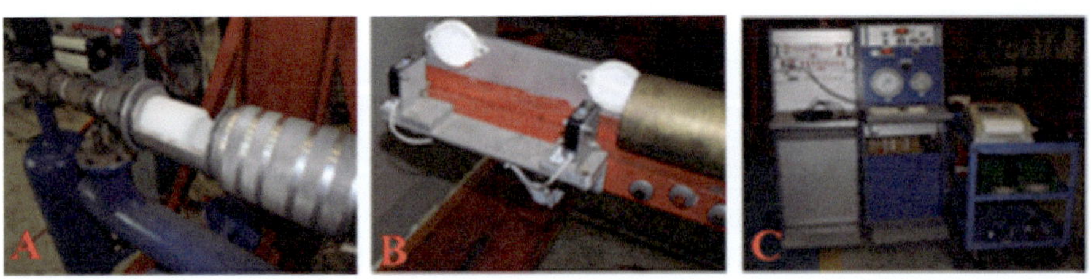

Fig. 10.19 Photos of the sabot in the gun (**a**), photocell station for hailstone velocity (**b**), and measuring and test equipment (**c**)

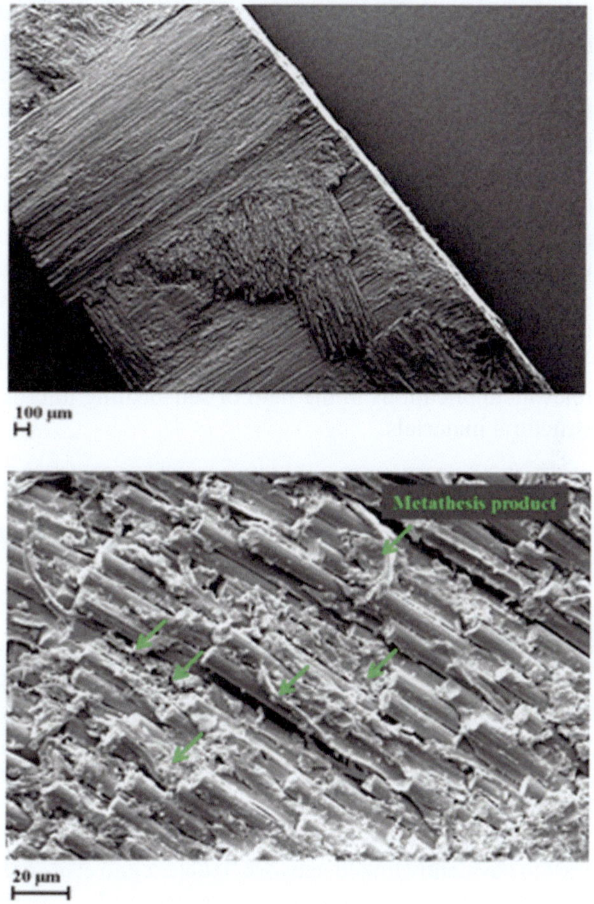

Fig. 10.20 FESEM of fracture surface of the delaminated panels

amines (e.g., DDS) after curing cycles up to 180 °C as those planned for high structural materials (aeronautical, naval, infrastructural, etc.).

One of the most relevant results concerning the use of $HG2_{Ph\text{-}Mes}$ is the possibility to reduce its percentage from 5 wt/wt% to 0.5 wt/wt% in self-healing resins. The stability at high temperatures in presence of aromatic primary amines allows to use $HG2_{Ph\text{-}Mes}$ in epoxy resins aimed at building load-bearing structures with the relevant advantage to reduce its percentage of 90% and hence the cost related to the self-healing materials of about 79% (compared to the material containing 5 wt/wt% of HG1).

The possibility to use the initiator solubilized in the form of molecular complex inside epoxy mixtures also allows high performance in the self-healing function (the ROMP reaction is activated whenever the healing agent contacts the matrix where the damaged capsule is embedded and not only when the healing agent contacts the initiator particles).

It is worth noting that, in the field of structural composites, solutions able to overcome shortcomings related to the microencapsulated systems will also allow to fulfill other requirements of structural materials without additional costs. For instance, the inclusion of microcapsules in CFRCs (self-healing aeronautical laminates) aimed at imparting self-healing ability is of relevant benefit also for the very appreciable damping performance of these kinds of self-healing composites compared to standard CFRC composites (Viscardi et al. 2017).

This peculiar property of the microencapsulated systems further allows to reduce the cost related to the noise/vibration control of aircraft structures. The numerical analysis performed by means of ACTRAN® solver confirmed an enhanced performance of the innovative material design, in terms of structural vibration as well as of sound pressure level inside the cabin, motivating the development of advanced technologies up to industrial scale in future industrial activities.

The synthetized initiator can pave the way toward the rational design of practical relevant applications in the field of self-healing microencapsulated multifunctional structural materials.

2 Enabling Self-Healing Properties Toward the Functional Materials of the Future

Recently, alternative strategies to the microencapsulated systems have been proposed. In this new approach, chain dynamics and principles of polymeric physics are an important aspect for the design of self-heling structural polymer. In order to design materials with dynamic properties, a reversible bonding system is required to enable a self-healing material, including aspects of supramolecular polymer chemistry. Thus, non-covalent bonds such as hydrogen bonds and π–π stacking are crucial elements on which the possibility of multiple healing cycles can be considered (in contrast to covalently linked networks) (Guadagno et al. 2017c, 2019; Herbst et al. 2013).

The field of supramolecular chemistry utilizes multiple, reversible, and, in certain cases, cooperative intermolecular interactions to create new materials with unique properties and functionalities (Faghihnejad et al. 2014). The non-covalent intermolecular interactions typically utilized in the design of supramolecular materials and polymers include hydrophobic (Annable et al. 1996; Tam et al. 1998), hydrogen bonding (Sijbesma et al. 1997; Sivakova et al. 2005; Cordier et al. 2008), metal–ligand (Beck and Rowan 2003; Weng et al. 2006), and ionic interactions (Eisenberg et al. 1990). Over the last two decades, many synthesis strategies have been developed to design new supramolecular polymers with unique characteristics such as enhanced bulk properties (i.e., plateau modulus, tensile modulus), self-healing capability, stimulus responsiveness, and the ability to assemble into well-defined nanostructures. In this field, self-healing polymer materials or composites have attracted considerable attention over the past decade due to their controllable and reversible molecular interactions, interesting mechanical properties, and potential applications. Many conventional healing approaches used in thermoplastic polymers and thermoset composites such as microencapsulation and thermally reversible cross-links (covalent bonds) require treatments at high temperature (i.e., high energy input) which are not required using this approach.

Recently, Guadagno et al. (2017c) applied such strategy to develop self-healing aeronautical resins by introducing nanometric flexible cages of hybrid materials into the host structure matrix. These nanometric flexible cages have been found to be able to enhance other properties such as flame resistance, mechanical properties, and thermal stability (Guadagno et al. 2017c).

Furthermore, hydrogen bonding moieties can be covalently attached to specific nanofillers to activate self-healing mechanisms integrated in multifunctional materials. It is worth noting that the main challenge to overcome in applying this type of mechanisms is the limited dynamic properties of thermosetting matrix segments. Recent developments have highlighted that this mechanism can be integrated in epoxy mixture (able to reduce moisture content) solidified with primary aromatic amines and therefore characterized by high values in the glass transition temperature and storage modulus. The presence of GPOSS in the nanofilled epoxy matrix strongly increases flame resistance and allows a decrease in the resin viscosity. This is a desirable effect as it counterbalances the increase in the viscosity due to the addition of nanofillers in the epoxy formulation. The increase in viscosity is an effect which must be carefully considered because it causes several processing problems such as difficulties in the dispersion stage of the nanofiller and/or filtration during the infusion process for the manufacturing of carbon fiber-reinforced composites. Dynamic mechanical results have shown that sample containing carbon nanoparticles tends to create a phase with increased mobility of the chains. This particular phase arrangement is most likely responsible to better promote reversible hydrogen bonds determined by interaction between epoxy resin and nanocages of POSS compounds. The presence of a more mobile phase in the multifunctional formulation can be advantageously exploited to enhance the self-healing efficiency up to 400% and then to strongly reduce the fatigue crack growth rate of CFR laminates impregnated with multifunctional formulations (Kadlec et al. 2016).

Mechanisms based on hydrogen bonding were also used by Guadagno et al. (2019). They developed structural self-healing resins based on dynamic hydrogen bonds capable to overcome many current industrial limitations. Hydrogen bonding moieties, such as barbiturate and thymine, able to act as reversible healing elements by their simultaneous donor and acceptor character, were covalently linked to multiwalled carbon nanotubes (MWCNTs) to formulate auto-repair nanocomposites (see Fig. 10.21).

The so-functionalized MWCNTs, embedded in a rubber-toughened epoxy formulation, lead to reversible MWCNT bridges through the matrix due to strong attractive interactions between the rubber phase, finely dispersed in the matrix, and MWCNT walls. Healing efficiencies have been assessed for the nanocharged epoxy formulation loaded with 0.5 wt/wt% of MWCNTs decorated with barbituric acid and thymine groups. For both functional groups, healing efficiencies higher than 50% have been found. Dynamic mechanical analysis (DMA) evidences an enhancement in epoxy chain movements due to micro/nanodomains of the rubber phase enabling self-healing behavior by recovering the critical fracture load. Results from this study may promote the wide development of safe and cost-efficient self-healing composites in aeronautical, automotive, civil engineering, and wind power industries as alternative to the microencapsulated systems.

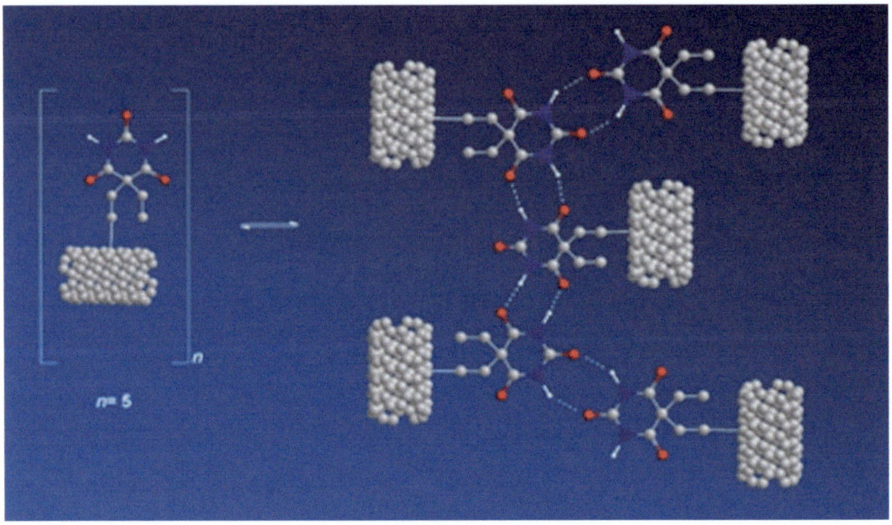

Fig. 10.21 Example of barbiturate-modified MWCNTs for reversible hydrogen bonds

3 Conclusion

The design of self-healing, structural, and functional material is a work which sounds like magic! Such material can drive the future changes, revolutionizing the idea of the material scientists on the vastness of application and bringing a new dimension in the future technologies. Never is this so true as now. This is the time where the nanoengineering offers the very real promise of a veritable cornucopia of enabling new smart materials, devices, and products. The human mission is to take the benefits of the challenges ahead in recent developments of the nanoscience which promise environmental sustainability in competitiveness and high levels of human safety. The objective of many industrial applications is set in this scenario. The development of structural, functional, and self-healing materials can revolutionize multiple industries: automotive, aeronautical, electronic, naval, etc. Until now, the traditional approach to the development of structural material is to address the load-carrying function and other functional requirements separately, resulting in a suboptimal load-bearing material with add-on attachments which perform the nonstructural functions with the penalty of added weight. Recently, however, there has been increased interest in the development of load-bearing materials and structures which have integral non-load-bearing functions, guided by recent discoveries about nanofillers and nanotechnology that can help to project materials working as self-healing multifunctional systems. This chapter discussed various developments that lead in this direction. The technological benefits of such systems have begun to be identified for targeted applications, and demonstrators are under construction for a wide range of applications from automotive, naval, space, and aerospace to civil engineering and domestic products.

Acknowledgments This project has received funding from the European Union's Horizon 2020 research and innovation program under grant agreement no. 760940.

References

Amendola E, Dello Iacono S, Pastore A, Curcio M, Giordano M, Iadonisi A (2015) Epoxy thermosets with self-healing ability. J Mater Sci Chem Eng 3:162–167

Annable T, Buscall R, Ettelaie R (1996) Network formation and its consequences for the physical behaviour of associating polymers in solution. Colloids Surf A Physicochem Eng Asp 112:97–116

Beck JB, Rowan SJ (2003) Multistimuli, multiresponsive metallo-supramolecular polymers. J Am Chem Soc 125:13922–13923

Binder WH (2013) In: Binder WH (ed) Self-healing polymers. From principles to applications. Wiley-VCH Verlag GmbH & Co. KGaA, Weinheim, pp 440

Brown EN, Sottos NR, White SR (2002) Fracture testing of a self-healing polymer composite. Exp Mech 42:372–379

Brown EN, Kessler MR, Sottos NR, White SR (2003) In situ poly(urea-formaldehyde) microencapsulation of dicyclopentadiene. J Microencapsul 20:719–730

Brown EN, White SR, Sottos NR (2004) Microcapsule induced toughening in a self-healing polymer composite. J Mater Sci 39:1703–1710

Brown EN, White SR, Sottos NR (2005a) Retardation and repair of fatigue cracks in a microcapsule toughened epoxy composite—part I: manual infiltration. Compos Sci Technol 65:2466–2473

Brown EN, White SR, Sottos NR (2005b) Retardation and repair of fatigue cracks in a microcapsule toughened epoxy composite—part II: in situ self-healing. Compos Sci Technol 65:2474–2480

Cho SH, White SR, Braun PV (2009) Self-healing polymer coatings. Adv Mater 21:645–649

Chowdhury RA, Hosur MV, Nuruddin M, Tcherbi-Narteh A, Kumar A, Boddu V, Jeelani S (2015) Self-healing epoxy composites: preparation, characterization and healing performance. J Mater Res Technol 4:33–43

Chung K, Lee S, Park M, Yoo P, Hong Y (2015) Preparation and characterization of microcapsule-containing self-healing asphalt. J Ind Eng Chem 29:330–337

Cordier P, Tournilhac F, Soulie-Ziakovic C, Leibler L (2008) Self-healing and thermoreversible rubber from supramolecular assembly. Nature 451:977–980

Dry C (2007) Multiple function, self-repairing composites with special adhesives. Int Patent WO/2007/005657

Eigler S, Hirsch A (2014) Chemistry with graphene and graphene oxide—challenges for synthetic chemists. Angew Chem Int Ed Engl 53:7720–7738

Eigler S, Grimm S, Hof F, Hirsch A (2013) Graphene oxide: a stable carbon framework for functionalization. J Mater Chem A 1:11559–11562

Eisenberg A, Hird B, Moore RB (1990) A new multiplet-cluster model for the morphology of random ionomers. Macromolecules 23:4098–4107

Everitt DT, Luterbacher R, Coope TS, Trask RS, Wass DF, Bond IP (2015) Optimisation of epoxy blends for use in extrinsic self-healing fibre-reinforced composites. Polymer 69:283–292

Faghihnejad A, Feldman KE, Yu J, Tirrell MV, Israelachvili JN, Hawker CJ, Kramer EJ, Zeng H (2014) Adhesion and surface interactions of a self-healing polymer with multiple hydrogen-bonding groups. Adv Funct Mater 24:2322–2333

Georgakilas V, Otyepka M, Bourlinos AB, Chandra V, Kim N, Kemp KC, Hobza P, Zboril R, Kim KS (2012) Functionalization of graphene: covalent and non-covalent approaches, derivatives and applications. Chem Rev 112:6156–6214

Gu X, Qi W, Xu X, Sun Z, Zhang L, Liu W, Pan X, Su D (2014) Covalently functionalized carbon nanotube supported Pd nanoparticles for catalytic reduction of 4-nitrophenol. Nanoscale 6:6609–6616

Guadagno L, Raimondo M (2012) Chapter 14: Use of FTIR analysis to control the self-healing functionality of epoxy resins. In: Theophile T (ed) Infrared spectroscopy—materials science, engineering and technology. InTech, Rijeka, pp 285–300

Guadagno L, Longo P, Raimondo M, Naddeo C, Mariconda A, Sorrentino A, Vittoria V, Iannuzzo G, Russo S (2010) Cure behavior and mechanical properties of structural self-healing epoxy resins. J Polym Sci B Polym Phys 48:2413–2423

Guadagno L, Longo P, Raimondo M, Naddeo C, Mariconda A, Vittoria V, Iannuzzo G, Russo S (2011) Use of Hoveyda–Grubbs' second generation catalyst in self-healing epoxy mixtures. Compos Part B 42:296–301

Guadagno L, Raimondo M, Naddeo C, Longo P (2013) Chapter 17: Application of self-healing materials in aerospace engineering. In: Binder WH (ed) Self-healing polymers: from principles to applications. Wiley-VCH Verlag, Weinheim, pp 401–412

Guadagno L, Raimondo M, Naddeo C, Longo P, Mariconda A (2014a) Self-healing materials for structural applications. Polym Eng Sci 54:777–784

Guadagno L, Raimondo M, Naddeo C, Longo P, Mariconda A, Binder WH (2014b) Healing efficiency and dynamic mechanical properties of self-healing epoxy systems. Smart Mater Struct 23:045001. (11 pp)

Guadagno L, Raimondo M, Vertuccio L, Mauro M, Guerra G, Lafdi K, De Vivo B, Lamberti P, Spinelli G, Tucci V (2015) Optimization of graphene-based materials outperforming host epoxy matrices. RSC Adv 5:36969–36978

Guadagno L, Raimondo M, Vietri U, Naddeo C, Stojanovic A, Sorrentino A, Binder WH (2016) Evaluation of the mechanical properties of microcapsule-based self-healing composites. Int J Aeros Eng 2016: Article ID 7817962. (10 pp)

Guadagno L, Mariconda A, Agovino A, Raimondo M, Longo P (2017a) Protection of graphene supported ROMP catalyst through polymeric globular shell in self-healing materials. Compos Part B 116:352–360

Guadagno L, Longo P, Mariconda A, Calabrese E, Raimondo M, Naddeo C, Vertuccio L, Russo S, Iannuzzo G (2017b) Grubbs-Hoveyda type catalyst for metathesis reactions in highly reactive environments. European Patent Application EP3141303 (A1)——2017-03-15

Guadagno L, Naddeo C, Raimondo M, Barra G, Vertuccio L, Sorrentino A, Binder WH, Kadlec M (2017c) Development of self-healing multifunctional materials. Compos Part B 128:30–38

Guadagno L, Vertuccio L, Naddeo C, Calabrese E, Barra G, Raimondo M, Sorrentino A, Binder WH, Michael P, Rana S (2019) Self-healing epoxy nanocomposites via reversible hydrogen bonding. Compos Part B 157:1–13

Herbst F, Döhler D, Michael P, Binder WH (2013) Self-healing polymers via supramolecular forces. Macromol Rapid Commun 34:203–220

Jin H, Hart KR, Coppola AM, Gergely RC, Moore JS, Sottos NR, White SR (2013) Chapter 15: Self-healing epoxies and their composites. In: Binder WH (ed) Self-healing polymers. From principles to applications. Wiley-VCH Verlag GmbH & Co. KGaA, Weinheim, pp 361–380

Jones AS, Rule JD, Moore JS, White SR, Sottos NR (2006) Catalyst morphology and dissolution kinetics of self-healing polymers. Chem Mater 18:1312–1317

Jones AS, Rule JD, Moore JS, Sottos NR, White SR (2007) Life extension of self-healing polymers with rapidly growing fatigue cracks. J R Soc Interface 4:395–403

Kadlec M, Nováková L, Mlch I, Guadagno L (2016) Fatigue delamination of a carbon fabric/epoxy laminate with carbon nanotubes. Compos Sci Technol 131:32–39

Kessler MR, White SR (2001) Self-activated healing of delamination damage in woven composites. Compos A: Appl Sci Manuf 32:683–699

Kessler MR, Sottos NR, White SR (2003) Self-healing structural composite materials. Compos A: Appl Sci Manuf 34:743–753

Liu W, Liu Y, Wang R (2011) MD simulation of single-wall carbon nanotubes employed as container in self-healing materials. Polym Polym Compos 19:333–338

Longo P, Mariconda A, Calabrese E, Raimondo M, Naddeo C, Vertuccio L, Russo S, Iannuzzo G, Guadagno L (2017) Development of a new stable ruthenium initiator suitably designed for self-repairing applications in high reactive environments. J Ind Eng Chem 54:234–251

Mariconda A, Longo P, Agovino A, Guadagno L, Sorrentino A, Raimondo M (2015) Synthesis of ruthenium catalysts functionalized graphene oxide for self-healing applications. Polymer 69:330–342
Motuku M, Vaidya UK, Janowski GM (1999) Parametric studies on self-repairing approaches for resin infused composites subjected to low velocity impact. Smart Mater Struct 8:623–638
Nia AS, Rana S, Döhler D, Osim W, Binder WH (2015) Nanocomposites via a direct graphene-promoted "click"-reaction. Polymer 79:21–28
Osim W, Stojanovic A, Akbarzadeh J, Peterlik H, Binder WH (2013) Surface modification of MoS_2 nanoparticles with ionic liquid-ligands: towards highly dispersed nanoparticles. Chem Commun 49:9311–9313
Raimondo M, Guadagno L (2012) Healing efficiency of epoxy-based materials for structural application. AIP Conf Proc 1459:223–225
Raimondo M, Guadagno L (2013) Healing efficiency of epoxy-based materials for structural applications. Polym Compos 34:1525–1532
Raimondo M, Longo P, Mariconda A, Guadagno L (2015) Healing agent for the activation of self-healing function at low temperature. Adv Compos Mater 24:519–529
Raimondo M, De Nicola F, Volponi R, Binder W, Michael P, Russo S, Guadagno L (2016) Self-repairing CFRPs targeted towards structural aerospace applications. Int J Struct Integr 7:656–670
Rule JD, Brown EN, Sottos NR, White SR, Moore JS (2005a) Wax-protected catalyst microspheres for efficient self-healing materials. Adv Mater 17:205–208
Rule JD, Sottos NR, White SR, Moore JS (2005b) The chemistry of self-healing polymers. Educ Chem 42:130–132
Sahoo NG, Rana S, Cho JW, Li L, Chan SH (2010) Polymer nanocomposites based on functionalized carbon nanotubes. Prog Polym Sci 35:837–867
Sarkar S, Moser ML, Tian X, Zhang X, Al-Hadeethi YF, Haddon RC (2014) Metals on graphene and carbon nanotube surfaces: from mobile atoms to atomtronics to bulk metals to clusters and catalysts. Chem Mater 26:184–195
Scheuermann GM, Rumi L, Steurer P, Bannwarth W, Mülhaupt R (2009) Palladium nanoparticles on graphite oxide and its functionalized graphene derivatives as highly active catalysts for the Suzuki−Miyaura coupling reaction. J Am Chem Soc 131:8262–8270
Shaygan NA, Rana S, Döhler D, Jirsa F, Meister A, Guadagno L, Koslowski E, Bron M, Binder WH (2015) Carbon-supported copper nanomaterials: recyclable catalysts for Huisgen [3+2] cycloaddition reactions. Chem—A Eur J 21:10763–10770
Sijbesma RP, Beijer FH, Brunsveld L, Folmer BJ, Hirschberg JH, Lange RF, Lowe JK, Meijer EW (1997) Reversible polymers formed from self-complementary monomers using quadruple hydrogen bonding. Science 278:1601–1604
Sivakova S, Bohnsack DA, Mackay ME, Suwanmala P, Rowan SJ (2005) Utilization of a combination of weak hydrogen-bonding interactions and phase segregation to yield highly thermosensitive supramolecular polymers. J Am Chem Soc 127:18202–18211
Tam KC, Jenkins RD, Winnik MA, Bassett DR (1998) A structural model of hydrophobically modified urethane−ethoxylate (HEUR) associative polymers in shear flows. Macromolecules 31:4149–4159
Toohey KS, Sottos NR, Lewis JA, Moore JS, White SR (2007) Self-healing materials with microvascular networks. Nat Mater 6:581–585
Toohey KS, Sottos NR, White SR (2009a) Characterization of microvascular-based self-healing coatings. Exp Mech 49:707–717
Toohey KS, Hansen CJ, Lewis JA, White SR, Sottos NR (2009b) Delivery of two-part self-healing chemistry via microvascular networks. Adv Funct Mater 19:1399–1405
van der Zwaag S (2007) In: van der Zwaag S (ed) Self healing materials: an alternative approach to 20 centuries of materials science, Springer Series in Materials, vol 100. Springer, Dordrecht, pp 1–18
Viscardi M, Arena M, Barra G, Guadagno L (2017) Smart carbon-epoxy laminate with high dissipation properties for vibro-acoustic optimization in the turboprop aircraft. Int J Mech 11:51–57

Weng W, Beck JB, Jamieson AM, Rowan SJ (2006) Understanding the mechanism of gelation and stimuli-responsive nature of a class of metallo-supramolecular gels. J Am Chem Soc 128:11663–11672
White SR, Sottos NR, Geubelle PH, Moore JS, Kessler MR, Sriram SR, Brown EN, Viswanathan S (2001) Autonomic healing of polymer composites. Nature 409:794–797
Wilson GO, Moore JS, White SR, Sottos NR, Andersson HM (2008a) Autonomic healing of epoxy vinyl esters via ring opening metathesis polymerization. Adv Funct Mater 18:44–52
Wilson GO, Caruso MM, Reimer NT, White SR, Sottos NR, Moore JS (2008b) Evaluation of ruthenium catalysts for ring-opening metathesis polymerization-based self-healing applications. Chem Mater 20:3288–3297. (Supporting Information)
Wu DY, Meure S, Solomon D (2008) Self-healing polymeric materials: a review of recent developments. Prog Polym Sci 33:479–522
Yang S, Bachman RE, Feng X, Müllen K (2013) Use of organic precursors and graphenes in the controlled synthesis of carbon-containing nanomaterials for energy storage and conversion. Acc Chem Res 46:116–128

Chapter 11
Laser Joining Processes for Lightweight Aircraft Structures

Peer Woizeschke

1 Introduction

Through its precise application, both spatially and temporally, with a specific power density and intensity distribution onto the workpiece surface, laser beams enable a wide variety of methods for the processing of materials (see Fig. 11.1 upper picture series). Depending on the specific process, the laser power, spot size, intensity profile, exposure time, or the process speed of the laser beam in relation to the workpiece varies.

Generally speaking, a distinction is made between the use of laser pulses and the application of a continuous-wave laser beam. The spot size of the laser beam on the workpiece surface can typically vary from a few micrometers to a number of millimeters. In addition to round spots, a number of other spot geometries can be achieved by shaping the beam. The intensity distribution within the laser beam also varies depending on the technology used, from a Gaussian distribution ("Gaussian mode") to a smooth power density in the focus of the laser beam, such as in the "top hat" beam profile, to multi-focus variants with two or more intensity maxima in the beam profile, a broad spectrum can be achieved (see Fig. 11.2). In addition to changes to the beam profile, the laser beam can also be modulated in both space and time by integrating optical scanner systems (1D, 2D, or 3D). For example, 1D beam oscillation is typically carried out transversely to the process direction (Schultz et al. 2014a) but in rarer cases also longitudinally (Woizeschke et al. 2015).

An essential aspect in all laser material processing is the degree to which the intensity of the laser beam works on the material, as different modes of laser processing are possible (see Fig. 11.1 lower picture series). In general, the laser beam heats the surface, whereby absorption depends, among other things, upon the angle of incidence, the polarization, and the wave length of the laser beam as well as on

P. Woizeschke (✉)
BIAS—Bremer Institut für angewandte Strahltechnik GmbH, Bremen, Germany
e-mail: woizeschke@bias.de

S. Pantelakis, K. Tserpes (eds.), *Revolutionizing Aircraft Materials and Processes*, https://doi.org/10.1007/978-3-030-35346-9_11

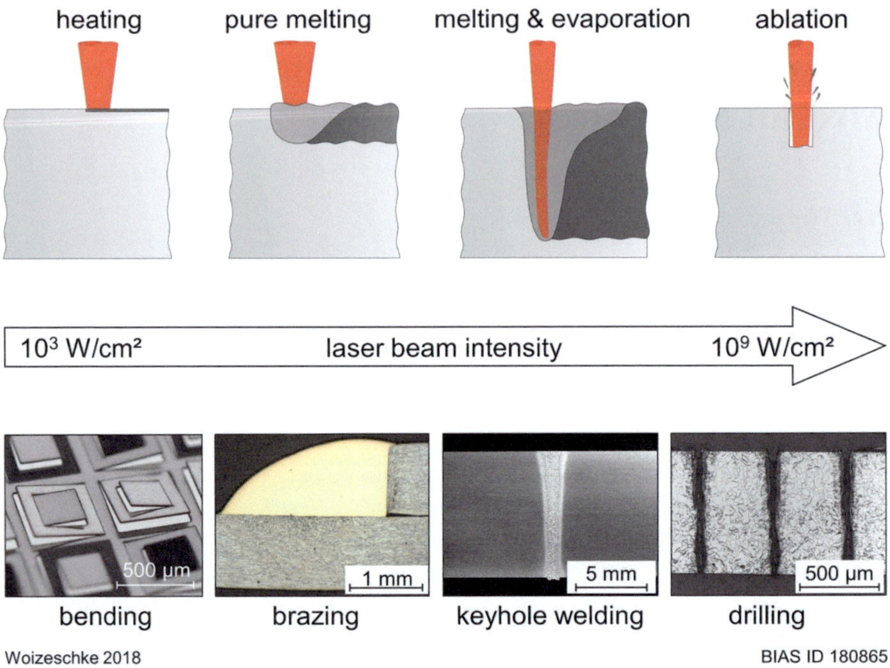

Fig. 11.1 Range of laser material processing with respect to the applied beam intensity on the workpiece

the material and surface composition (Dausinger 1995). With an increase in power, the actual local energy input onto a workpiece surface does not simply increase linearly with the locally applied power density because, on the one hand, the material-dependent absorption can change with increasing temperature and, on the other hand, the workpiece surface can undergo phase changes. After a certain energy input, there is first a melting of the surface, which influences the degree of absorption independently from the temperature increase. After a certain so-called threshold intensity, an evaporation of the irradiated materials occurs; this can form a vapor channel in the material, which tends to significantly increase absorption. In addition to a flatter angle of the laser beam hitting the surface within the so-called vapor capillary (also known as the keyhole), the significantly higher absorption rate can be traced back to the multiple reflections of the laser beam within the capillary. Depending on the laser source and the material in the keyhole, further absorption mechanisms rely on the formation of metal vapor and plasma in and over the capillary (Zhao and DebRoy 2003). With a sufficiently high intensity of the laser beam, the metal vapor from the capillary or the surrounding gas can be ionized. For CO_2 lasers (carbon dioxide laser sources with a 10.6 µm wavelength), plasma absorption plays an important role (Katayama et al. 2012). For material processing with a solid-state laser (wavelength around 1 µm), the plasma absorption coefficient is around 100 times less (Miyamoto 1997), so that absorption in the plasma can be

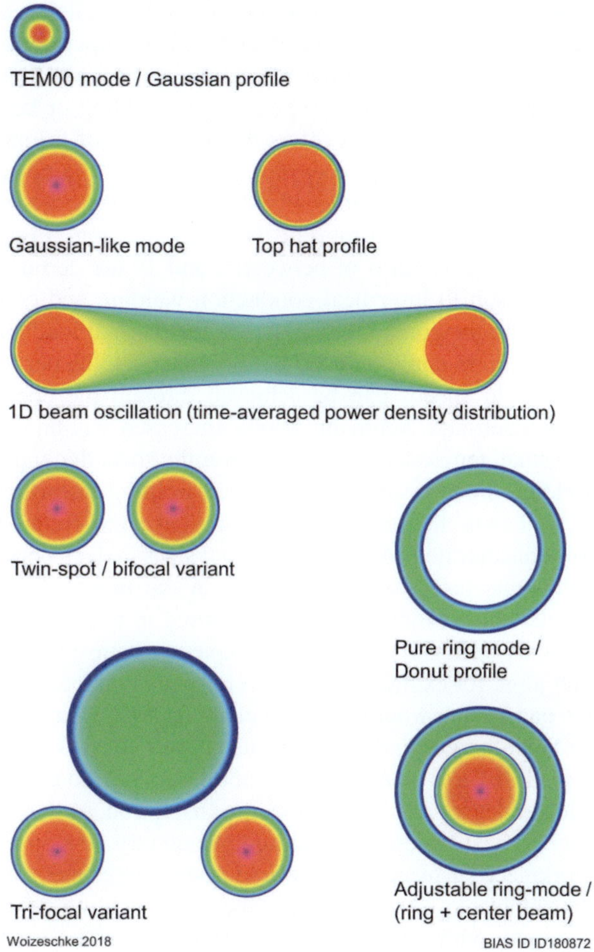

Fig. 11.2 Selection of possible laser beam intensity distributions for material processing

disregarded (Martin et al. 2001). Current solid-state laser sources are proving to be a cost-effective energy source due to their high effectiveness and an efficient energy conversion rate in comparison to other laser sources and other welding methods (Vollertsen 2009). A high beam intensity can be achieved through a high laser power and/or a high beam quality (good focusing ability) to produce small focus diameter.

While welding without capillary formation is referred to as a heat conduction process, the deep penetration welding effect (the keyhole formation) refers to processes with the creation of the above-described vapor capillaries. Depending on the laser beam and process parameters, the vapor capillary (the keyhole) changes its shape. At a simplified description, the vapor capillary during laser beam welding appears to take on the shape of a finger drawing through the material as if through a liquid. The shape of the vapor capillary (curvature and depth) is affected in particular by the process speed of the laser beam (Jin and Li 2004). Due to the evaporation processes and metal vapor flows within the capillary, this is a highly dynamic process which can lead to different kinds of imperfections and defects (Matsunawa et al. 2000). Laser welding that employs the formation of such a vapor capillary is known as laser beam deep penetration welding or keyhole welding. Compared to many other methods, it offers the advantage of bringing the energy deep into the material of the workpiece, meaning that, for example, narrow weld seams can be achieved that are significantly deeper than they are wide, which is referred to as having a high aspect ratio of the seam (see, e.g., Fig. 11.1, picture 3 of the lower series). Aspect ratios of between 1 and 10 are certainly possible and achieved. In comparison to laser heat conduction welding and conventional welding, the heat input into the component, and thus the thermal load and any related weld defaults, can be significantly reduced for the same welding depth (Dilthey 2006). In addition, high weld speeds can be achieved through the deep penetration welding effect (Vollertsen and Neumann 2009). Increased welding speeds, reduced heat inputs, and small temperature gradients over the sheet thickness due to the deep penetration welding mode resulted in decreased welding distortion in the macro (Radaj 1992) as well as in the micro range (Woizeschke and Vollertsen 2014). A disadvantage of this concentrated energy input—spot sizes in deep penetration welding lie in case of solid-state laser sources typically in the range of 15–600 μm—includes a high degree of positioning accuracy required in relation to the laser beam on the workpiece as well as between the workpieces themselves. When joining two sheets in a butt joint, for example, care must be taken that the laser beam does not primarily hit the gap, which would lead to weld defects or even weld failure.

With the currently primary used laser sources, which have wavelengths in the region of 1 μm (near-infrared), aluminum has, in comparison to steel and titanium, a lower degree of material-specific absorption, so that for heat conduction processes such as heat conduction welding, which rely on a simple Fresnel absorption on the workpiece surface, high-powered laser sources are necessary. Prior to the existence of suitable laser power units, the laser welding of aluminum was achieved, for example, with a coating that increased the absorption rate (Arata and Miyamoto 1974). A further possibility lays in the use of active process gases, which enabled the laser welding of aluminum through an exothermic reaction, even for lower laser beam intensities achieved. The achievable increase of the energy input was, however, also accompanied by negative consequences for the weld seam properties (Kou 1987). The breakthrough in laser beam welding of aluminum didn't occur until the 1980s, when higher laser beam powers became available (today, fiber lasers allow outputs of over 100 kW (Katayama et al. 2015)). In the 1990s and 2000s, laser beam welding became established for the joining of aluminum-based materials (Heider 1994), and their applications can be found in countless industries, such as in the manufacture of automobiles (Schubert et al. 1997), aircraft (Heimerdinger 2003), and ships (Russell 1997) as well as in power plant construction (Jones et al. 2000) and raw material production (Ono et al. 2001).

Due to the lower absorption of the laser beam in the wavelength range of solid-state lasers (rod, fiber, and disk laser), for aluminum alloys, the threshold intensity for keyhole formation is, with typical values of around 5×10^6 W/cm^2, higher than that of steel (~10^6 W/cm^2). Special demands made on the process are the result of the specific material properties of aluminum (Wilden and Neumann 2010). Aluminum materials form a stable oxide layer on the surface and demonstrate a rapid decrease in the solubility of hydrogen during the transition from liquid to solid states and also have a high ability to conduct heat. The low kinematic viscosity and the comparatively large interval between the solidus and evaporation temperatures of aluminum alloys lead to a larger melt pool as well as increased melt pool and capillary dynamics during deep penetration welding. However, in addition to the abovementioned problem of the gap, there are further challenges in the form of process defects such as pores and spatters. A distinction is made between the so-called metallurgical hydrogen-induced pores and process pores. Process pores can be primarily traced back to instabilities of the capillary (Seto et al. 2001). Remedial action regarding process pores is possible with, for example, the use of double focus techniques (Hohenberger 2003). The high level of melt pool dynamics is also the reason for the lower surface quality of the deep welded seams in comparison to heat conduction welding. Particularly in the use of interesting aluminum alloys, there is a high susceptibility to hot cracking. The occurrence of hot cracks is an essential challenge in the laser welding of aluminum alloys, which can be compensated for by the addition of alloy elements into the melt pool via a filler wire (Berkmanns et al. 1992), as is known from arc welding (Schoer 1980). With the important 6xxx series of alloys, the susceptibility to hot cracking can, for example, be significantly reduced through a consistent distribution of silicon with contents of over 2 wt.% in the melt pool (Ostermann 1998). The addition of silicon can occur during the process through the use of aluminum-silicon (AlSi) filler wires (typically silicon-containing aluminum alloys with silicon contents between 5 wt.% and 12 wt.%), as is done in aircraft manufacture, among others (Schumacher 2002). A significant reduction in the susceptibility to hot cracking can be achieved for aluminum alloys that are prone to hot cracking through the addition of grain-refining elements or compounds such as titanium (Ti), titanium diboride (TiB_2), or zircon (Zr) in the melt pool (Heimerdinger 2003), through the changing of the aluminum alloy to be welded to aluminum-magnesium-scandium (AlMgSc) or aluminum-copper-lithium-scandium (AlCuLiSc) alloys, which were specifically developed for thermal manufacturing processes (Palm 2000), and through the use of coatings to add hot-crack-reducing alloy elements (Heß et al. 2011).

In the field of lightweight construction, especially, two approaches are currently being followed, namely the multi-material design for the suitable combination of different materials and the additive manufacturing of suitable structures; these are covered by G. Lampeas in another chapter of this book. The material mix in aircraft manufacture encompasses in particular aluminum and titanium alloys as well as carbon fiber-reinforced plastics (CFRP); thus, processes for joining these materials are crucial for aircraft manufacture in order to make full use of the material-specific advantages of mixed construction. In the laser welding of aluminum alloys, as covered in this chapter, as well as for the laser welding of titanium alloys, which is not covered here (see, e.g., Bergmann 2004), laser material processing methods are more suited for replacing, or also complementing or improving, joining processes for the production of such mixed joints. According to the Sub-Platform Joining of the European Technology Platform Manufuture (Allison and Scudamore 2014), joining processes for multi-material joints can generally be divided into three categories. Martinsen et al. (2015) expanded these definitions in a review paper with a fourth category for hybrid approaches, so that mechanical, chemical, thermal, and hybrid joining processes can be distinguished. The following description of processes that compete with laser joining for the joining of aluminum and titanium makes use of these categories to give an overview. Depending on the method used, the force transmission between the joining partners occurs via an adhesive bond/a substance joining, a form closure (positive-locking joint), or a force closure.

Methods that use thermal energy to form a joint are assigned to the category of thermal joining. Documented thermal joining methods for aluminum-titanium joints include, in addition to laser processes (Kreimeyer et al. 2001), arc welding (Wilden et al. 2007a), electron-beam welding (Skoda et al. 1996), vacuum brazing (Takemoto et al. 1990), friction welding (Fuji et al. 1995), friction stir welding (Dressler et al. 2009), electromagnetic resistance welding (Marya et al. 2005), resistance spot welding (Nesterov et al. 1990), ultrasonic welding (Zhu et al. 2012), diffusion bonding (Wilden and Bergmann 2004), spark plasma sintering (Miriyev et al. 2016), explosive bonding (Kahraman et al. 2007), and impact bonding (Wang et al. 2016), whereby each reference provides an example of the respective method. Some of the processes are suitable for the formation of continuous weld seams, while others lead to spot-shaped, round, or rectangular joining zones. Common to all thermal processes is the rule that the joining of aluminum and titanium goes hand in hand with the formation of an intermetallic phase at the boundary and thus also includes the formation of a substance-to-substance joint. However, the methods differ significantly in their process temperatures and durations, which range from temperatures well below the melting point of aluminum and process durations in hours, such as in diffusion bonding (see, e.g., Wilden and Bergmann 2004), to temperatures above the evaporation point of titanium, in the case of deep penetration welding with vapor capillary formation within the titanium joining partner, and joining speeds in multiple meters per minute (see, e.g., Kreimeyer et al. 2005).

Mechanical joining creates a joint via a mechanical process. The result is, in contrast to thermal methods, a positive locking or force closure between the aluminum and titanium joining partners. To this category belong not only conventional bolting, screwing, and riveting but also such approaches as clinching (He et al. 2015) or self-piercing riveting (Zhang et al. 2016). A joint is chemically formed when a chemical reaction is responsible for the bonding. Therefore, the adhesive bonding of aluminum and titanium is referred to as chemical joining, and the transmission of force is determined by the bond between the adhesive and the joining partner as well as by the cohesion within the adhesive itself. At the boundary between the adhesive and the joining partner, micro- (see, e.g., Habenicht 2009) or nano-interlocking (see, e.g., Zimmermann et al. 2012) mechanisms can also play a role. Hybrid processes comprise the interactions of at least two joining processes. For the material combination of aluminum-titanium, methods used include gas tungsten arc welding assisted hybrid friction stir welding (Bang et al. 2013), ultrasonic-assisted brazing (Ma et al. 2011), and fiber laser-cold metal transfer arc hybrid welding (Gao et al. 2014). With regard to the specific alloys and material thicknesses, studies on individual joining processes often involve various joining partner combinations and types of joint. In general, these include different versions of lap or butt joints (rarely T-joints), whereby lap joints occur more frequently. For butt joints, in addition to differences in the transmissible force, the connections differ in the thickening in the joining zone (e.g., 180% in Kempa (2014) or 170% in Schumacher et al. (2007)), affecting the assembly space and the mass of the joint. It is similar for lap joints, whereby the overlapping of the joining partners always leads to a thickening in the joining area, giving the structure additional material and weight.

With an aluminum-titanium combination, there are numerous intermetallic phases. In comparison to the titanium-rich phase TiAl (see, e.g., Yamaguchi et al. 2000) and Ti_3Al (see, e.g., Semiatin et al. 1998), which as titanium aluminides form their own material class for high-temperature applications, the aluminum-rich phase Al_3Ti, a tri-aluminide, displays on the one hand a lower density and a higher oxidation resistance but on the other hand is an extremely brittle material as it only has four slip planes (Yamaguchi et al. 1987). Due to the significant difference between the thermal expansion coefficients of aluminum and titanium, thermal stresses occur in the bimetal bond during the cooling of an aluminum-titanium joint, which can under certain circumstances lead to a joint failure even without external loading (Majumdar et al. 1997). This risk increases with the degree of phase formation, whereby particularly the melting of both materials is critical, as excessive phase forming takes place in the weld area due to increased diffusion and the occurrence of mixing; see, e.g., Tomashchuk et al. (2015). In 1997, Majumdar et al. showed that in laser beam welding of aluminum and titanium, an intermediate niobium layer is required in order to achieve a crack-free joint in the case of the melting of both joining partners (Majumdar et al. 1997). Through the use of a pulsed laser beam with adjusted beam modulation, Wilden et al. achieved success with a crack-free formable weld joint, although its connective strength is not known (Wilden et al. 2007b). Furthermore, Katayama produced an article on the combination of, among others, aluminum and titanium and titled it "The unweldables" (Katayama 2013) as no conventional welding technique could create a suitable connection between these material pairings. In the melt joining of aluminum and titanium via a laser beam, an approach became established in the research whereby only the aluminum joining partner is locally melted in the joint zone, while the titanium remains in a solid state, allowing the formation of intermetallic phases to be suitably limited. This principle was already used in the mid-1990s by Skoda et al. for joining aluminum and titanium using an electron beam (Skoda et al. 1996). Based on this approach, in the field of laser beam joining, previous studies have focused on, in addition to aluminum and titanium (Schubert et al. 1997), particularly the joining of aluminum and steel (e.g., Radscheit 1997). Such a joint is often referred to as a firmly bonded joint with a double character, as there is a weld structure on the aluminum side and a brazed joint on the titanium side. As this approach uses neither classic welding nor classic brazing, various terms have been invented or modified, such the German terms "Schwöten" and "Schlöten" as well as "combined welding-brazing," "weld brazing," or "braze welding." These terms can frequently be traced back to studies on aluminum-steel material combinations as well as aluminum-titanium. In order to prevent misunderstandings and contradictions, this chapter simply refers to this method as joining.

In the field of lightweight structures, the application of laser processes is not limited to the welding of similar materials or the joining of dissimilar metals like aluminum and titanium. Laser processes are also used to pre-treat surfaces prior to the adhesive bonding or to realize concepts for lightweight hybrid aluminum-CFRP joints by integrating transition structures made of titanium which are currently under research.

In addition to material-specific lightweight construction using lightweight metals such as aluminum, titanium, their combination, and a mix with fiber-reinforced plastics, which offers the potential to reduce fuel consumption through a reduction in weight with the aim of achieving CO_2 reduction goals by 2050, a further laser-based manufacturing concept shall be demonstrated using an aerodynamic approach (see an overview of aerodynamic technologies to improve aircraft performance in Abbas et al. (2013), whereby theoretically high savings potentials can be reached if implemented properly: the reduction in the air resistance at the wings and vertical stabilizers of aircraft using hybrid laminar flow control (HLFC) (Schrauf 2005).

The long-established laser cutting of metallic sheets, and recently also of CFRP, as well as laser brazing, laser beam melting (or other laser-based processes) for additive manufacturing, laser structuring, or laser drilling are not further covered in this chapter, nor is the labelling of cables or marking of components using laser beams, which have also been in use for a number of years. These processes are only briefly mentioned at this point for the sake of completeness.

1.1 Laser Welding of Aluminum Alloys

The introduction to this chapter also provided an introduction to the laser processing of materials as well as the laser beam welding of aluminum. Laser beam deep penetration welding in particular has a high application potential due to its high degree of absorption. It allows the creation of deep yet narrow weld seams with high welding speeds, meaning that the heat input, and thus the heat effect on the component, remains low despite the high heat conductivity of aluminum alloys. Vollertsen et al. compared weld depths per kilowatt laser output power according to the welding speed for various laser beam sources with different beam qualities, indicated by the beam parameter product, on the example of the aluminum alloy EN AW-6082 (see Fig. 11.3 according to Vollertsen and Neumann (2009)). With increasing beam quality (a smaller beam parameter product), the weld depth increased. The deepest welds were achieved by a single-mode fiber laser, which had a Gaussian beam profile (TEM00 mode) and the smallest focal diameter.

Standing against the high potential of using laser beams for the welding of aluminum, there are, however, a series of challenges. This subsection will therefore

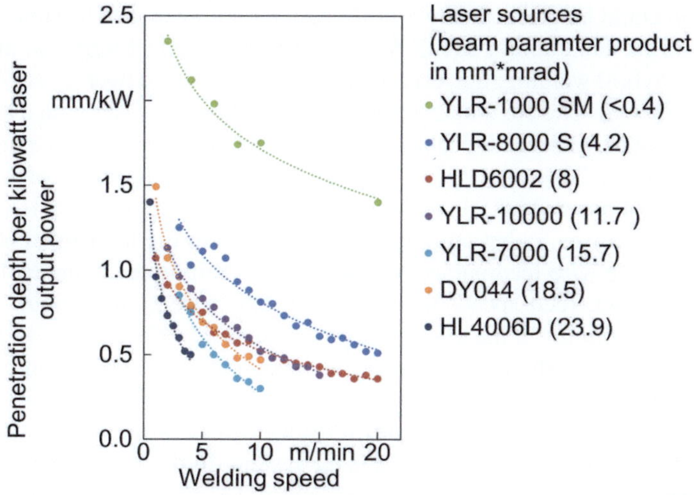

Fig. 11.3 Comparison of the penetration depth per kilowatt laser output power for laser sources with different beam qualities (beam parameter products) and max powers according to Vollertsen and Neumann (2009)

address the specific challenges of this process and the current solution approaches from research and development. Hereby, the increase in the gap bridging ability, the increase in the seam surface quality, the reduction in the susceptibility to hot cracking, the prevention of spatters and pores, and the increase in process reliability are highlighted. Especially, the laser welding of thin sheets up to thicknesses of a few millimeters is covered in this chapter.

Regarding the gap bridging ability, the challenge lies in the fact that the actual advantage of the laser beam, namely a precise and concentrated energy input with a limited local zone of effect, can quickly lead to problems in the laser welding of sheets in a butt joint with gap sizes greater than the technical zero gap due to the small laser spot. Typically, without filler materials, gaps of only around 10% of the sheet thickness are workable in order to prevent the too high amounts of transmission of the focused laser beam. Such a precise adjustment of the gap often takes a lot of effort, in particular for thinner sheets with welds that are multiple meters in length. A defocusing of the beam would increase the size of the laser spot on the workpiece; however, on the one hand, the efficiency would decrease significantly after dropping below the deep penetration welding threshold, and on the other hand, for larger gaps, this would in no way guarantee a complete joint between the edges of both materials. The use of a filler wire (Kappelsberger 1987), which also allows a suitable alloying, for example, of silicon to reduce the hot cracking susceptibility, is generally a proven approach in all forms of welding. In this way, for example, a 1-mm-wide gap was successfully bridged in the CO_2 laser welding of a 2-mm-thick steel sheet in 1999 (Sun and Kuo 1999). In a comparison of five different processes regarding the gap bridging ability during the welding of 2.1-mm-thick sheets of aluminum alloy AA 5182, it was shown that in laser welding without a filler wire, a

0.2 mm gap could be bridged, while bridging a 0.6 mm gap was feasible with cold wire (Aalderink and Pathiraj 2010). A gap of 1 mm could be bridged in the case of laser-MAG hybrid welding; however, this led to an increased heat input (Aalderink and Pathiraj 2010). The combination of a 1 kW single-mode laser with a high beam quality with an MSG process enabled, in the case of steel sheets, also a bridging of gaps that reached 50% of the sheet thickness at a welding speed of 4.5 m/min (Thomy and Vollertsen 2007). In general, laser welding with filler wire is, compared to the other processes, rather complex; as in addition to the laser process parameters, there are the wire feed parameters (leading or trailing feed, angle of the wire axis, impact point, feed speed, wire diameter, and wire composition). Even the slightest mispositioning of the filler wire can lead to weld defects or faults. Remedial measures could lie in the inclination of the beam axis (Beck 1996), in the double focus technique (Xie 2002), and in the dynamic power modulation (Geisel 2002). The spatial beam modulation, as is known from electron-beam welding (Farrel and Ferrario 1987), can be effected via a scanner system. In the narrow gap welding of comparatively thicker aluminum sheets in multilayer processes, beam oscillation has the result that any mispositioning of the wire perpendicular to the sheet becomes noncritical (Göbel et al. 2007).

In their recent work, Schultz et al. combined a filler wire feed with transversal beam oscillation during the joining of thin aluminum sheets. This process enabled constant gaps of 190% of the sheet thickness and increasing gaps of up to 300% to be reliably bridged (see Fig. 11.4 for experiments with different alloys and filler wires (Schultz et al. 2014a)). The effort in the preparation (clamping technology, positional tolerances, etc.) can thus be significantly reduced. Due to the continued focused laser beam and welding speeds up to 10 m/min, the process efficiency also remains high (Schultz and Seefeld 2015). A benchmark test for the evaluation of the gap bridging ability was introduced in Woizeschke et al. (2016a).

The quality of a laser-welded seam is determined by the occurrence of seam defects, faults, and imperfections. For the user, the surface quality and sufficient strength are decisive criteria; therefore, spatters, pores, cracks, and inhomogeneities

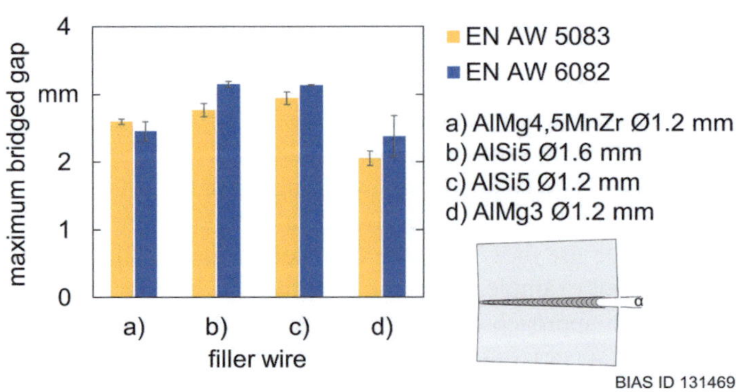

Fig. 11.4 Bridgeability in laser beam welding with focused beam, wire feed, and beam oscillation for a selection of aluminum alloys and filler wires according to Schultz et al. (2014a)

in the weld and along its length should be avoided as far as possible. A crack in the weld or in the heat-affected zone represents an extremely critical weld defect (Dilthey 2005). In the formation of cracks, a distinction must be made between cold cracks and hot cracks. Cold cracks are formed in the solidified seam or the heat-affected zone and bear typically rugged fracture areas. In contrast, hot cracks form in fluid or mushy conditions between the solidus and liquidus temperatures of the alloy, whereby the cracks—caused by the surface tension—are rounded at least at the microscopic level (see Fig. 11.5 (right)). In Fig. 11.5 (left), the grains can be clearly seen; hot cracks always run in an intergranular or interdendritic direction. There is a differentiation among hot cracks between solidification cracks, which occur within the seam (often in the middle), and liquation cracks, which can occur in the (partially melted zone of the) heat-affected zone or, in multilayer processes, in the lower layers. The following considers solidification cracks. Particularly in the case of the alloys of interest to lightweight construction, namely 2xxx (AlCu), 6xxx (AlMgSi), and 7xxx (AlZnMg and AlZnMgCu), there is a high susceptibility to hot cracking. The combination of deep and narrow welds and high welding speeds leads in deep penetration laser welding using modern fiber and disk lasers to the possible occurrence of high-temperature gradients and cooling rates during the solidification of the melt pool. This results in a mostly stalklike grain structure that grows from the melt line to the center of the seam. If the remainder of the melt in the center of the weld can no longer compensate for the increasing shrinkage of the weld metal with backfeed, a hot crack forms toward the end of the solidification process (Cross 2005).

The most commonly used method for the reduction of the susceptibility to hot cracking is influencing the alloy composition in the weld metal through the addition of suitable filler materials (typically in the form of a wire). For example, for the welding of 6xxx alloys, achieving/adjusting silicon contents in the melt to just over 2% is recommended (Schulze 2010). Ploshikhin showed that the susceptibility to hot cracking in the laser welding of the alloy AA6056 depends on the silicon content, as this influences the bearable elongation (measured using the distance from the weld seam to the non-clamped edge of the sheet) in the two-phase area of the

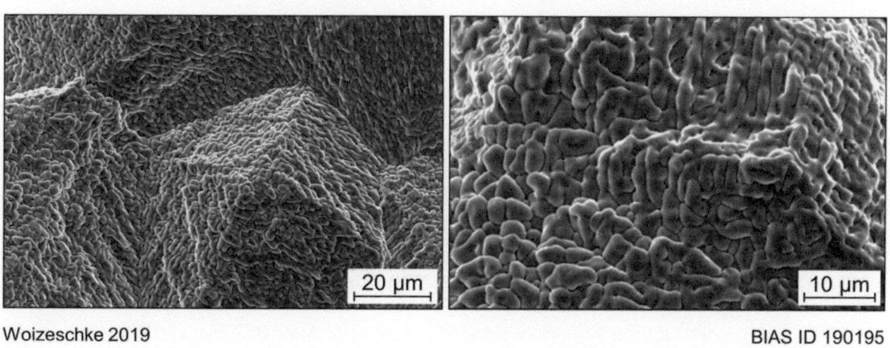

Fig. 11.5 Typical surface of a hot crack

partially solidified melt (Ploshikhin et al. 2004). The critical aspect regarding the configuration of the silicon content in the melt is especially a uniform homogenous distribution of the silicon in the weld metal via an AlSi wire. Effects on the dispersion in the melt pool at constant laser process parameters stem from, among others, the wire position and the wire feed speed (Pretorius et al. 2004). Gatzen further showed that the distribution of elements in the melting pool can be influenced by magnetically induced flow modulation when silicon is introduced (Gatzen 2014). The frequency was identified as an essential parameter, whereby a change of 5 Hz already has significant effects.

A refining of the structure toward finer, particularly equiaxed dendritic grains can increase the strength (Arata et al. 1973), the ductility (Janaki Ram et al. 2003), and the fracture resistance of the weld metal (Schempp et al. 2013) and thus improve the mechanical properties. The grain refining, that is, the reduction of the grain size in the resulting structure, is a proven measure against hot cracking. In the field of casting, in addition to such approaches as stirring (Fonseca de Arruda and Prates de Campos 1983) and vibrating (Cahoon et al. 1992), the addition of grain refiners (McCartney 1989) has also been investigated. Subsequently, the approach was transferred to arc joining processes; see (Brock and Aidun 1995) for gas metal arc welding (GMAW) and (Schempp et al. 2012) for gas tungsten arc welding (GTA), as well as laser welding (Tang and Vollertsen 2014). Tang et al. showed for the welding of 6xxx alloys with a disk laser that the grain size significantly decreased with the addition of titanium/boron (Ti/B in a 5:1 ratio) (see Fig. 11.6 (Tang 2014)), while a proportion of equiaxed dendritic grains in the weld metal of 100% is achievable (Tang and Vollertsen 2014). At higher weld speeds, more grain refiners tend to be needed in order to achieve a comparable grain refining (Tang and Vollertsen

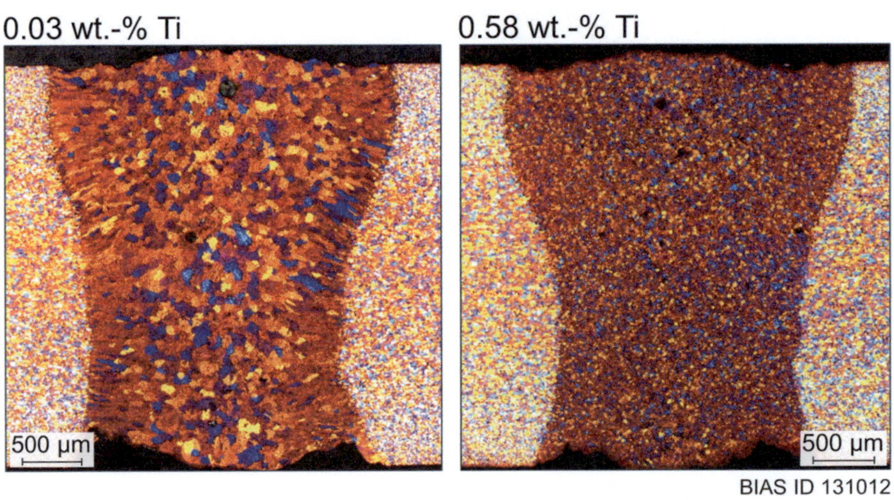

Fig. 11.6 Seam cross-sections of laser-welded aluminum EN AW-6082 sheets (left) without adding grain-refining elements during the process and (right) with the addition of titanium/boron (Ti/B in a 5:1 ratio) according to Tang (2014)

2014). It was shown that the susceptibility to hot cracking can be reduced through a reduction in the grain size or also through an increase in the proportion of equiaxed dendritic grains to a minimum level. A further reduction in grain size led once more to an increase in the susceptibility to hot cracking (Tang and Vollertsen 2014). Further grain-refining elements are, for example, scandium (Dev et al. 2007) and zircon (Matsuda et al. 1986). As with casting, vibrations induced during laser beam deep penetration welding can also lead to grain refining or also the increase in the proportion of equiaxed dendritic grains—despite the high solidification speeds (Woizeschke et al. 2017). This approach is currently being researched (Radel 2018). Initial results also show a reduction in the susceptibility to hot cracking (Radel and Woizeschke 2018).

In addition to the metallurgical and structure-specific influencing factors, the strain field plays a particularly decisive role in the formation of hot cracks. The temperature distribution has a decisive influence not only on solidification but also on the thermomechanical processes. The occurrence of hot cracks depends on the tensile stresses affecting the film created by the remainder of the melt in the partially solidified area of the weld metal. In addition to the shrinkage during solidification, the dead weight and/or external loads of the structure, as well as thermally induced mechanical processes, can influence the strain field (Hilbinger 2001). Any remelting of tack welds or weld seams can result in additional internal stresses. Generally speaking, there are various (thermo)mechanical approaches for the prevention of hot cracking. In addition to the use of synchronous pressing rollers that were laterally offset from the seam (Liu et al. 1996), Yang et al. developed a hot cracking mitigation technique by introducing a trailing heat sink during welding (Yang et al. 2000), and Ploshikhin et al. investigated using at least a second defocused laser beam (or multi-beam welding) during laser beam welding to induce compressive stress (Ploshikhin et al. 2007). Scanners that allow a beam oscillation can also have a positive influence on the temperature gradient, whereby the decrease in the susceptibility to hot cracking can be traced back to a reduction in the growth of dendrites (Kutsuna et al. 2002). A special feature can be seen in the use of double-sided twin beam lasers, which tend to be employed for the welding of stringers to the fuselage skin in aircraft manufacture. Gruss showed that in this case, a shift of both laser beams toward each other, whereby two separate melt pools are formed instead of one, causes the structural rigidity to increase, thereby decreasing the susceptibility to mid-weld cracking (Gruss 2008). Furthermore, Gruss et al. showed that the recesses in the stringer as well as the welding sequence represent starting points that influence the susceptibility to hot cracking (Gruss et al. 2008). In contrast to continuous-wave welding, pulse welding, in particular of thinner materials, enables the temperature field to be influenced through pulse shaping (temporally changing the power). Zhang et al. used this method to reduce the hot crack susceptibility in pulse welding by lowering the cooling rates (Zhang et al. 2008). Beyond pulse shaping, two-beam processes for the laser pulse welding of aluminum alloys are currently being researched and developed with the aim of increasing the weld penetration depth (e.g., Zhang et al. 2008) and the welding speed (e.g., Mathers 2002), improving seam quality (e.g., Chen and Molian 2007), reducing the number of pores (Bergmann et al. 2013), and decreasing the generation of cracks (e.g., Nakashiba et al. 2011). For example, Bergmann et al. showed an 80% increase in the full penetration welding speed for the pulse welding of 0.5 mm-thin sheets of the aluminum alloy AA 5754 with an Nd:YAG primary laser (pulse peak power 2.3 kW, pulse duration 5 ms, diameter 400 μm) and an overlaid secondary diode laser (laser power 150 W, spot diameter 2 mm) (Bergmann et al. 2015). The process window of laser power and pulse length could be significantly increased for process results between "no penetration" and "poor seam quality," even at 100 W (Bergmann et al. 2015). It was also shown that the susceptibility to hot cracking can be significantly reduced without the need for a filler wire (Bergmann et al. 2015).

Pores and spatters are two significant and undesirable phenomena in the deep penetration laser welding of aluminum. Pores can occur as micropores or macropores. They can be traced back to the solubility of elements or to the process behavior; hence, these are also referred to as process pores. Process pores are in particular caused by the strong dynamics of the vapor capillary, which are characteristic for this process keyhole welding. Such pores are of course not desirable as they diminish the weld seam quality. Spatters, on the other hand, represent a loss of material and can result in a weld with blowouts, blowholes, underfills, undercuts, and craters. Kaplan and Powell categorized different types of spatters (Kaplan and Powell 2011). Spatters can also adhere to the surface and subsequently may need to be removed in an additional process step. During welding with a CO_2 laser, Seto et al. observed a fluctuating keyhole shape during the welding processes with the aid of X-ray transmission imaging (Seto et al. 2001). The capillary fluctuated dynamically in the direction of depth, and bubbles continuously formed at the base of the capillary. Some of these escaped through the keyhole itself or through the surface of the melt pool. Others remained in the weld metal until it solidified and thus became pores in the weld seam. The formation of bubbles or pores is generally attributed to capillary collapse (Lin et al. 2017), that is, the encircling and closing off of a lower capillary area as well as the bulging of the capillary rear wall or its base. In addition to capillary dynamics, converging melt pool flows, and buoyancy can also lead to a mitigation of bubbles (Shi and Wan 2016). The formation of spatters during deep penetration laser welding is directly associated with a high level of process dynamics. In order for a spatter to form, a local quantity of melt must contain sufficient kinetic energy to overcome the surface tension and exit the melt pool (Kaplan and Powell 2011), whereby there is a broad spectrum regarding the spatter number, size, and release speed as well as the type and location of its formation (Kaplan and Powell 2011). Spatters can be released from the capillary wall or the melt pool and are influenced by such factors as melt pool waves, capillary wall fluctuations, migrating steps at the capillary front, and shear forces induced by the metal vapor. Measures to counteract the formation of pores and spatters will particularly lead to a process stabilization. Thus, a reduction in the ambient pressure leads to a more stable process behavior (e.g., Katayama et al. 2012). For example, an ambient pressure of 0.1 kPa reduces the formation of spatters and pores in comparison to normal pressure, despite a greater weld penetration depth (Katayama et al. 2012). Using an appropriate filler wire (Schubert et al. 1998) or an electromagnetic field (Schneider et al. 2013) can also lead to a calming of the melt pool. Schneider et al. showed for AlMg3 sheets that an oscillating magnetic field can decrease the pore area in radiography testing by up to 80% (Schneider et al. 2013). It should be underlined that, in comparison to partial penetration welding, full penetration welding generally leads to significantly fewer pores (e.g., Seto et al. 2001). A positively reducing effect on the number of pores was achieved by Katayama et al. in the welding of AA 5083 by employing a directed steam of nitrogen (Katayama and Kawahito 2009). A number of studies have focused on the input of energy into the keyhole. These approaches were based on the influence of temporal and/or spatial variations in the laser parameters (e.g., laser power modulation, laser pulse, beam oscillation, angle of incidence, polarization, beam formation, and double or multiple foci). It appears that a too high intensity should be avoided (Volpp 2017). In general, a pulsing (Katayama et al. 1997) or a temporal laser power modulation is suitable (Tsukamoto et al. 2001). However, Klassen showed for the welding of EN AW-6013 that a stimulation with a constant frequency destabilizes the processes and thus recommended a low-frequency laser modulation with a varying frequency of between 25 and 50 Hz (Klassen 2000). Shaping a single beam to a caustic with multiple beam waists in the direction of the beam propagation can also influence the energy input and the formation of spatters. In regard to this, the studies by Volpp et al. showed that the number of spatters varies with the changes in the beam (Volpp et al. 2016). The use of dual spot welding can stabilize both the melt pool and the keyhole dynamics (Haboudou et al. 2003). Gref showed for the laser welding of aluminum that the capillary should be wide open so that the outflow of the metal vapor is not disturbed (Gref 2005).

Fetzer et al. investigated the keyhole geometry during the laser welding of aluminum with a welding speed of up to 50 m/min (Fetzer et al. 2018). The X-ray analysis showed that with speeds above approximately 30 m/min, there was significantly less bulging as well as less deep fluctuations in the capillary, which had greatly increased in diameter. The pore formation is negligible at such high speeds. A further increase in speed led to an opening up of the capillary into a wedge-like shape. Pore formation was no longer detectable, and the welding depth was extremely constant during the process (Fetzer et al. 2018). If the availability of high-powered lasers continues to expand in the future, then this represents a very promising approach.

In addition to the process pores described here, the issue of hydrogen porosity should also be mentioned. The strong temperature dependence of the solubility of hydrogen in the aluminum melt can lead to problems in laser beam deep penetration welding due to the significant temperature gradients and cooling speeds. There are multiple sources of hydrogen: the source material, the filler material, the presence of oxides and hydroxides or contaminants on the surface, the protective gas, or the ambient atmosphere. Higher hydrogen contents occur in particular in aluminum casting components, for example, in automobile manufacture. In addition to the avoidance or removal of possible sources of hydrogen, an increase in the solidification times is useful on the process side in order to provide more time for the hydrogen to leave the melt pool. Further defects to be considered under certain circumstances are the inclusion of oxides and the loss of some alloying elements with low evaporation temperatures by selective vaporization. Further details on these aspects, which cannot be further covered here, can be found in, for example, the overview by Cao et al. (2003).

An increasingly important criterion is the surface quality. In this regard, deep penetration welded seams are lagging significantly behind seams produced through heat conduction welding and especially laser brazing (Vollertsen et al. 2017). The latter tends to be used for areas that are later visible to the end user. In deep penetration laser welding, the process dynamics have a significant influence on the surface quality of the seam. Process calming approaches thus often lead to better surface qualities. The actual approach in so-called buttonhole welding attempts to combine deep penetration laser welding, with its good process efficiency (high levels of absorption), with a high gap bridging ability and a high surface quality of the seam in order to create seams with an acceptable visual quality. This process employs the targeted creation of a cavity ("hole") in the melt pool during laser welding with a filler wire feed and a laser beam that oscillates perpendicularly to the welding direction. Schultz et al. were already able to show robust process windows for pendulum frequencies of 250 Hz in the buttonhole welding of 1-mm-thick aluminum sheets of the alloy EN AW-6082 with a filler wire (Schultz et al. 2017a). The oscillation frequency affects the shape of the molten wire tip, such as the wire tip length and melting angle, and thus the buttonhole formation (Cho et al. 2018a). Cho et al. (2018b) showed that the cavity dampens the melt pool dynamics to such an extent that a highly dynamic area forms in front of the cavity while a calmer area appears behind it. This dampening effect of the buttonhole (see Fig. 11.7) is responsible for the high surface quality of the seams (Schultz and Woizeschke 2018). In combination with the high gap bridging ability of this method, a highly versatile and useful process will become available in the future. A further current approach for increased surface

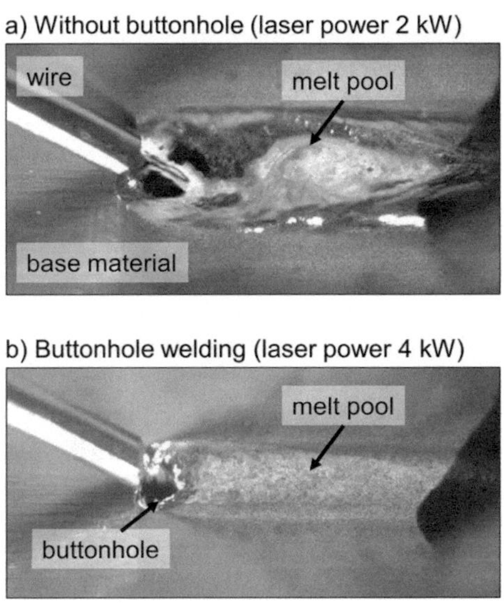

Fig. 11.7 Observation of the process behavior (**a**) without buttonhole (laser power of 2 kW) and (**b**) with existing buttonhole (laser power of 4 kW) according to Schultz and Woizeschke (2018)

quality in laser welding is the "adjustable ring-mode fiber" technology (Maina et al. 2018). This refers to the use of a high-powered laser with a beam profile consisting of a central partial beam and an encircling ring partial beam. While the power of the central beam determines the weld depth, the ring part is responsible in particular for the temperature distribution. Combined into a dual mode, Maina et al. were able to show that deep penetration welded seams could be created with a reduced roughness on the surface (Maina et al. 2018). So far, however, the achieved surface quality has not been able to match that of buttonhole welding.

The constant increase in laser power and beam quality has led, however, in the field of laser lenses to challenges in the form of a so-called focus shift, that is, a change in the focus position of the laser beam as well as depending on the conditions of the focus diameter (Reitemeyer et al. 2009). The focus shift depends on the lens system as well as the laser beam (lens material, lens thickness, lens coatings, lens wavelengths, laser power, beam quality, beam diameter, intensity distribution, beam-on-times, heat conduction, contaminants, etc.). With increasing laser power, a linearly increasing focal position shift is observed for clean lenses, whereby the focus diameter remains almost constant (Reitemeyer et al. 2009). In the case of a clouding of the lens (e.g., protective glass), the focus shift increases in conjunction with an increase in the diameter (Reitemeyer 2012). Reitemeyer therefore developed a laser head that actively compensated for the shift in the focus position (Reitemeyer 2012). The focus of a reference beam that runs through the whole processing head is captured by a camera so that the actual focus position shift can be determined online (Reitemeyer et al. 2010). With a collimator motor, in-line control is possible, allowing a constant focus position, and thus consistent weld seam properties, to be achieved (Reitemeyer 2012). In addition, compensating optical systems have been developed to reduce the thermal shift in high-power laser systems (thermal lensing compensation). The concept is based on a lens material, which has a negative temperature coefficient in the material refractive index; see, e.g., (Scaggs and Haas 2010). Besides focus shifting effects, a constant weld depth is not self-evident for deep penetration laser welding due to process variations and material inhomogeneities. Currently, the optical coherence tomography (OCT) is implemented in more and more laser material processes demonstrating its potential to resolve alignment issues (Bautze 2018) and control the weld depth in-line (Kogel-Hollacher et al. 2016). Mittelstädt et al. investigate the possibility of applying OCT systems in combination with beam oscillation for realizing the deep penetration laser welding of hidden T-joints; see preliminary results in Mittelstädt et al. (2018).

1.2 Laser Processes for Aluminum-Titanium(-CFRP) Joints

For laser-based joining of aluminum and titanium components for multi-material structures, deep penetration and heat conduction laser processes have been developed and analyzed. Figure 11.8 gives an overview of the concepts by schematic sketches. In deep penetration laser welding using focused high-performance laser

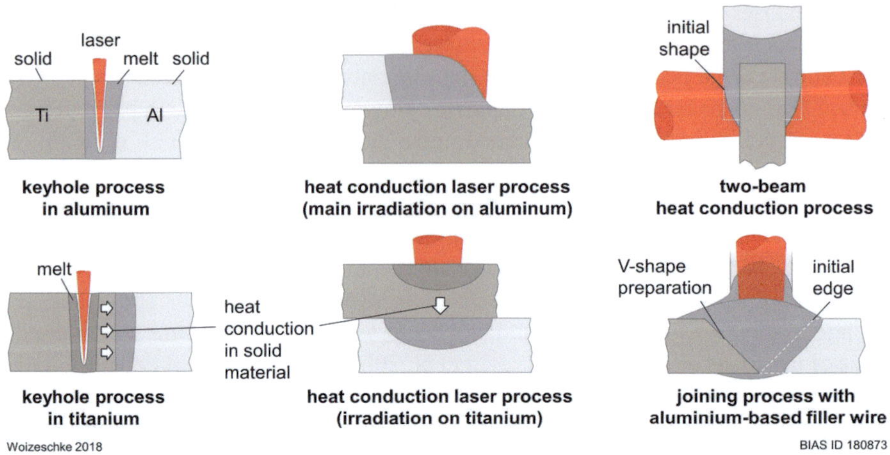

Fig. 11.8 Overview of concepts for laser joining of aluminum and titanium components based on heat conduction and deep penetration keyhole processes

sources, the laser beam can be positioned onto either the titanium or the aluminum joining partner. Kreimeyer et al. achieved a deep weld in titanium with the use of a titanium filler wire (grade 2 titanium) (Kreimeyer et al. 2005), while Song et al. produced deep penetration welds in aluminum without an additional material feed (Song et al. 2013). Song et al. furthermore determined the positioning of the laser beam on the aluminum where no melting of the titanium occurred. They did, however, underline that in the process window during the joining of aluminum and titanium using deep penetration welding with keyhole formation, an accuracy of 0.1 mm regarding the positioning of the laser beam to the interface is required along the entire length of the weld seam (Song et al. 2013). For the weld seam created by Kreimeyer et al., the 1-mm-thick titanium (Ti6Al4V) remained mostly in a solid state in the interface area to the 1.15-mm-thick aluminum (AA6016) (Kreimeyer et al. 2005). Only in the upper and lower parts of the weld seam could a mixing, i.e., multiphase areas comprising aluminum, titanium, and intermetallic phases, be observed (Kreimeyer et al. 2005). Casalino and Mortello confirmed the findings of Kreimeyer et al. for the combination of the aluminum alloy AA5754 and the titanium alloy Ti6Al4V and sheet thicknesses of 2 mm with a comparable process design (Casalino and Mortello 2016).

The characteristics of the heat-conduction-based laser beam joining process using a defocused laser beam are similar to those of heat conduction welding. According to the degree of absorption, a certain portion of the energy of the laser beam is absorbed by the workpiece surface and is transferred to the interior through both heat conduction and melt pool convection. Hereby, approaches exist that use either a single-sided or double-sided laser beam operation. The diameter of the laser beam onto the workpiece varies according to the type of process and typically lies within the range of a few millimeters. In addition to the round or elliptical—due to the angle of incidence of the laser beam—spot geometries on the workpiece surface, rectangular spots can also be used in order to control the energy input. Depending on the process used, the laser beam is completely positioned on the titanium, partially on the titanium and partially on the aluminum, or completely on the aluminum. While a variety of processes require the addition of material in the form of a filler wire, there are variations in the field of lap joints and double-sided joining that do not require additional material. It is the same for the use of flux, which is required in many, but not all, processes.

In 1997, Schubert et al. presented an aluminum-titanium lap joint whereby the laser beam heated the titanium in such a manner that the aluminum joining partner melted from the rear and, with the use of flux, a substance-to-substance joint was produced with the formation of intermetallic phases (Schubert et al. 1997). In 2001, Kreimeyer et al. reported the laser joining of aluminum (AlMg0.4Si1.2, AA6016) and titanium (Ti6Al4V) in a lap joint using a clamp head and without flux or filler material (Kreimeyer et al. 2001). Hereby, a 0.8-mm-thick titanium sheet overlapped a 1-mm-thick aluminum sheet. The laser was positioned so that the beam was mainly absorbed by the titanium. The aluminum melted completely in the joining zone and wetted the titanium sheet at its front face and overlying side faces. In 2003, the Airbus Deutschland GmbH registered a patent for a seat track using the hybrid construction method. In this approach, a titanium section was joined to an aluminum carrier in a butt weld using a thermal joining process (Zerner 2003); see a seat track prototype produced by BIAS GmbH according to Möller et al. (2012) in Fig. 11.9.

The patent application describes the process of joining, for example, in the manufacture of a seat track for an aircraft passenger seat, whereby a titanium component is laid into a groove at the front face of an aluminum component, which is locally thickened. Subsequently, a defocused laser beam melts the aluminum in this area. The technical realization of this process, with two conventional laser-processing optical systems running on both sides of the sample in combination with locally accompanying protective gas chambers, was demonstrated by Kreimeyer and Vollertsen (2005). The mechanical-technological properties of the joint between 2-mm-thick aluminum (AA 6056) and 1.8-mm-thick titanium (Ti6Al4V) were

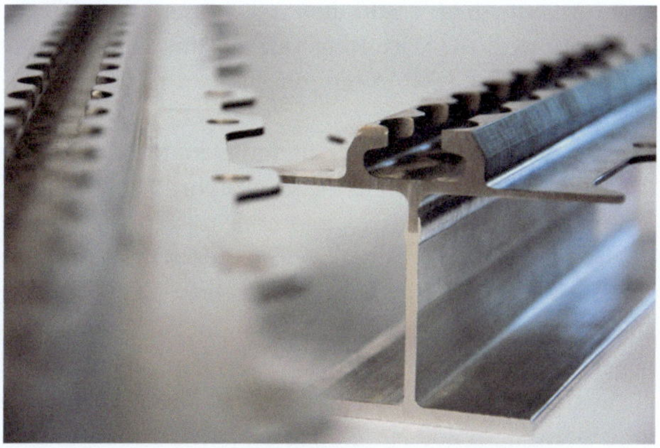

Fig. 11.9 Hybrid aluminum-titanium seat track joined by a double-sided laser beam process. (Photo: Woizeschke)

characterized in Kocik (2009). Regarding the weld reinforcement, it has been shown that the total thickness depends significantly on the energy input per unit length, whereby, in the parameter window under consideration, through weld seam bulging, a thickening of the seam areas by between 0.85 and 1.8 mm compared to the aluminum base sheet can occur. Chen et al. (2009, 2010) examined the joining of 1.5-mm-thick aluminum (aluminum-magnesium alloy 5A06 Al) and titanium of the same thickness (Ti6Al4V) using a one-sided laser process with a rectangular spot profile (2 mm × 4 mm) with an aluminum alloy AlSi12 filler material (outer diameter 2 mm) with a flux core. The edges of both joining partners were chamfered (=scarfed), producing a V-shaped groove prior to the joining. Weld seam reinforcements were formed due to the filler material, resulting in thickenings in the joining zone of between half (Chen et al. 2009) and one (Chen et al. 2010) times the thickness of the base sheet. Woizeschke distributed also a primary laser beam onto two processing heads for the simultaneous irradiation of the joining zone from both sides in order to significantly reduce the temperature gradients along the interface layer (Woizeschke 2017). In comparison to the previously described work by Kocik et al., the laser spots were positioned more on the aluminum. The process design allowed the examination, at constant process parameters, of the influence of the titanium edge geometry on the weld seam geometry, the interfacial layer, and the properties of the joint (Woizeschke and Vollertsen 2016).

In addition to sheet-sheet joint configurations, Möller et al. introduced a wire concept whereby titanium wire loops were used as transition elements between aluminum sheets and CFRP laminates. Using a thermal laser-based (double-sided) joining process, the wire loops were joined to the aluminum joining partner, while the connection to the CFRP was achieved through an interlocking connection of the carbon fiber rovings, which were subsequently infiltrated with a hardening resin system. A further concept for aluminum-CFRP joints using a titanium transition structure (Woizeschke and Wottschel 2013) refers back to the basic concept from a patent by Herrmann et al. (2008), whereby thin titanium sheets or foils served as transition elements (Woizeschke and Vollertsen 2015). Within the CFRP laminates, these comparatively thin titanium sheets stepwise replaced the CFRP layers toward the edge of the component, so that at the CFRP component a titanium laminate edge was formed, which could then in a subsequent step be laser joined to an aluminum sheet. As with the wire loop variant, this is a butt joint configuration between the aluminum and CFRP joining partners; a prototype (Woizeschke 2017) is shown in Fig. 11.10.

One essential parameter of multi-material structures regarding applications in the future is the strength of such hybrid aluminum-titanium(-CFRP) joints. The aforementioned processes and concepts for laser beam joining of aluminum and titanium using deep penetration welds within one of the joining partners showed conclusively that joint strengths of over 200 N/mm^2 can be achieved under quasi-static tensile loading if the energy input has been appropriately selected. In the deep penetration welding works reviewed here, the interface area of the joining partner always ran perpendicular to the sheet surface. Kreimeyer et al. derived an average joint strength of 200 N/mm^2 for the combination of 1-mm-thick aluminum AA 6016

Fig. 11.10 Prototype of a slim integral joining concept for lightweight aluminum-CFRP joints by generating a titanium laminate edge at the CFRP part to enable the laser-based joining to the aluminum sheet (Woizeschke 2017)

and 1.15-mm-thick titanium Ti6Al4V (Kreimeyer et al. 2005). Song et al. (2013) found a maximum joint strength of 203 ± 27 N/mm^2 for a joint between aluminum AA6061-T6 and titanium Ti6Al4V (each with a sheet thickness of 2 mm) with a 1.0 mm distance between the laser beam axis and the interface area. For the combination of the titanium alloy T40 (grade 2 titanium) and the aluminum alloy AA 5754, the experiments by Casalino et al. found, regarding the influence of the energy per unit length, in the case of 1.5-mm-thick sheets a maximum joint strength of 191 N/mm^2 at 50 J/mm and a constant distance of 0.75 mm (Casalino et al. 2015). With increasing energy per unit length, the joint strength decreased. In a further work, Casalino and Mortello investigated the combination of the same aluminum alloy AA5754 with the titanium alloy Ti6Al4V, whereby an average joint strength of 210 N/mm^2 was achieved (value taken from a diagram) at a sheet thickness of the joining partners of 2 mm (Casalino and Mortello 2016).

For laser beam joining using a defocused beam or two defocused beams, the following joint strengths have been reported from the studies reviewed here, depending on the process and combination used. Joint strength refers here to the maximum tensile strength in regard to the aluminum base sheet thickness (outside the overlap or the thickening due to the weld seam reinforcement in the joining zone). Experiments by Schubert et al. on lap joints determined joint strengths between 164 and 237 N/mm^2, whereby it is not known which alloys were used (Schubert et al. 1997). The average value was 200 ± 22 N/mm^2. The lap joints between aluminum AA6016 and titanium Ti6Al4V tested by Kreimeyer et al. resulted in joint strengths between 192 and 226 N/mm^2 with an average of 220 N/mm^2 (Kreimeyer et al. 2001). With simultaneously formed double-sided butt joints between aluminum AA 6056 and titanium Ti6Al4V1 (base sheet thicknesses: 2 and 1.8 mm) with suitable process parameters, the joint fractures are typically in the heat-affected aluminum outside of the thickened joining zone area (Kocik et al. 2006). Schumacher et al. found for the aluminum tempers T6 and T4 average joint strengths of 236 ± 24 N/mm^2 and 242 ± 6 N/mm^2, respectively (Schumacher et al. 2007). Due to the location of the fracture in the heat-affected zone of the base sheet, which is thinner than the joining zone, the maximum force transfer of the weld seam is unknown. In the study by Y. Chen et al., the frontal face alignment of the titanium component was not perpendicular to the sheet surface, but a 45° scarfing at the titanium edge was present (Chen et al. 2009). For the tensile samples, the process-dependent weld reinforcements were removed prior to the tensile test. For these modified one-sided scarfed samples, a maximum average tensile strength of 290 N/mm^2 was determined for a minimal energy per unit length for the series of experiments. The fracture occurred within the aluminum weld metal. The joint strength related to the aluminum base sheet thickness as well as investigations into samples with weld reinforcement were not documented.

In the case of the double-sided joining of aluminum-titanium butt joints, the homogenous interface layer properties achieved by Woizeschke through the advanced process design enabled an analytical prediction of the influence of the titanium sheet edge geometry on the resulting joint strength using a model based on the material and geometry parameters (Woizeschke and Vollertsen 2016), thus facilitating a targeted design of the titanium edge geometry. Hereby, a double-sided 15° bevel on the titanium edge, i.e., a 30° titanium wedge shape, enabled the highest force transfer in tensile tests (Woizeschke 2017). These joints consisting of 4-mm-thick aluminum (EN AW-6082 T6) and 2.6-mm-thick titanium (Ti6Al4V) always experienced a ductile failure in the heat-affected zone of the aluminum (Woizeschke 2017). In comparison to the previous state of the art, the thickness of the seam was here the same as that of the aluminum sheet, i.e., there was no thickening in the joining zone.

Möller et al. and Woizeschke et al. were the first to use a (double-sided) laser beam joining process for the joining of aluminum sheets with titanium wire loops (Möller et al. 2010) as well as titanium laminates (Woizeschke and Wottschel 2013) to realize the integral transition to CFRP components. While the wire loops resulted in relatively low joint strengths due to the limitation in the achievable load-bearing titanium cross-section, even with an improved double-rowed configuration of the connection (see Schumacher et al. 2014), the titanium laminate offered sufficient potential regarding the load-bearing cross-section of the laminate that the location of the failure could be shifted into the aluminum joining partner (Woizeschke and Wottschel 2013). In Woizeschke and Vollertsen (2018), it was shown that a failure in the heat-affected zone of the aluminum could be achieved in case of aluminum-titanium subjoints (without CFRP) as long as a defect-free wetting and an infiltration of the interspaces in the laminate by aluminum had occurred. In this case, the laminate behaves in a quasi-monolithic manner, much like a single sheet (Woizeschke and Vollertsen 2018).

In case of the aluminum-titanium-CFRP concept, a 30° wedge shape of the titanium laminate edge consisting of five-layered (5 × 0.5 mm) titanium laminate as a transition structure, an average force transmission of 628 N per millimeter of seam length was achieved (Woizeschke 2017). However, the investigations of Woizeschke also showed that the reproducible manufacture of joints capable of withstanding high levels of stress between titanium-CFRP hybrid laminates and aluminum sheets remains a challenge. The identified cause-effect relationships between the laminate-specific factors and the joint strength that need to be considered were summarized in a flow diagram in Woizeschke and Vollertsen (2018).

For such joining concepts with titanium transition structures as well as for adhesive bonding of hybrid titanium-CFRP joints, the bonding strength of this hybrid laminate is essential for the application. The challenge hereby lies especially on creating a sufficient aging resistance of the joints. A novel environmentally friendly and economical laser-based pretreatment process for the titanium surfaces as alternative to wet-chemical or mechanical processes was developed and investigated by Specht et al. (2014). The laser-based formation of nanostructured titanium-oxide (TiO_2) layers (see Fig. 11.11) allows the production of hydrothermally stable bonds. The process is based on the generation of TiO_2 nanoparticles in an oxygen-containing atmosphere by laser-induced material evaporation and condensation. The layer is formed by scanning the surfaces with overlapping laser pulses (Specht 2015). In addition, CFRP parts can also be pre-treated by laser processes in order to improve the bond strength of metal-composite joints. Reitz et al. used IR laser (and UV laser) prior to an adhesive bonding process (Reitz et al. 2017). Two opposing effects were identified: strength increasing surface activation and reducing weakening of the interface between fibers and matrix near the surface (Reitz et al. 2017).

In sum, laser-based joining processes show high potential for lightweight hybrid multi-material structures by enabling slim integral aluminum-titanium(-CFRP) joints with high strengths and ductile failure behaviors with failure locations in the heat-affected zone.

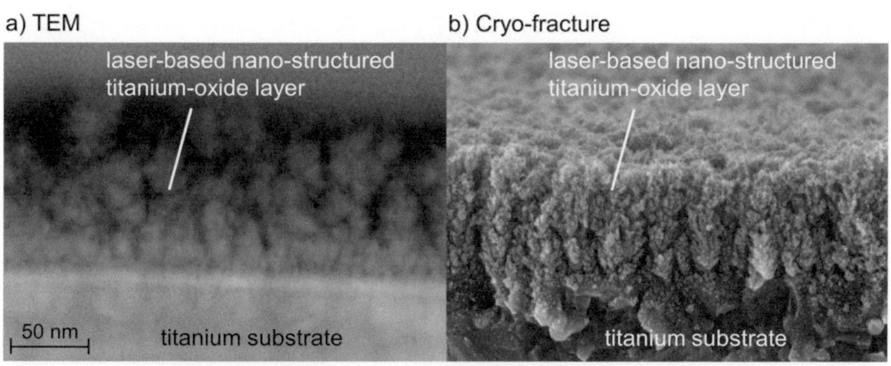

Fig. 11.11 Laser-based formation of nanostructured titanium-oxide layers to improve the bond strength of metal-composite joints: (**a**) transmission electron micrograph and (**b**) image of the cryo-fractured layer, according to (Specht 2015)

1.3 Laser Processes for HLFC Structures

Hybrid laminar flow control (HLFC) enables the expansion of the area with laminar air flow at the wing area through suction of the air via perforated fuselage skin (Young et al. 2001), in particular at the wing and vertical stabilizers (see Fig. 11.12). The production of HLFC structures requires the combination of various laser processes with completely different parameters (see Fig. 11.13 left): (1) the perforating (drilling) of the fuselage skin, (2) the direct one-sided deep penetration laser welding of stringers to the thin fuselage skin with a defined and low weld depth for the reinforcement and creation of separate pressure chambers, and (3) the laser beam straightening through laser bending to create the required aerodynamic outer contour of the wing without dents, steps, or waves (Woizeschke et al. 2016b). The laser is the key enabling technology for this concept (Woizeschke et al. 2016b). Laser beams with very different intensities and spot sizes on the workpiece are used. While the percussion drilling—more than half a million drill holes per hour in this case (Schultz et al. 2017b)—combines a high power density on the spot with short durations of the laser beam pulses so that at the workpiece a removal of material occurs (at least partially) through ablation (Messaoudia et al. 2015), the one-sided laser welding of the stringers is realized through a keyhole process (Schultz et al. 2014b), and the laser straightening occurs through a heat conduction process by using the same but defocused continuous-wave laser beam (Reitemeyer et al. 2013). The production of such an advanced lightweight aircraft structure (Fig. 11.13 (right)) demonstrates the technical opportunities of laser material processing.

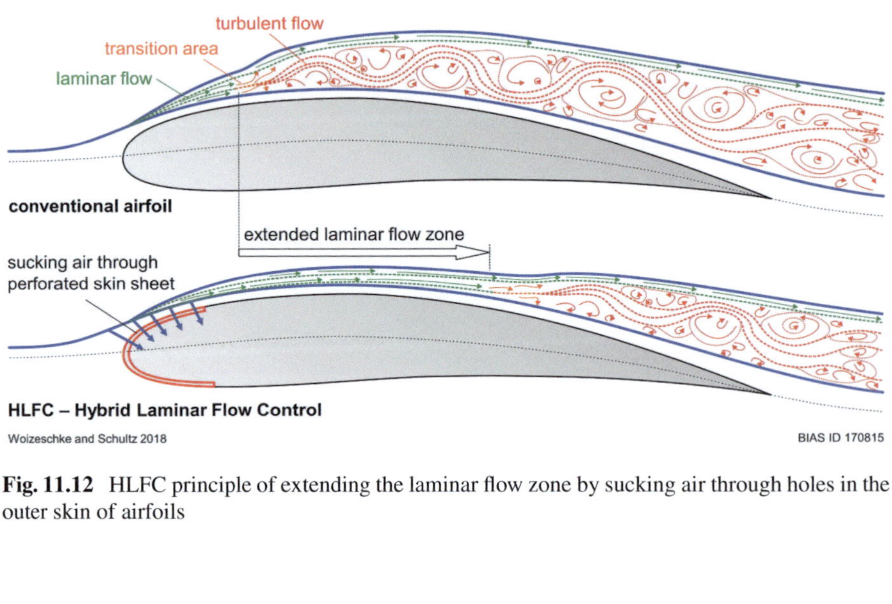

Fig. 11.12 HLFC principle of extending the laminar flow zone by sucking air through holes in the outer skin of airfoils

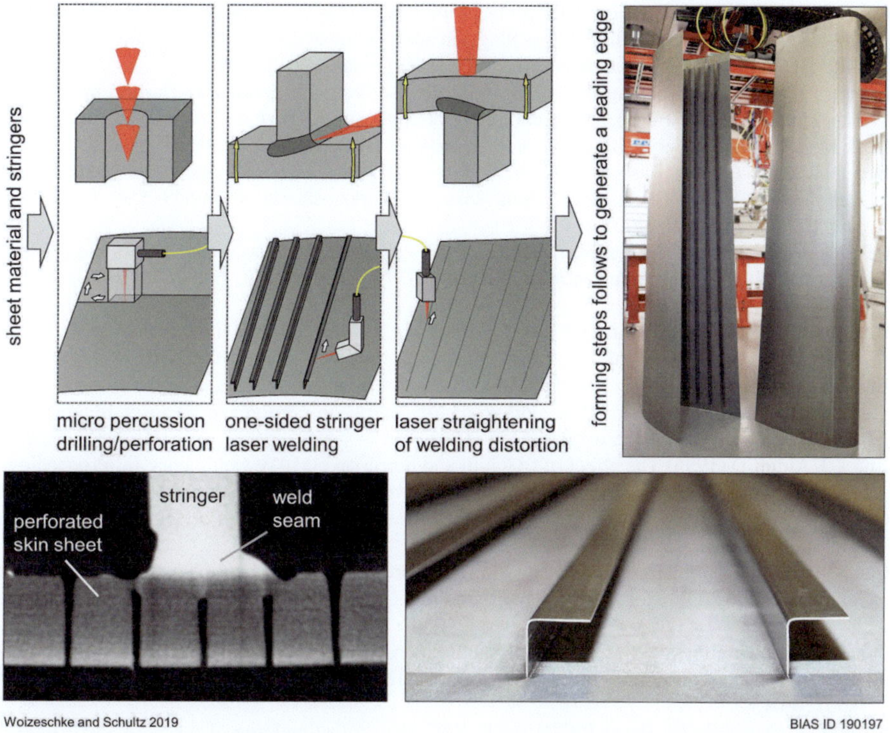

Fig. 11.13 Manufacturing procedure of HLFC structures in the form of a combination of three different laser processes: laser micro drilling, laser deep penetration welding, and laser straightening; computed tomography image (bottom left), welded stringers on a perforated skin sheet (bottom right), and demonstrators (top right)

Acknowledgments The author would like to thank his current and former colleagues at the BIAS Institute in Bremen for their many years of cooperation and various support.

References

Aalderink BJ, Pathiraj B (2010) Seam gap bridging of laser based processes for the welding of aluminium sheets for industrial applications. Int J Adv Manuf Technol 48:143–154. https://doi.org/10.1007/s00170-009-2270-x

Abbas A, de Vicente J, Valero E (2013) Aerodynamic technologies to improve aircraft performance. Aerosp Sci Technol 28(1):100–132. https://doi.org/10.1016/j.ast.2012.10.008

Allison A, Scudamore R (2014) European Strategic Research Agenda: Joining (prepared by the Joining Sub-Platform)

Arata Y, Miyamoto I (1974) Some fundamental properties of high power laser beam as a heat source. Trans JWRI 3:1–20

Arata Y, Matsuda F, Mukae S et al (1973) Effect of weld solidification mode on tensile properties of aluminum weld metal. Trans JWRI 2(2):184–190

Bang H, Bang H, Song H et al (2013) Joint properties of dissimilar Al6061-T6 aluminum alloy/Ti–6%Al–4%V titanium alloy by gas tungsten arc welding assisted hybrid friction stir welding. Mater Des 51(0):544–551. https://doi.org/10.1016/j.matdes.2013.04.057
Bautze T (2018) Seam tracking with OCT and beam oscillation for automotive laser remote welding applications. In: European automotive laser applications (EALA)
Beck M (1996) Modellierung des Lasertiefschweißens. B. G. Teubner, Stuttgart
Bergmann JP (2004) Laserstrahlschweißen von Titanwerkstoffen unter Berücksichtigung des Einflusses des Sauerstoffes. Mat.-wiss. u. Werkstofftech. 35(9):543–556. https://doi.org/10.1002/mawe.200400776
Bergmann JP, Bielenin M, Stambke M et al (2013) Effects of diode laser superposition on pulsed laser welding of aluminum. Lasers Manuf (LiM) (41):180–189. https://doi.org/10.1016/j.phpro.2013.03.068
Bergmann JP, Bielenin M, Feustel T (2015) Aluminum welding by combining a diode laser with a pulsed Nd: YAG laser. Weld World 59(2):307–315. https://doi.org/10.1007/s40194-014-0218-8
Berkmanns J, Behler K, Beyer E (1992) Der Laser—ein neues Werkzeug zum Schweißen von Aluminiumwerkstoffen. In: Proceedings of 4th European conference on laser treatment of materials ECLAT '92, Oberursel, pp 151–156
Brock J, Aidun D (1995) The effect of titanium and boron on GMAW weldments. In: Proceedings of the 6th international conference on aluminum weldments, pp 343–356
Cahoon JR, Tandon KN, Chaturvedi MC (1992) Effect of gravity level on grain refinement in aluminum alloys. Metall Trans A 23:3399–3404
Cao X, Wallace W, Immarigeon J-P et al (2003) Research and progress in laser welding of wrought aluminum alloys. II. Metallurgical microstructures, defects, and mechanical properties. Mater Manuf Process 18(1):23–49. https://doi.org/10.1081/AMP-120017587
Casalino G, Mortello M (2016) Modeling and experimental analysis of fiber laser offset welding of Al-Ti butt joints. Int J Adv Manuf Technol 83:89–98. https://doi.org/10.1007/s00170-015-7562-8
Casalino G, Mortello M, Peyre P (2015) Yb–YAG laser offset welding of AA5754 and T40 butt joint. J Mater Process Technol 223:139–149. https://doi.org/10.1016/j.jmatprotec.2015.04.003
Chen W, Molian P (2007) Dual-beam laser welding of ultra-thin AA 5052-H19 aluminum. Int J Adv Manuf Technol 39:889–897. https://doi.org/10.1007/s00170-007-1278-3
Chen Y, Chen S, Li L (2009) Effects of heat input on microstructure and mechanical property of Al/Ti joints by rectangular spot laser welding-brazing method. Int J Adv Manuf Technol 44(3–4):265–272. https://doi.org/10.1007/s00170-008-1837-2
Chen S, Li L, Chen Y et al (2010) Si diffusion behavior during laser welding-brazing of Al alloy and Ti alloy with Al-12Si filler wire. Trans Nonferrous Metals Soc China 20(1):64–70. https://doi.org/10.1016/S1003-6326(09)60098-4
Cho W-I, Schultz V, Woizeschke P (2018a) Numerical study of the effect of the oscillation frequency in buttonhole welding. J Mater Process Technol 261:202–212. https://doi.org/10.1016/j.jmatprotec.2018.05.024
Cho W-I, Woizeschke P, Schultz V (2018b) Simulation of molten pool dynamics and stability analysis in laser buttonhole welding. Procedia CIRP 74:687–690. https://doi.org/10.1016/j.procir.2018.08.042
Cross CE (2005) On the origin of weld solidification cracking. In: Böllinghaus T, Herold H (eds) Hot cracking phenomena in welds. Springer, Berlin, pp 3–18
Dausinger F (1995) Strahlwerkzeug Laser: Energiekopplung und Prozesseffektivität. Zugl.: Stuttgart, Univ., Habil.-Schr. Laser in der Materialbearbeitung. Teubner, Stuttgart
Dev S, Stuart AA, Kumaar RRD et al (2007) Effect of scandium additions on microstructure and mechanical properties of Al–Zn–Mg alloy welds. Mater Sci Eng A 467(1–2):132–138. https://doi.org/10.1016/j.msea.2007.02.080
Dilthey U (2005) Schweißtechnische Fertigungsverfahren 2: Verhalten der Werkstoffe beim Schweißen, 3., bearbeitete Auflage. VDI-Buch. Springer, Berlin
Dilthey U (2006) Schweißtechnische Fertigungsverfahren 1: Schweiß- und Schneidtechnologien, 3., bearbeitete Auflage. VDI-Buch. Springer, Berlin

Dressler U, Biallas G, Alfaro Mercado U (2009) Friction stir welding of titanium alloy TiAl6V4 to aluminium alloy AA2024-T3. Mater Sci Eng A 526(1–2):113–117. https://doi.org/10.1016/j.msea.2009.07.006

Farrel WJ, Ferrario JD (1987) A computer-controlled, wide-bandwidth deflection system for electron beam welding and heat treating. Weld J 10:41–49

Fetzer F, Hagenlocher C, Weber R (2018) High power, high speed, high quality. LTJ 15(3):28–31. https://doi.org/10.1002/latj.201800017

Fonseca de Arruda AC, Prates de Campos Filho M (1983) Rheocasting techniques applied to grain refinement of aluminum alloys. In: Proceedings of the conference on solidification technology in the foundry and cast house, Coventry (UK), pp 143–146

Fuji A, Ameyama K, North TH (1995) Influence of silicon in aluminium on the mechanical properties of titanium/aluminium friction joints. J Mater Sci 30

Gao M, Chen C, Gu Y et al (2014) Microstructure and tensile behavior of laser arc hybrid welded dissimilar Al and Ti alloys. Materials 7(3):1590–1602. https://doi.org/10.3390/ma7031590

Gatzen M (2014) Durchmischung beim Laserstrahltiefschweißen unter dem Einfluss niederfrequenter Magnetfelder. Dissertation. Strahltechnik, Bd. 55. BIAS Verlag, Bremen

Geisel M (2002) Prozeßkontrolle und -steuerung beim Laserstrahlschweißen mit den Methoden der nichtlinearen Dynamik. Meisenbach Bamberg

Göbel G, Brenner B, Beyer E (2007) New application possibilities for fiber laser welding. In: Lu Y (ed) 26th International congress on applications of lasers & electro-optics (ICALEO 2007), Orlando, pp 102–108

Gref W (2005) Laserstrahlschweißen von Aluminiumwerkstoffen mit der Fokusmatrixtechnik. Laser in der Materialbearbeitung. Utz, München

Gruss H (2008) Schweißgerechte Struktur- und Prozessstrategien im Flugzeugbau. Dissertation

Gruss H, Herold H, Streitenberger M (2008) Verbesserung des Strukturverhaltens laserstrahlgeschweißter Haut-Stringer-Verbindungen. In: 6. Laser-Anwenderforum Bremen, BIAS-Verlag

Habenicht G (2009) Kleben: Grundlagen, Technologien, Anwendungen, 6., aktualisierte Aufl. VDI. Springer, Berlin

Haboudou A, Peyre P, Vannes AB et al (2003) Reduction of porosity content generated during Nd:YAG laser welding of A356 and AA5083 aluminium alloys. Mater Sci Eng A 363(1):40–52. https://doi.org/10.1016/S0921-5093(03)00637-3

He X, Zhang Y, Xing B et al (2015) Mechanical properties of extensible die clinched joints in titanium sheet materials. Mater Des 71:26–35. https://doi.org/10.1016/j.matdes.2015.01.005

Heider P (1994) Lasergerechte Konstruktion und lasergerechte Fertigungsmittel zum Schweissen grossformatiger Aluminium-Strukturbauteile, Als Ms. gedr. Fortschrittberichte VDI : Reihe 2, Fertigungstechnik, Nr. 326. VDI-Verl., Düsseldorf

Heimerdinger C (2003) Laserstrahlschweißen von Aluminiumlegierungen für die Luftfahrt. Zugl.: Stuttgart, Univ., Diss., 2003. Laser in der Materialbearbeitung. Utz, München

Herrmann A, Schiebel P, Hoffmeister C et al (2008) Verbindungen zwischen einem monolithischen Metallbauteil und einem endlosfaserverstärkten Laminatbauteil sowie Verfahren zur Herstellung derselben (DE102008047333B4)

Heß A, Bassi C, Schellinger F (2011) Heißrissfreies Remotelaserstrahlfügen von 6xxxer Aluminiumwerkstoffen. Laser Mgazin 2:30–31

Hilbinger RM (2001) Heißrissbildung beim Schweißen von Aluminium in Blechrandlage. Zugl.: Bayreuth, Univ., Diss., 2001. Institut für Materialforschung—Bayreuth, vol 9. Utz Wiss, München

Hohenberger B (2003) Laserstrahlschweissen mit Nd:Yag-Doppelfokustechnik: Steigerung von Prozeßstabilität, Flexibilität und verfügbarer Strahlleistung. Zugl.: Stuttgart, Univ., Diss., 2002. Laser in der Materialbearbeitung. Utz, München

Janaki Ram GD, Mitra TK, Shankar V et al (2003) Microstructural refinement through inoculation of type 7020 Al–Zn–Mg alloy welds and its effect on hot cracking and tensile properties. J Mater Process Technol 142(1):174–181. https://doi.org/10.1016/S0924-0136(03)00574-0

Jin X, Li L (2004) An experimental study on the keyhole shapes in laser deep penetration welding. Opt Lasers Eng 41(5):779–790. https://doi.org/10.1016/S0143-8166(03)00034-4

Jones L, Alfile J-P, Aubert P et al (2000) Advanced cutting, welding and inspection methods for vacuum vessel assembly and maintenance. Fusion Eng Des 51–52:985–991. https://doi.org/10.1016/S0920-3796(00)00412-9

Kahraman N, Gulenc B, Findik F (2007) Corrosion and mechanical-microstructural aspects of dissimilar joints of Ti-6Al-4V and Al plates. Int J Impact Eng 34(8):1423–1432. https://doi.org/10.1016/j.ijimpeng.2006.08.003

Kaplan AFH, Powell J (2011) Spatter in laser welding. J Laser Appl 23(3):32005. https://doi.org/10.2351/1.3597830

Kappelsberger E (1987) Werkstückspezifische Bedingungen beim Laserschweißen: Laser/Optoelektronik in der Technik/Laser/Optoelectronics in Engineering

Katayama S (2013) The unweldables. Laser Community 2:24–25

Katayama S, Kawahito Y (2009) Elucidation of phenomena in high-power fiber laser welding and development of prevention procedures of welding defects. In: Gapontsev DV, Kliner DA, Dawson JW et al (eds) SPIE LASE: lasers and applications in science and engineering. SPIE, 71951R

Katayama S, Seto N, Kim JD et al (1997) Formation mechanism and reduction method of porosity in laser welding of stainless steel. In: Proceedings of the 16th international congress on applications of lasers and electro-optics (ICALEO), pp 83–92

Katayama S, Kawahito Y, Mizutani M (2012) Latest progress in performance and understanding of laser welding. Phys Procedia 39:8–16. https://doi.org/10.1016/j.phpro.2012.10.008

Katayama S, Mizutani M, Kawahito Y et al (2015) Fundamental research of 100 kW fiber laser welding technology. In: Proceedings of the lasers in manufacturing conference (LiM), #342

Kempa S (2014) Potentiale und Grenzen beim thermischen Fügen von Multimaterial-Verbindungen. Schweißtechnik Soudure 2014(1):36–39

Klassen M (2000) Prozeßdynamik und resultierende Prozeßinstabilitäten beim Laserstrahlschweißen von Aluminium. Dissertation, Universiät Bremen

Kocik R (2009) Analyse und Bewertung der mechanisch-technologischen Eigenschaften von geschweissten Mischverbindungen aus Aluminium und Titan. Forschungsberichte aus der Stiftung Institut für Werkstofftechnik, Bremen, Bd. 46. Shaker, Aachen

Kocik R, Vugrin T, Seefeld T (2006) Laserstrahlschweißen im Flugzeugbau: Stand und künftige Anwendungen. In: 5. Laser-Anwenderforum Bremen, BIAS-Verlag

Kogel-Hollacher M, Schoenleber M, Bautze T et al (2016) Measurement and closed-loop control of the penetration depth in laser materials processing. In: 9th international conference on photonic technologies (LANE)

Kou S (1987) Welding metallurgy. A Wiley Interscience publication, Wiley, New York

Kreimeyer M, Vollertsen F (2005) Processing titanium-aluminum hybrid joints for aircraft applications. In: Proceedings of the third international WLT-conference on lasers in manufacturing 2005, LiM2005, Munich

Kreimeyer M, Wagner F, Zerner I et al (2001) Laser beam joining of aluminium with titanium with the use of an adapted working head. In: DVS-Berichte, vol 212. Verlag für Schweißen und verwandte Verfahren, DVS-Verlag, pp 317–321

Kreimeyer M, Wagner F, Vollertsen F (2005) Laser processing of aluminum-titanium-tailored blanks. Opt Lasers Eng 43(9):1021–1035. https://doi.org/10.1016/j.optlaseng.2004.07.005

Kutsuna M, Shido K, Okada T (2002) Fan shaped cracking test of aluminum alloys in laser welding. In: Miyamoto I, Kobayashi KF, Sugioka K et al (eds) LAMP 2002: international congress on laser advanced materials processing. SPIE, p 230

Lin R, Wang H-P, Lu F et al (2017) Numerical study of keyhole dynamics and keyhole-induced porosity formation in remote laser welding of Al alloys. Int J Heat Mass Transf 108:244–256. https://doi.org/10.1016/j.ijheatmasstransfer.2016.12.019

Liu W, Tian X, Zhang X (1996) Preventing weld hot cracking by synchronous rolling during welding. Weld J 75(9):297.s–304.s

Ma Z, Zhao W, Yan J et al (2011) Interfacial reaction of intermetallic compounds of ultrasonic-assisted brazed joints between dissimilar alloys of Ti6Al4V and Al4Cu1Mg. Ultrason Sonochem 18(5):1062–1067. https://doi.org/10.1016/j.ultsonch.2011.03.025

Maina MR, Okamoto Y, Okada A et al (2018) High surface quality welding of aluminum using adjustable ring-mode fiber laser. J Mater Process Technol 258:180–188. https://doi.org/10.1016/j.jmatprotec.2018.03.030
Majumdar B, Galun R, Weisheit A et al (1997) Formation of a crack-free joint between Ti alloy and Al alloy by using a high-power CO_2 laser. J Mater Sci 32:6191–6200
Martin B, Loredo A, Pilloz M et al (2001) Characterisation of cw Nd: YAG laser keyhole dynamics. Opt Laser Technol 33(4):201–207. https://doi.org/10.1016/S0030-3992(01)00014-7
Martinsen K, Hu SJ, Carlson BE (2015) Joining of dissimilar materials. CIRP Ann Manuf Technol 64(2):679–699. https://doi.org/10.1016/j.cirp.2015.05.006
Marya M, Marya S, Priem D (2005) On the characteristics of electromagnetic welds between aluminium and other metals and alloys. Weld World 49(5/6):74–84
Mathers G (2002) The welding of aluminium and its alloys. CRC Press, Boca Raton
Matsuda F, Nakata K, Nishio Y et al (1986) Effect of zirconium addition on improvement of solidification crack susceptibility of Al-Zn-Mg alloy weld—fundamental research on solidification crack susceptibility of Al-Zn-Mg alloy weld (Report 1). Q J Jpn Weld Soc 4(1):115–120. https://doi.org/10.2207/qjjws.4.115
Matsunawa A, Seto N, Kim J-D et al (eds) (2000) Dynamics of keyhole and molten pool in high-power CO_2 laser welding, 3888 IS
McCartney DG (1989) Grain refining of aluminium and its alloys using inoculants. Int Mater Rev 34(1):247–260. https://doi.org/10.1179/imr.1989.34.1.247
Messaoudia H, Mehrafsun S, Schrauf G et al (2015) Highly reproducible laser micro drilling of titanium-based HLFC sections. In: Lasers in manufacturing, Munich, Germany
Miriyev A, Levy A, Kalabukhov S et al (2016) Interface evolution and shear strength of Al/Ti bimetals processed by a spark plasma sintering (SPS) apparatus. J Alloys Compd 678:329–336. https://doi.org/10.1016/j.jallcom.2016.03.137
Mittelstädt C, Seefeld T, Woizeschke P et al (2018) Laser welding of hidden T-joints with lateral beam oscillation. Procedia CIRP 74:456–460. https://doi.org/10.1016/j.procir.2018.08.151
Miyamoto (1997) Laser welding (2)—welding of thick plate. In: Proceedings of the 41st laser materials processing conference, pp 21–34
Möller F, Thomy C, Vollertsen F et al (2010) Novel method for joining CFRP to aluminium. In: Laser assisted net shape engineering 6, Proceedings of the LANE 2010, Part 2 5, pp 37–45. https://doi.org/10.1016/j.phpro.2010.08.027
Möller F, Thomy C, Vollertsen F (2012) Joining of titanium-aluminium seat tracks for aircraft applications—system technology and joint properties. Weld World 56(3–4):108–114. https://doi.org/10.1007/BF03321341
Nakashiba S, Okamoto Y, Sakagawa T et al (2011) Welding characteristics of aluminum alloy by pulsed Nd:YAG laser with pre- and post-irradiation of superposed continuous diode laser. In: Proceedings of international congress on applications of lasers and electro optics (ICALEO), #M205
Nesterov AF, Trubitsin AP, Prokhorov AN (1990) Special features of resistance welding titanium to aluminium. Weld Int 4(6):464–466. https://doi.org/10.1080/09507119009447761
Ono M, Shiozaki T, Shinbo Y et al (2001) Development of high power laser pipe welding process. Q J Jpn Weld Soc 19(2):233–240. https://doi.org/10.2207/qjjws.19.233
Ostermann F (1998) Anwendungstechnologie Aluminium. VDI-Buch. Springer, Berlin
Palm F (2000) Bericht aus der Werkstoff-Forschung der DaimlerChryslerAG in Ottobrunn zum Schweißen im Flugzeugbau. In: DVS-Berichte, vol 208. DVS-Media, pp 53–58
Ploshikhin V, Prikhodovsky A, Makhutin M et al (2004) Rechnergestützte Entwicklung der Verfahren zum rissfreien Laserstrahlschweißen. NMB Bayreuth
Ploshikhin V, Prikhodovski A, Ilin A et al (2007) Computer aided development of the crack-free laser welding processes. Key Eng Mat 353–358:1984–1994. https://doi.org/10.4028/www.scientific.net/KEM.353-358.1984
Pretorius T, Kreimeyer M, Sepold G et al (2004) Simulation of laser beam welded aluminium joints. In: Proceedings of 14th international conference on computer technology in welding and manufacturing

Radaj D (1992) Heat effects of welding: temperature field, residual stress, distortion. Springer, Berlin
Radel T (2018) Mechanical manipulation of solidification during laser beam welding of aluminum. Weld World 62(1):29–38. https://doi.org/10.1007/s40194-017-0530-1
Radel T, Woizeschke P (2018) Reduction of hot cracking susceptibility during laser welding of aluminum by vibrations. Weld World 41(1):164. https://doi.org/10.1007/s40194-018-00680-2
Radscheit CR (1997) Laserstrahlfügen von Aluminium mit Stahl. Dissertation. Strahltechnik, Bd. 4. BIAS-Verlag, Bremen
Reitemeyer D (2012) Stabilisierung der Fokuslage beim Schweißen mit Faser- und Scheibenlasern. Zugl.: Bremen, Univ., Dissertation. Strahltechnik, Bd. 49. BIAS-Verlag, Bremen
Reitemeyer D, Seefeld T, Vollertsen F et al (2009) Influences on the laser induced focus shift in high power fiber laser welding. Lasers Manuf (LiM):293–298
Reitemeyer D, Seefeld T, Vollertsen F (2010) Online focus shift measurement in high power fiber laser welding. Phys Procedia 5:455–463. https://doi.org/10.1016/j.phpro.2010.08.073
Reitemeyer D, Schultz V, Syassen F et al (2013) Laser welding of large scale stainless steel aircraft structures. Lasers Manuf (LiM) 41:106–111. https://doi.org/10.1016/j.phpro.2013.03.057
Reitz V, Meinhard D, Ruck S et al (2017) A comparison of IR- and UV-laser pretreatment to increase the bonding strength of adhesively joined aluminum/CFRP components. Compos A Appl Sci Manuf 96:18–27. https://doi.org/10.1016/j.compositesa.2017.02.014
Russell JD (1997) Application of laser welding in shipyards. In: Beckmann LHJF (ed) Lasers and optics in manufacturing III. SPIE, pp 174–183
Scaggs M, Haas G (2010). Thermal lensing compensation objective for high power lasers. International Congress on Applications of Lasers & Electro-Optics (ICALEO2010), pp 1511–1517. https://doi.org/10.2351/1.5062011
Schempp P, Cross CE, Schwenk C et al (2012) Influence of Ti and B additions on grain size and weldability of aluminium alloy 6082. Weld World 56(9):95–104. https://doi.org/10.1007/BF03321385
Schempp P, Cross CE, Häcker R et al (2013) Influence of grain size on mechanical properties of aluminium GTA weld metal. Weld World 57(3):293–304. https://doi.org/10.1007/s40194-013-0026-6
Schneider A, Avilov V, Gumenyuk A et al (2013) Laser beam welding of aluminum alloys under the influence of an electromagnetic field. Lasers Manuf (LiM) (41):4–11. https://doi.org/10.1016/j.phpro.2013.03.045
Schoer H (1980) Über das Schweißrissigkeitsverhalten von Aluminiumwerkstoffen. Metall 34(6):546–551
Schrauf G (2005) Status and perspectives of laminar flow. Aeronaut J 109(1102):639–644. https://doi.org/10.1017/S000192400000097X
Schubert E, Zerner I, Sepold G et al (1997) Laser beam joining of material combinations for automotive applications. In: Beckmann LHJF (ed) Lasers and optics in manufacturing III. SPIE, pp 212–221
Schubert E, Klassen M, Skupin J et al (1998) Effect of filler wire on process stability in laser beam welding of aluminium-alloys. In: Proceedings of the 6th international conference on welding and melting by electron and laser beams (CISFFEL), pp 195–204
Schultz V, Seefeld T (2015) Schlussbericht - Strahlmodulation beim Schweißen mit hoch fokussierenden Festkörperlasern mit Zusatzwerkstoff (Spaltüberbrückbarkeit), IGF-Vorhaben 17.558N
Schultz V, Woizeschke P (2018) High seam surface quality in keyhole laser welding: buttonhole welding. J Manuf Mater Process (JMMP) 2(4):78. https://doi.org/10.3390/jmmp2040078
Schultz V, Seefeld T, Vollertsen F (2014a) Gap bridging ability in laser beam welding of thin aluminum sheets. Phys Procedia 56:545–553. https://doi.org/10.1016/j.phpro.2014.08.037. 8th International conference on laser assisted net shape engineering LANE 2014
Schultz V, Thomy C, Vollertsen F et al (2014b) Development of a laser welding and straightening process for aircraft structures for hybrid laminar flow control. In: IIW annual assembly: Doc. no. IV-1180-14

Schultz V, Cho W-I, Woizeschke P et al (2017a) Laser deep penetration weld seams with high surface quality. In: Overmeyer L, Reisgen U, Ostendorf A, Schmidt M (eds) Lasers in manufacturing (LiM2017). USB stick

Schultz V, Stephen A, Kalms M et al (2017b) Laser ermöglichen die Fertigung kraftstoffsparender Flugzeugstrukturen. Laser Magazin 5:23–24

Schulze G (2010) Die Metallurgie des Schweissens: Eisenwerkstoffe—nichteisenmetallische Werkstoffe, 4., neu bearbeitete Aufl. VDI-Buch. Springer, Heidelberg

Schumacher J (2002) Erfahrungen bei der Serieneinführung für Laserstrahlschweißen im Flugzeugbau. In: 4. Laser-Anwenderforum. BIAS Verlag, Bremen, pp 247–256

Schumacher J, Irretier A, Kocik R et al (2007) Investigation of laser-beam joined titanium-aluminum hybrid structures. In: Applied production technology APT'07, pp 149–160

Schumacher J, Clausen B, Zoch H-W (2014) Strength and failure behaviour of carbon fibre reinforced plastics (CFRP)-aluminium seam structures. Mat.-wiss. u. Werkstofftech 45(12):1108–1115. https://doi.org/10.1002/mawe.201400359

Semiatin SL, Seetharaman V, Weiss I (1998) Hot workability of titanium and titanium aluminide alloys—an overview. Mater Sci Eng A 243(1-2):1–24. https://doi.org/10.1016/S0921-5093(97)00776-4

Seto N, Katayama S, Matsunawa A (2001) Porosity formation mechanism and suppression procedure in laser welding of aluminium alloys. Weld Int 15(3):191–202. https://doi.org/10.1080/09507110109549341

Shi P, Wan P (2016) Numerical simulation of formation process of keyhole-induced pore for laser deep penetration welding. In: International conference on advanced electronic science and technology (AEST), pp 368–373

Skoda P, Dujak J, Michalicka P (op. 1996) Creation of heterogeneous weld joints of titanium- and aluminium-based materials by electron beam welding. In: Welding science & technology: Japan—Slovak welding symposium: proceedings of the international welding conference, 5–7 March 1996. Faculty of Metallurgy, Department of Materials Science, Košice, pp 157–160

Song Z, Nakata K, Wu A et al (2013) Interfacial microstructure and mechanical property of Ti6Al4V/A6061 dissimilar joint by direct laser brazing without filler metal and groove. Mater Sci Eng A 560(0):111–120. https://doi.org/10.1016/j.msea.2012.09.044

Specht U (2015) Untersuchungen zur Entstehung laserinduzierter nanoporöser Strukturen auf Titanoberflächen für langzeitstabile Klebungen. Dissertation. University of Bremen

Specht U, Ihde J, Mayer B (2014) Laser induced nano-porous Ti–O-layers for durable titanium adhesive bonding. Mat.-wiss. u. Werkstofftech 45(12):1116–1122. https://doi.org/10.1002/mawe.201400360

Sun Z, Kuo M (1999) Bridging the joint gap with wire feed laser welding. J Mater Process Technol 87(1):213–222. https://doi.org/10.1016/S0924-0136(98)00346-X

Takemoto T, Nakamura H, Okamoto I (1990) Aluminum brazing filler metals for making aluminum to titanium joints in a vacuum (physics, process, instrument & measurement). Trans JWRI 19(1):39–44

Tang Z (2014) Heißrissvermeidung beim Schweißen von Aluminiumlegierungen mit einem Scheibenlaser. Dissertation. Strahltechnik, Bd. 53. BIAS Verlag

Tang Z, Vollertsen F (2014) Influence of grain refinement on hot cracking in laser welding of aluminum. Weld World 58(3):355–366. https://doi.org/10.1007/s40194-014-0121-3

Thomy C, Vollertsen F (2007) Hybrid Welding of thin sheet material with single-mode fibre laser. In: Proceedings of the IIW International Institute of Welding Commission XII intermediate meeting, IIW Doc. Nr. XII-1912-07 (CD)

Tomashchuk I, Sallamand P, Cicala E et al (2015) Direct keyhole laser welding of aluminum alloy AA5754 to titanium alloy Ti6Al4V. J Mater Process Technol 217:96–104. https://doi.org/10.1016/j.jmatprotec.2014.10.025

Tsukamoto S, Kawaguchi I, Honda H et al (2001) Suppression of porosity using pulse modulation of laser power in 20kW laser welding. In: Proceedings of the 20th international congress on applications of lasers and electro-optics (ICALEO), pp D607–D615

Vollertsen F (2009) Properties and prospects of high brightness solid state lasers. LTJ 6(5):27–31. https://doi.org/10.1002/latj.200990071
Vollertsen F, Neumann S (2009) High brightness solid state laser: development and application. In: Proceedings of the 5th international congress on laser advanced material processing LAMP09, #317
Vollertsen F, Woizeschke P, Schultz V et al (2017) Developments for laser joining with high-quality seam surfaces. Lightw Des Worldw 10(5):6–13. https://doi.org/10.1007/s41777-017-0041-1
Volpp J (2017) Dynamik und Stabilität der Dampfkapillare beim Laserstrahltiefschweißen. Dissertation. Strahltechnik, Bd. 63. BIAS Verlag
Volpp J, Srowig J, Vollertsen F (2016) Spatters during laser deep penetration welding with a bifocal optic. AMR 1140:123–129. https://doi.org/10.4028/www.scientific.net/AMR.1140.123
Wang H, Vivek A, Wang Y et al (2016) Laser impact welding application in joining aluminum to titanium. J Laser Appl 28(3):32002. https://doi.org/10.2351/1.4946887
Wilden J, Bergmann JP (2004) Manufacturing of titanium/aluminium and titanium/steel joints by means of diffusion welding. Weld Cut 3:285–290
Wilden J, Neumann T (2010) Moderne Strahlquellen im Einsatz – Welche metallurgischen Ansätze ergeben sich? In: Strahlschweißen von Aluminium. DVS-Media, Düsseldorf, pp 39–44
Wilden J, Bergmann JP, Reich S et al (2007a) Grundlegende Untersuchungen zum flussmittelfreien Löten von Leichtmetall-Material-Mix Konstruktionen
Wilden J, Bergmann JP, Holtz R et al (2007b) Einsatz von gepulsten Nd:YAG-Lasern für das Fügen von Werkstoffen und Werkstoffkombinationen mit anspruchsvollen Eigenschaften
Woizeschke P (2017) Eigenschaften laserstrahlgefügter Mischverbindungen aus Aluminium und Titan in Abhängigkeit der Kantengeometrie und Halbzeugstruktur. Dissertation, University of Bremen, Strahltechnik, Bd. 65, Bremen, urn:nbn:de:gbv:46-00106005-12
Woizeschke P, Vollertsen F (2014) Distortion effects in deep penetration laser micro welding: challenges and results. Lasers Eng 28(5-6):337–359
Woizeschke P, Vollertsen F (2015) Fracture analysis of competing failure modes of aluminum-CFRP joints using three-layer titanium laminates as transition. J Mater Eng Perform 24(9):3558–3572. https://doi.org/10.1007/s11665-015-1638-3
Woizeschke P, Vollertsen F (2016) A strength-model for laser joined hybrid aluminum–titanium transition structures. CIRP Ann Manuf Technol 65(1):241–244. https://doi.org/10.1016/j.cirp.2016.04.027
Woizeschke P, Vollertsen F (2018) Joining of aluminum to titanium sheets/ laminates for aluminum-CFRP transitions. In: Herrmann AS (ed) Outcome of the research unit FOR1224: CFRP-Al transition structures in lightweight constructions: "Schwarz-Silber". BoD, Norderstedt, pp 31–76
Woizeschke P, Wottschel V (2013) Recent developments for laser beam joining of CFRP-aluminum structures. materials science engineering, symposium B6—hybrid structures. Procedia Mater Sci 2:250–258. https://doi.org/10.1016/j.mspro.2013.02.031
Woizeschke P, Mosgowoi E, Vollertsen F (2015) Decreasing pore formation in multiple-sheet laser joining with interfacial polymeric contaminations. Weld World 59(5):683–692. https://doi.org/10.1007/s40194-015-0244-1
Woizeschke P, Kügler H, Vollertsen F (2016a) Bewerten und Vergleichen der Spaltüberbrückbarkeit unterschiedlicher thermischer Fügeverfahren—Der Benchmark-Spalt. Der Praktiker 8:356–358
Woizeschke P, Schultz V, Messaoudia H et al (2016b) Laser processing of lightweight aircraft structures. (Keynote). In: Proceedings of the 84th laser materials processing conference, Japan Laser Processing Society (JLPS), pp 21–26
Woizeschke P, Radel T, Nicolay P et al (2017) Laser deep penetration welding of an aluminum alloy with simultaneously applied vibrations. Lasers Manuf Mater Process 4(1):1–12. https://doi.org/10.1007/s40516-016-0032-9
Xie J (2002) Dual beam laser welding. Weld J 81(10):223–230
Yamaguchi M, Umakoshi Y, Yamane T (1987) Plastic deformation of the intermetallic compound Al3Ti. Philos Mag A 55(3):301–315. https://doi.org/10.1080/01418618708209869

Yamaguchi M, Inui H, Ito K (2000) High-temperature structural intermetallics. Acta Mater 48(1):307–322. https://doi.org/10.1016/S1359-6454(99)00301-8

Yang YP, Dong P, Zhang J et al (2000) A hot-cracking mitigation technique for welding high-strength aluminum alloy. Weld Res Suppl 9:17

Young TM, Humphreys B, Fielding JP (2001) Investigation of hybrid laminar flow control (HLFC) surfaces. Aircr Des 4(2–3):127–146. https://doi.org/10.1016/S1369-8869(01)00010-6

Zerner I (2003) Sitzschiene(DE10360807B4). Accessed 18 Oct 2016

Zhang J, Weckman DC, Zhou Y (2008) Effects of temporal pulse shaping on cracking susceptibility of 6061-T6 aluminum Nd:YAG laser welds. Weld Res 87:18–30

Zhang X, He X, Xing B et al (2016) Influence of heat treatment on fatigue performances for self-piercing riveting similar and dissimilar titanium, aluminium and copper alloys. Mater Des 97:108–117. https://doi.org/10.1016/j.matdes.2016.02.075

Zhao H, DebRoy T (2003) Macroporosity free aluminum alloy weldments through numerical simulation of keyhole mode laser welding. J Appl Phys 93(12):10089–10096. https://doi.org/10.1063/1.1573732

Zhu Z, Lee K, Wang X (2012) Ultrasonic welding of dissimilar metals, AA6061 and Ti6Al4V. Int J Adv Manuf Technol 59(5–8):569–574. https://doi.org/10.1007/s00170-011-3534-9

Zimmermann S, Specht U, Spieß L et al (2012) Improved adhesion at titanium surfaces via laser-induced surface oxidation and roughening. Mater Sci Eng A 558(0):755–760. https://doi.org/10.1016/j.msea.2012.08.101

Chapter 12
Adhesive Bonding of Aircraft Structures

Konstantinos Tserpes

1 Introduction

Joints are the most critical parts in aircraft structures. Most structural failures in aircrafts have started from joints. The ability of the aeronautical engineers to apply new conceptual designs and to use new materials depends highly on the ability to efficiently design joints. Adhesive bonding finds an increasing use in aircrafts in the last 20 years which is proportional to the increasing use of carbon-fiber-reinforced plastic (CFRP) materials. The reason for the increasing use of adhesive bonding is the numerous advantages they offer over conventional mechanical joining techniques such as:

- Ability to join dissimilar materials such as composites with metals and thermosets with thermoplastics
- Fast and cheap joining process
- Sealing property (adhesive fills gaps and voids)
- Aerodynamically smoother surfaces
- Eliminates galvanic corrosion
- Uniform distribution of mechanical stress
- Better vibration damping behavior
- Better fatigue resistance

On the other hand, adhesive bonding possesses many disadvantages which limit its use and, in some cases, prohibit certification such as:

- Joint strength depends highly on the surface preparation of the adherends and the presence of defects in the bondline.

K. Tserpes (✉)
Laboratory of Technology and Strength of Materials, Department of Mechanical Engineering and Aeronautics, University of Patras, Patras, Greece
e-mail: kitserpes@upatras.gr

S. Pantelakis, K. Tserpes (eds.), *Revolutionizing Aircraft Materials and Processes*, https://doi.org/10.1007/978-3-030-35346-9_12

- Need for redesign of the parts to be assembled.
- Difficult disassembly of the joined parts.
- Durability limitations.

In the present chapter, the evolution of adhesive bonding technology in aircraft structures is described for both metallic and composite structures. The chapter is organized as follows. In Sect. 2, the evolution of use of adhesive bonding in aircraft structures is briefly described. Section 3 lists the basic adhesive materials along with their main properties and application areas. Section 4 presents the basics of the bonding process by focusing on surface preparation and quality assurance. Section 5 discusses the main defects that may appear in the bondline and describes the conventional and the extended NDT methods applied to adhesive joints. Section 6 discusses the basics of the mechanical behavior of bonded joints and the main destructive characterization methods. Finally, Sect. 7 discusses some design aspects of bonded joints and presents the status of the research conducted in Europe toward certification of adhesive bonding for primary composite structures.

Although adhesive bonding is extensively used also for patch repairing of aircraft structures, the present chapter is concentrated on joining applications since the repair applications are extensively discussed in Chap. 13 of the book.

2 The Evolution of Adhesive Bonding in Aircraft Structures

The historical record of the use of adhesive bonding in aircrafts until 1991 is given in Table 12.1 (Higgins 2000). The first use of adhesives is traced back in 1945 for DeHavilland Dove aircraft in which the Redux 775 adhesive was used in liquid and powder form. The development was continued for DeHavilland aircrafts for the same adhesive in the same form and for Fokker aircrafts for the same adhesive in liquid form, powder form, and film form. In these aircrafts, adhesive bonding was used for making primary aircraft structural parts. In 1963, hot cure epoxy adhesives began to be used in Boeing aircraft 727, while in 1972, the first Airbus aircraft, the A300, also used the hot cure epoxy adhesive Cytec FM123-2.

Since 1963, adhesive bonding finds an increasing use in the assembly of secondary aircraft structural parts showing a parallel course with the development of composite materials. The photo of Fig. 12.1 showing the structural parts of the Airbus A380 containing bonded joints is indicative of the present status of the extensive use of adhesive bonding in airliners. A characteristic example revealing the potential of adhesive bonding technology is the experimental Lockheed Martin X-55 Advanced Composite Cargo Aircraft (ACCA). The fuselage of the MX-55 consists of two large composite parts (upper and lower skins) bonded to circular frames.

Table 12.1 Historical record of the use of adhesives in airliners from 1945 to 1991 (Higgins 2000)

Aircraft	Year of first flight	Adhesive
Liquid and powder system		
DeHavilland Dove	1945	Redux 775
Vickers Viscount	1948	Redux 775
DeHavilland Heron	1950	Redux 775
DeHavilland Comet	1951	Redux 775
Fokker F27 Friendship	1955	Redux 775
Vickers Vanguard	1959	Redux 775
Vickers VC-10	1962	Redux 775
Hawker Siddeley Trident	1962	Redux 775
Fokker F28 Friendship	1967	Redux 775
Fokker 50 TurboProp	1985	Redux 775
Fokker 100 Regional Jet	1988	Redux 775
Film system		
DeHavilland 125	1962	Redux 775
Hawker Siddeley Trident	1962	Redux 775
Hawker Siddeley 146	1975	Redux 775
British Aerospace RJ series	1989	Redux 775
Fokker 50 Turboprop	1995	Redux 775
Fokker 100 Regional Jet	1995	Redux 775
Hot cure epoxy		
Boeing 727	1963	Cytec FM1000
Boeing 737	1967	Cytec FM1000
BAe Jetstream 31	1967	Cytec FM1000
BAe Jetstream 31	1982	Hexcel Redux 308A
BAe Jetstream 31	1991	3M AF163-2
BAe Jetstream 41	1991	3M AF163-2
SAAB 340	1983	Cytec FM73
Airbus A300	1972	Cytec FM123-2
Airbus A300	1972	Cytec FM123-5
Airbus A300	1972	3M AF126
Airbus A300	1982	Cytec FM73
Airbus A300	1991	3M AF163-2
Airbus A310	1982	Cytec FM73
Airbus A310	1991	3M AF163-2

3 Adhesive Materials

An adhesive is considered as *structural* when it withstands at least a stress of 6.9 MPa in a single-lap shear test. However, for aeronautical applications, additional criteria are applied as adhesives function in specialized conditions. In aircrafts, adhesives are applied in the form of either a paste or a film. Paste adhesives with high viscosity tend to form thicker bondlines and, therefore, fill and bridge

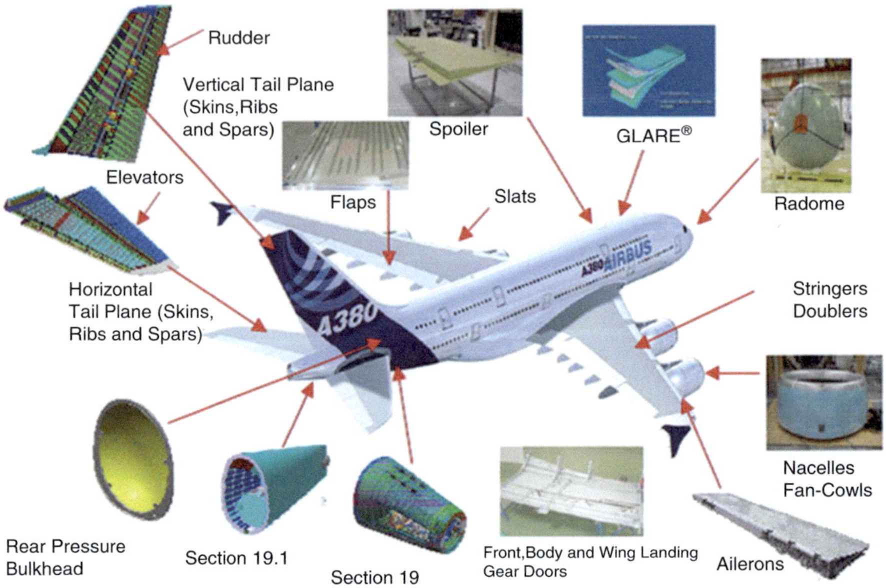

Fig. 12.1 Photo showing the structural parts of the Airbus A380 aircraft which contain bonded joints

gaps between bonded surfaces and can offer significantly greater elongation and impact resistance. On the other hand, film adhesives are used for bonding structural parts of large areas. Paste adhesives are more suitable for metallic adherends while film adhesives for composite adherends. When film adhesives are applied to "green" (B-staged) composites and co-cured, the adhesive dissolves molecules into the adherend, forming covalent chemical bonds that chemically lock the two together. Furthermore, adhesives also can attach a cured part, such as a stringer, to an uncured part. They also are used in a secondary bonding process, where two cured composite parts, or a composite and metal part, are joined.

In the following, a list of the basic adhesive materials together with their main properties and applications is given:

Anaerobics They cure when in contact with metal, and the air is excluded. They are often known as "locking compounds," being used to secure, seal, and retain turned, threaded, or similarly close-fitting parts. They are based on synthetic acrylic resins.

Cyanoacrylates They cure through reaction with moisture held on the surface to be bonded. They are suited to small plastic parts and to rubber. They are a special type of acrylic resin.

Toughened acrylics Toughened acrylics are fast curing and offer high strength and toughness. Both one-part and two-part systems are available. They tolerate minimal surface preparation and bond well to a wide range of materials.

Polyurethanes They are chemically reactive formulations which may be one-part or two-part systems and are usually fast curing. They provide strong impact-resistant joints and have better low-temperature strength than any other adhesive. Polyurethanes are useful for bonding glass-fiber-reinforced plastics.

Silicones Silicones are not very strong adhesives but are known for their flexibility and high temperature resistance. They are available in one- or two-part forms. The latter functions like the two-part epoxies and the former like the one-part polyurethanes. When the one-part adhesives cure, they liberate either alcohol or acetic acid (the familiar smell of vinegar). They are often used as sealants. Their adhesion to surfaces is only fair, but like their flexibility, their durability is excellent.

Phenolics Phenolics were the first adhesives for metals and have a long history of successful use for joining metal to metal and metal to wood. They require heat and pressure for the curing process.

Polyimides Polyimides are mainly used in applications which exploit their ability to withstand temperature up to 350 °C. They are available as liquids or films, and although they have good strength retention at high temperature, they have the disadvantage of being expensive and difficult to handle.

Hot melts Hot melts are based on modern thermoplastics and are used for fast assembly of structures designed to be only lightly loaded.

Plastisols Plastisols are modified PVC dispersions which require heat to harden. The resultant joints are often resilient and tough. Rubber adhesives are based on solutions of latexes and solidify through loss of the solvent medium. They are not suitable for sustained loadings.

Pressure-sensitive adhesives Pressure-sensitive adhesives are suitable for use as tapes and labels and, although they do not solidify, are often able to withstand adverse environments. This type of adhesive is not suitable for sustained loadings.

Epoxies They consist of an epoxy resin plus a hardener. They allow great versatility in formulation since there are many resins and many different hardeners. Epoxy adhesives can be used to join most materials. Epoxies have good strength, do not produce volatiles during curing, and have low shrinkage. However, they can have low peel strength and flexibility and are brittle. Epoxy adhesives are available in one-part, two-part, and film form and produce extremely strong durable bonds with most materials. Epoxy adhesives dominate adhesive bonding in aircraft structures.

By exploiting the achievements of nanotechnology, and especially the extraordinary physical properties of carbon nanotubes, researchers are working on the development of nanofilled adhesives with enhanced mechanical, thermal, and electrical properties. The first experimental results from this investigation are very encouraging (Wernik and Meguid 2014; Zielecki et al. 2017); however, much remains to be done, especially in the processing of the materials, for the specific technology to enter production.

4 Bonding Process

The bonding process comprises the steps of surface treatment, curing, and quality assurance. All operations must be performed by qualified personnel. All applications of adhesive are conducted in controlled areas which do not contain dust, oily vapors, aerosols, or other contaminants deleterious to bonding. The temperature in the controlled areas must range from 18.3 to 29.4 °C, and the relative humidity must be lower than 60%. The equipment used for the curing of the adhesive includes presses, bond forms, autoclaves, ovens, and pressure equipment. The details and procedures to be followed are included in the process specification. Each bonding tool shall be qualified prior to production.

4.1 Processing and Equipment

Bulk adhesives such as paste adhesives are the simplest and most reproducible adhesives to apply. They produce heavy coatings that fill voids, bridge gaps, or seal joints. However, the process is limited to high-viscosity materials. They can be in the form of either a high-viscosity extrudable liquid or a trowelable mastic. These systems can be troweled on or extruded through a caulking gun.

Film adhesives offer a clean, hazard-free operation with minimum waste and excellent control of film thickness. However, the method is generally limited to parts with flat surfaces or simple curves. Optimum bond strength requires curing under heat and pressure, which may involve considerable equipment and floor space, particularly for large parts.

The use of dry adhesive films is progressing more rapidly than the other forms because of their following advantages (Petrie 1999):

1. High repeatability—no mixing or metering, constant thickness.
2. Easy to handle—low equipment cost, relatively hardware-free, and clean operating.
3. Very little waste—preforms can be cut to size.
4. Excellent physical properties—wide variety of adhesive types available.

Pressure devices should be designed to maintain constant pressure on the bond during the entire cure cycle. They must compensate for bondline thickness reduction from adhesive flow or thermal expansion of assembly parts. Large bonded areas, such as aircraft parts, are usually cured in an autoclave. The parts are mated first and covered with a rubber blanket to provide uniform pressure distribution. The assembly is then placed in an autoclave, which can be pressurized and heated. Vacuum-bagging techniques can be a less expensive method of applying pressure to large parts. A film or plastic bag is used to enclose the assembly, and the edges of the film are sealed airtight. A vacuum is drawn on the bag, enabling atmospheric pressure to force the adherends together. Vacuum bags are especially effective on

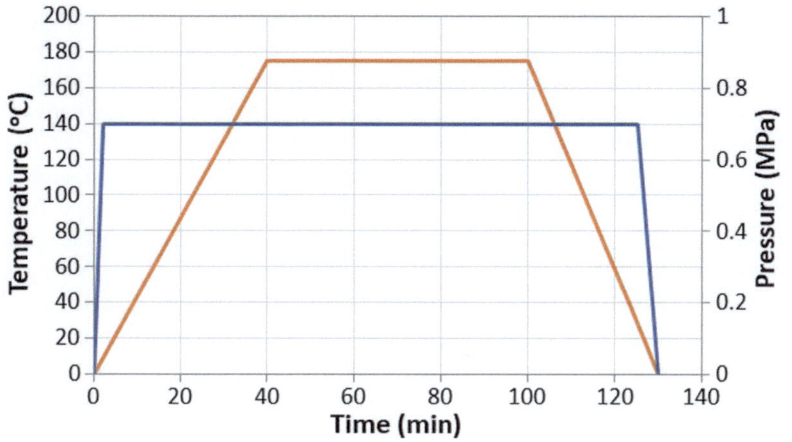

Fig. 12.2 Typical curing cycle of the FM 300 adhesive

large areas because the size is not limited by pressure equipment. Pressures, of course, are limited to atmospheric pressure. Many structural adhesives require heat as well as pressure to cure. Even with conventional room-temperature curing systems, most often the strongest bonds are achieved by an elevated temperature cure. With many adhesives, trade-offs between cure times and temperature are permissible. Generally, the manufacturer will recommend a certain curing schedule for optimum properties. Figure 12.2 shows a typical curing cycle of the widely used adhesive FM 300 of Cytec.

4.2 Surface Treatment

The surface treatment determines the strength of the bonded joint to a large extent. Provided the chosen adhesive can withstand the loads, the life of the bond is directly proportional to the degree of surface treatment. Surface treatment differs for the different adherends (Davis 2011).

Aluminum surfaces are usually prepared for adhesive bonding in aerospace applications either by etching or by anodization in acid solutions. The two widely used treatments are phosphoric acid anodization and chromic acid anodization. In addition, sol–gel or grit blasting/silane coupling agent treatments are promising as environmentally acceptable processes that can be used for repair.

The desire to use adhesively bonded titanium structures at elevated temperatures has been a driving force in the development of high-temperature adhesives (those curing at temperatures up to 400 °C). Like Al, durable surface preparations for titanium can be achieved by forming oxides in anodizing and/or etching solutions. For high-temperature applications, alternative surface treatments are necessary. Sol–gel processes similar to those for aluminum are the most promising. Another high-temperature-compatible surface treatment is a microrough titanium coating deposited by plasma spraying.

In the manufacturing of CFRP parts, peel plies, release films, and mold release agents are used for easy de-molding and surface consistence. The residues of films and plies on the CFRP surfaces have been found to be responsible for bondline failures and insufficient mechanical performance. To deal with this, manual grinding processes have been introduced in aerospace industry for surface treatment before bonding. These processes can remove different amounts of surface contamination and are still standard treatment procedures today. Manual grinding procedures depend to a high extent on the worker's skills; they are very difficult to control, particularly in the case of complex parts; they are very time-consuming and create a lot of grinding dust emissions in the workshop. For introduction of automated adhesive bonding processes, alternative surface treatments are needed to enable a high, reproducible quality standard. Therefore, alternative CFRP surface treatments like atmospheric plasma, low-pressure plasma, automated grit blasting, and laser treatment are meanwhile available.

The use of atmospheric plasma as an alternative treatment method for adhesive bonding is meanwhile state-of-the-art for plastics and metals. For CFRP bonding, the objective is to remove or activate surface contaminations coming from peel plies, release films, and release agents, used in the fabrication for composites (Fig. 12.3).

Plasma as an ionized gas can be created thermally or by an electrical field, among others (Wachinger et al. 2009). In a low-pressure plasma (LPP) equipment, the electrons are accelerated in an electrical field. The energy created inside the ionized gas

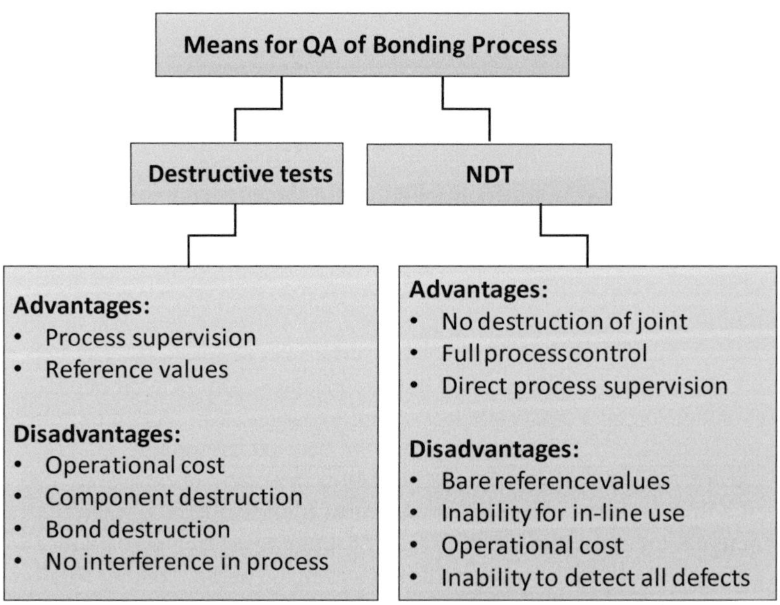

Fig. 12.3 Quality assurance concept in industrial process (Michaloudaki 2005)

can break any chemical bond of organic molecules. The temperature of the treated material however remains at room temperature. The process is flexible and can be configured through variation of the process parameters such as gas flow, gas composition, pressure, and power. In contrast to atmospheric plasma, a plasma chamber and vacuum equipment are necessary.

A meanwhile qualified procedure in the aerospace is the use of a carbon dioxide laser for the removal/activation of release agent residues on CFRP surfaces before painting (Wachinger et al. 2009). The application of the same laser technique, as a surface treatment method for structural adhesive bonding, is still under investigation. The degree of treatment (removal of surface contaminations like siloxanes and hydrocarbons, removal of resin system layer like epoxies, and attack of the carbon fibers) depends very much on the laser systems and treatment conditions.

Grit blasting applied as surface treatment of CFRP for subsequent structural adhesive bonding shall evenly abrade the surface comparable to the grinding (sanding) procedure on flat surfaces, thereby offering higher feasibility for automation and applicability on surfaces with considerable waviness in the submillimeter range without severe damage of the composites fibers (Wachinger et al. 2009).

4.3 Quality Assurance

According to DIN EN ISO 9001, adhesive bonding technology is a "special process," i.e., the quality of the bonded joints must be assured by use of specific methods and procedures. Quality assurance comprises suitable process monitoring, destructive tests on random specimens, and also nondestructive testing (NDT). The advantages and disadvantages of destructive and nondestructive testing methods are listed in Fig. 12.3 (Michaloudaki 2005).

In the present chapter, two sections are devoted to the two characterization approaches: Sect. 5 to the nondestructive methods and Sect. 6 to the destructive methods.

5 NDT Characterization

5.1 Current Practice

5.1.1 Pre-bond NDT

After surface treatment, the wettability of the adherends' surface is usually tested through measurements of the contact angle. This test involves the wetting of clean surfaces and the drop of water on them. Afterward, the spread of a liquid drop of known volume is measured by a transparent gauge placed over the drop. A more sophisticated device is the Fokker contamination tester which uses an oscillating probe to measure the electron emission energy of the surface (Adams and Drinkwater 1997). This method is not reliable which brings the need for better surface treatment and for developing new surface analysis methods.

5.1.2 After-Bond NDT

Conventional NDT methods for the quality assessment of the bondline comprise (Adams and Drinkwater 1997) conventional ultrasonics, oblique incidence ultrasonics, lamb waves, sonic vibrations, spectroscopic methods, acoustic emission, thermal methods, radiography, and optical holography. In recent years, advanced techniques such as the neutron radiography and X-ray computed tomography are also used. The latter two methods are not considered the most suitable for industrial use because their cost and test time are not often justifiable by their sensitivity.

Ultrasonics is the most widely used after-bond NDT method. Ultrasound inspection methods for bonds can be classified into two categories: bulk wave-based and lamb wave-based methods. Typical applications are the "pulse-echo" and the "through-transmission" techniques. The properties of the wave propagating through the specimens are monitored. Defects may affect these properties. For example, voids can scatter the wave, reducing the amplitude of the signal reaching the receiver. Delaminations of reasonable size, i.e., comparable to the size of the ultrasonic beam, can even stop the wave creating a strong ultrasound contrast.

The "pulse-echo" and the "through-transmission" techniques are commonly used for performing C-scans. C-scans are performed by measuring point by point the whole area to be inspected. The final result is an overview of how the wave properties change across the inspected region, which can be correlated to the specimen properties. Tserpes et al. (2011a, b, 2012) have detected defects in a Pi-shaped bonded joint using C-scan (Fig. 12.4) and correlated the C-scan signal with a stiff-

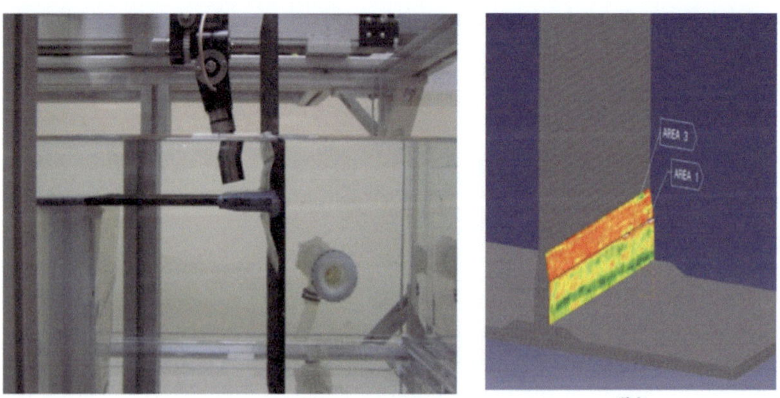

(a) (b)

Fig. 12.4 (**a**) C-scan experiment on a Pi-shaped bonded joint on the go; (**b**) C-scan image of two scan areas ness degradation scheme of the adhesive. Most common techniques use time of arrival, phase speed, and signal amplitude. Although time-consuming, C-scans are the most common and reliable technique used in the industry.

The industrial potential of ultrasound has been considerably enhanced in recent years due to the development of novel algorithms for analysis and new test equipment such as phased array sensors and equipment for portable testing such as the Fokker Bondtester 90 which is a highly flexible tool used in laboratories as well as in the field. The Fokker Bondtester 90 uses a measuring technique based on the ultrasonic resonance-impedance principle.

5.2 Extended NDT

In recent years, extended NDT (ENDT) techniques both for pre-bond and after-bond testing of bonded joints are under development. In this area, significant progress has been made by European projects ENCOMB (2014) and ComBoNDT (2015).

The objective of the ENCOMB project was the development and adaptation of ENDT methods for pre- and post-bond inspection of CFRP aircraft structural components. State-of-the-art ENDT techniques have been screened, and the most suitable ones have been taken forward for development and adaptation. As part of this process, five contamination scenarios have been identified as of primary importance for aircraft manufacturers, namely, release agent, moisture, Skydrol, thermal degradation, and imperfect curing of the adhesive. For the characterization of CFRP adherend surfaces, 14 ENDT technologies have been tested (Table 12.2). The results have shown that (1) the hydraulic fluid/water contamination is detectable by seven technologies, (2) release agent contamination by four technologies, (3) moisture in CFRP adherends by seven technologies, and (4) heat damage of adherends by four technologies. Several techniques show good results for the detection of different contamination levels. For the characterization of CFRP adhesive bonds, nine ENDT technologies have been tested (Table 12.2). The results show that (1) weak bond due to release agent is detectable by five technologies, (2) weak bond due to moisture by three technologies, and (3) poor curing of the adhesive by four technologies. Several techniques show good results for the detection of different contamination levels. A summary of the results of each ENDT method can be found in ENCOMB (2014).

The objective of the ComBoNDT project was to enhance the most promising quality assurance concepts of ENCOMB and advance them to a technology readiness level (TRL) of five to six. The ComBoNDT project was successful, but no details will be given here since most of its results are not published yet. The main conclusion of ComBoNDT is that many technologies have managed to reach the target TRL; however, a lot of work still has to be done in order for the methods to meet the requirements of the industry.

Table 12.2 The ENDT techniques investigated in the ENCOMB project

Pre-bond NDT testing techniques	Post-bond NDT testing techniques
Optically stimulated electron emission (OSEE)	Nonlinear ultrasound
Laser scanning vibrometry	Laser Shock Adhesion Test (LASAT)
Infrared (IR) spectroscopy	THz/GHz reflectometry
Aerosol wetting test	Laser ultrasound
Portable handheld FTIR spectroscopy	Laser scanning vibrometry
Laser-induced breakdown spectroscopy	Electromechanical impedance
THz/GHz polarization-frequency reflectometry (PFR)	Ultrasonic frequency analysis
THz/GHz quasi-optical method of internal reflection reflectometry	Industrial laser ultrasound
Optical fiber sensors	Active thermography using optical excitation
Electronic nose technology (E-nose)	
Dual-band active thermography	
Laser-induced fluorescence	
THz technology	
Optical coherence tomography	

6 Destructive Characterization

Destructive characterization of adhesive joints is a difficult task because the involved failure mechanisms are complicated. The task is even more difficult for composite adherends since quite often failure in the bondline interacts with failure in the surface layers. The basic failure modes of adhesive joints are schematically described in Fig. 12.5. In most cases, the "desired" failure mode is cohesive failure although the failure mode is not always representative of the joint's strength.

Different types of destructive tests are used for characterizing the quality of bonded joints. A very good description of existing methods is given in da Silva et al. (2012). The most commonly used tests are those used in fracture mechanics for characterizing mode I, mode II, and mixed-mode fracture toughness of materials (Floros et al. 2015). The basic test used by the aircraft industry is the mode I fracture toughness test on double cantilever beam (DCB) specimens (Fig. 12.6). This choice is in contradiction with design requirements since bonded joints are mainly designed to be loaded in shear, and therefore, mode II and mode III fracture toughness tests should be conducted.

Despite the wide acceptance of fracture toughness tests, they are very expensive and time-consuming and, in some cases, such as defected bonded joints and adhesive CFRP joints, do not give representative results for the joint's strength since the crack is growing within a complex material system (failure jumping from one interface to the other and adherend failure). The problem is enhanced by the fact that the existing standards have been developed for fracture toughness tests of metals. In Markatos et al. (2013, 2014), Tserpes et al. (2014), and Pantelakis and Tserpes (2014), the authors have investigated the effect of different pre-bond contaminations

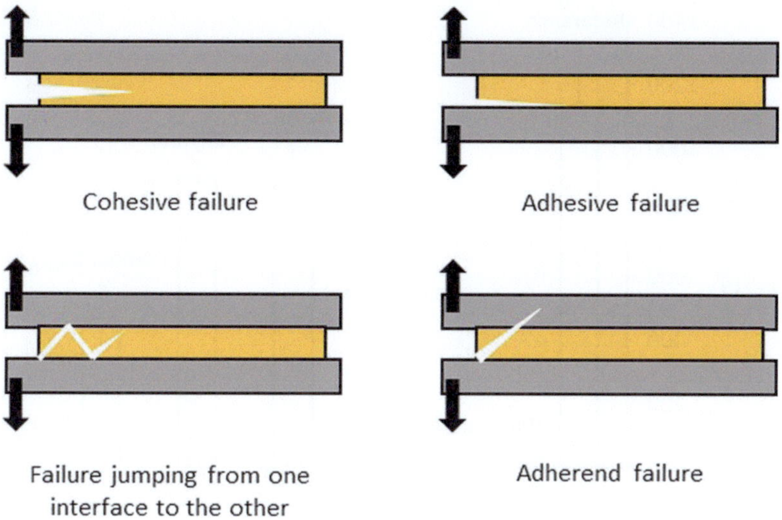

Fig. 12.5 Schematic representation of basic failure modes in bonded joints

Fig. 12.6 DCB specimen (adhesively bonded CFRP) during a Mode-I fracture toughness test

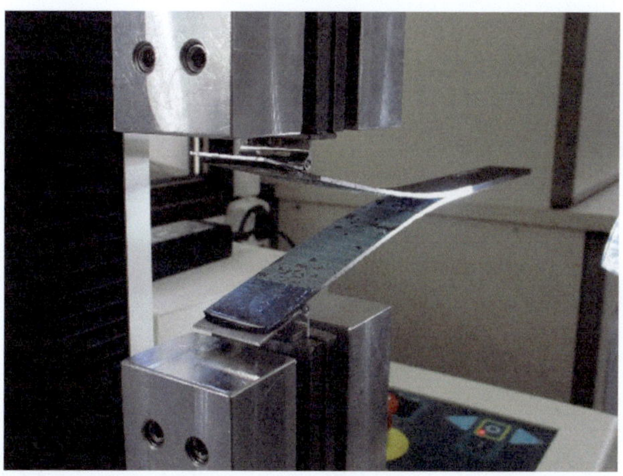

on the critical energy release rate G_{IC} of CFRP joints using DCB tests. As shown in Fig. 12.7, the DCB tests, although proved capable of capturing the effect of the different contaminations, show a large standard deviation, which in some cases might hinder the evaluation of the results. This is exactly the case for the recent investigation on the effect of deicing fluid as can be seen in Fig. 12.8 (Moutsompegka et al. 2017). Another important finding reported in Moutsompegka et al. (2017) is the much bigger effect of deicing fluid on the G_{IIC} than on G_{IC} (Fig. 12.9) which supports the need for re-examination of industry's decision to fully rely on DCB tests.

A faster and less expensive test is the single-lap shear (SLS) test. The SLS is mainly used for a fast and not a detailed evaluation of bonded joints since the devel-

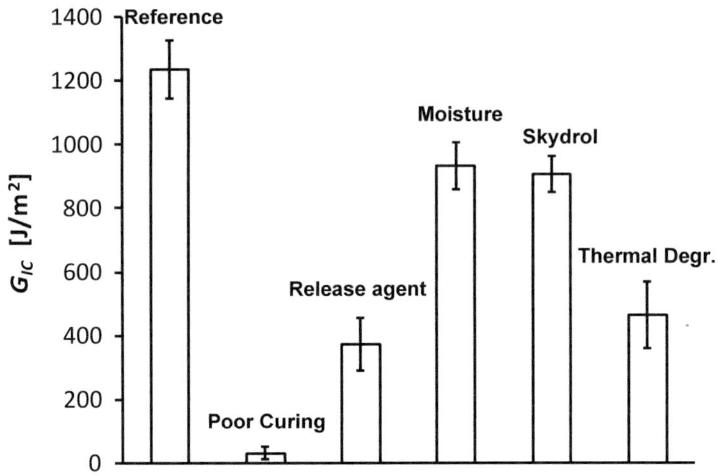

Fig. 12.7 Average crack initiation GIC values for the six contamination scenarios (Markatos et al. 2013)

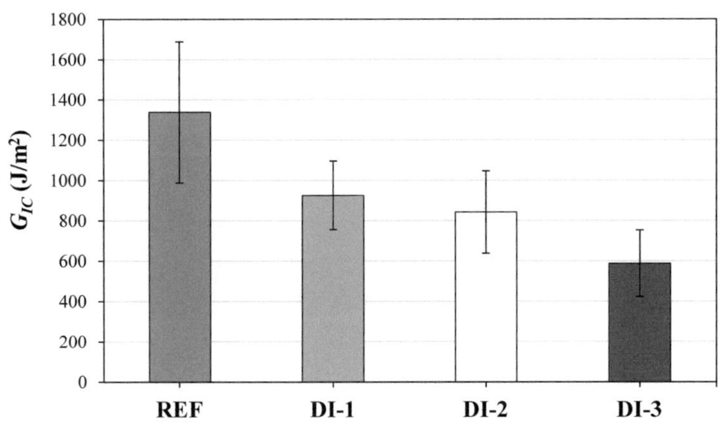

Fig. 12.8 Comparison of GIC values between the reference adhesive CFRP joint and the three levels of deicing contamination (Moutsompegka et al. 2017)

oped loading conditions are not representative of the loading conditions that appear in aircraft structures. For instance, the SLS test has been used with success in Katsiropoulos et al. (2012) for a first assessment of the differences between two adhesive materials under static and fatigue loading conditions.

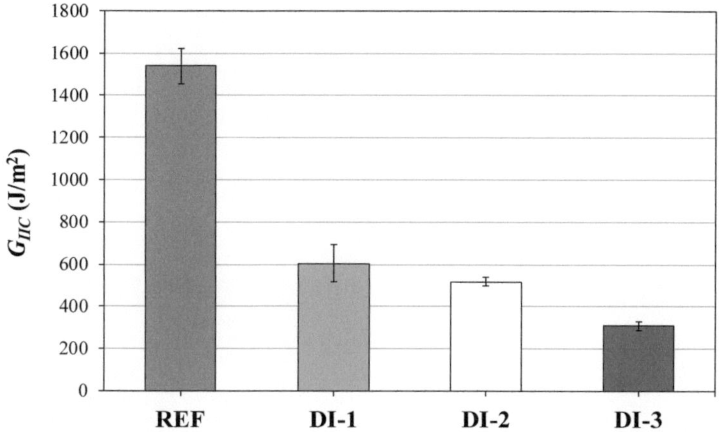

Fig. 12.9 Comparison of G_{IIC} values between the reference adhesive CFRP joint and the three levels of deicing contamination (Moutsompegka et al. 2017)

7 Design Aspects and Certification

7.1 Modular Joining Elements

Implementation of adhesive joints requires, in some cases, redesign of the structural parts to be assembled as existing designs were selected so as to implement bolted connections. The main objective of the redesign of the structural parts is the load to be transferred between the assembled parts mainly through shear. Therefore, large adhesive areas loaded in shear must be created. A very useful tool in the application of bonded joints is multifunctional profiles of different shapes which act as joining elements between the assembled parts. The multifunctional profiles are made of textile composite material and can be of different shapes as shown in Fig. 12.10. An example of a bonded joint between a composite skin and a beam realized by means of a Pi-shaped profile is illustrated in Fig. 12.11.

The design of multifunctional profiles to ensure their own integrity as well as the integrity of the bondline is a major challenge. Large progress in the area was done in the European project MOJO (2011). The main outcome of the MOJO project (MOJO 2011) is the numerical optimization of the geometry of the profiles (Tserpes and Koumpias 2015) and the redesign of the flap-track beam of the Airbus A400M which was originally made from state-of-the-art metallic materials and advanced exclusively by mechanical fasteners (Fig. 12.12) Tserpes et al. (2011a, b). The objective of the redesign process was to fully replace metallic materials by advanced composites with enhanced through-the-thickness strength and to partially replace mechanical fasteners by adhesive bonds implemented through joining profiles. The new flap-track beam has been subjected to mechanical tests using realistic static and fatigue loading conditions. The experimental results showed that the composite

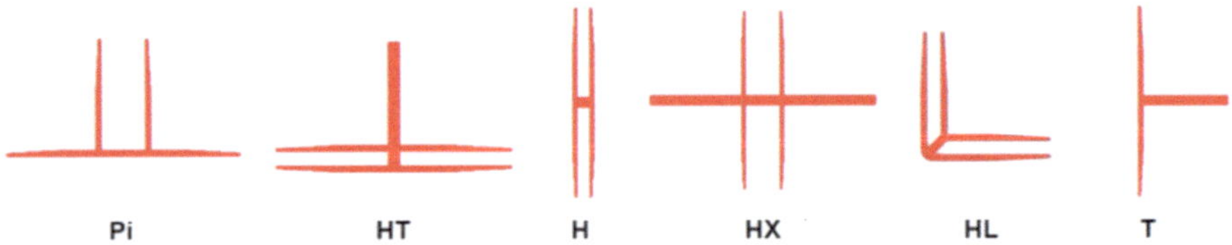

Fig. 12.10 Schematics of joining profiles

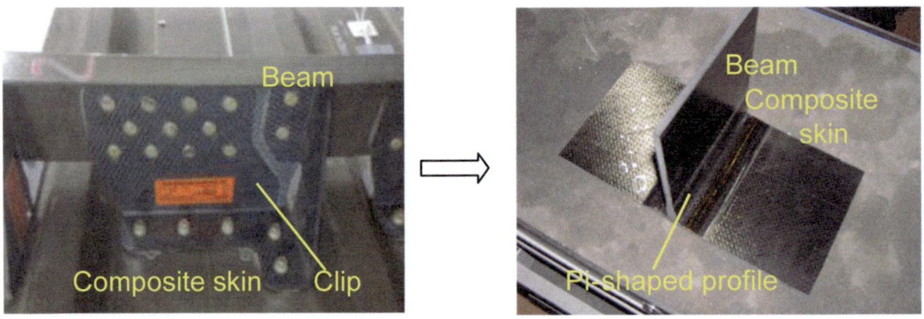

Fig. 12.11 Adhesive joining of a composite skin with a composite beam using a Pi-shaped profile

Fig. 12.12 The composite, partially-bonded, flap-track beam of the Airbus A400M (MOJO 2011)

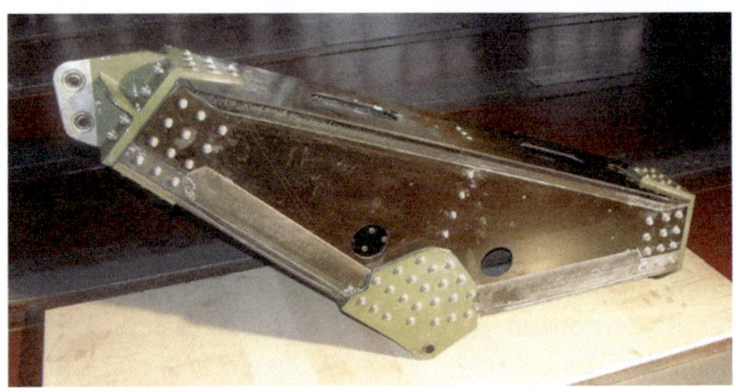

flap-track beam is capable to effectively carry the load which the original fastened metallic flap-track beam has been designed to carry. It is noted that the novel flap-track beam has won the JEC 2010 Process Innovation Award.

7.2 *Certification of Bonded Composite Primary Aircraft Structures*

While, for secondary structures, adhesive bonding is a common practice, certification rules that are applicable for primary bonded structures prevent the use of bolt-free bonded joints for primary structures, as a result of earlier experiences, where the interpretation of the rules led to in-service premature failure incidents on adhesively bonded joints (BOPACS 2017). However, it is well-known that the full cost and weight savings of composites cannot be realized until bonded joints can be certified without fasteners.

Regarding the existing certification requirements for bonded composite primary aircraft structures, the FAA's document 14 CFR § 23.573, which sets forth requirements for damage tolerance and fatigue evaluation of primary composite airframe structures, states that for any bonded joint, the failure of which would result in catastrophic loss of the airplane, the limit load capacity must be substantiated by one of the following methods:

(a) The maximum disbonds of each bonded joint consistent with the capability to withstand the loads must be determined by analysis, tests, or both. Disbonds of each bonded joint greater than this must be prevented by design features.
(b) Proof testing must be conducted on each production article that will apply the critical limit design load to each critical bonded joint.
(c) Repeatable and reliable nondestructive inspection techniques must be established that ensure the strength of each joint.

In the last 10 years, the aircraft industry is working intensively toward fulfillment of the Means of Comply. The work is performed in two axes: the development of ENDT methods capable of detecting kissing bonds which are not detectable at the moment by conventional NDT methods such as ultrasound and the development of design features capable of stopping crack growth in the bondline before becoming critical for the residual strength of the bonded part.

The progress in the area of the ENDT method development in Europe has been described in Sect. 5.2.

In the second axis, significant progress has been made through the European project BOPACS. The objectives of BOPACS were to develop new, efficient disbond-arresting design features and to gain a fundamental understanding of crack growth and crack arrestment in composite bonded joints. Figure 12.13 shows representative design features for the four main categories investigated in BOPACS by numerical simulation and mechanical tests. In Fig. 12.14, crack stopping due to corrugation in a DCB specimen is illustrated by means of the experimental and numerical load-displacement curves as well as by the predicted debonding progression. In general, the "through-the-thickness" features like the different types of bolts have given the best results followed by the "adhesive bondline architecturing" features where locally toughened adhesives were introduced in the bondline. The tests on a wide single-lap shear specimen showed that bolts used as crack stoppers are capable of slowing down the crack propagation speed of an initial damage in a high load transfer joint. The demonstrated crack-arresting capability for the tested configurations shows a very good performance which is well above the needed crack-arresting capability of typical aerospace configurations in terms of number of cycles to rupture with and without crack arrestors. Also, the residual strength after fatigue testing and arrested crack growth gives good confidence for limit load capability after crack arresting to fulfil today's certification requirements.

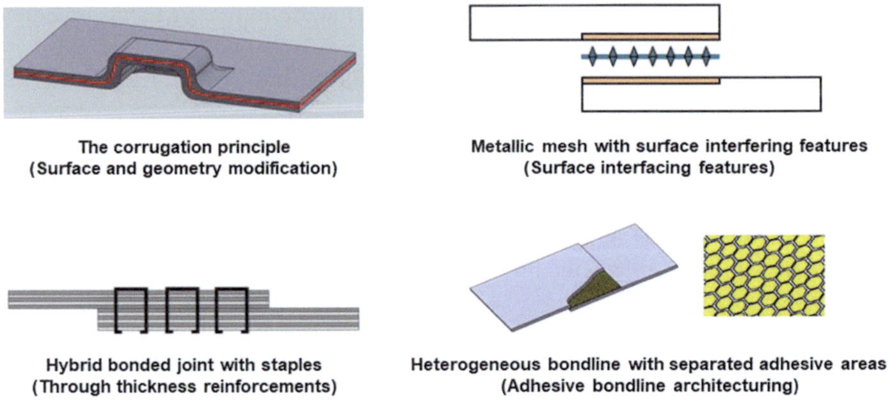

Fig. 12.13 The four categories of disbond-stopping design features investigated in the BOPACS project

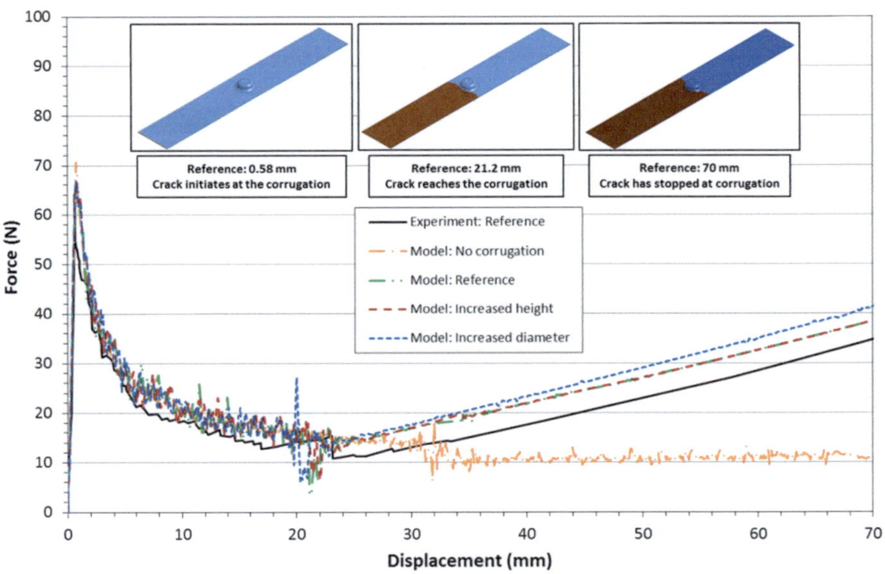

Fig. 12.14 Load-displacement curves and crack propagation in a DCB specimen with corrugation (Tserpes et al. 2016)

In conclusion, despite the significant progress that has been accomplished, we are still far from the certification of adhesive bonding for primary composite aircraft structures. Besides the fulfillment of the Means of Comply, progress needs to be made on other aspects such as the change in culture. For instance, there has to be assurance that we are capable to bond reliably and repeatably and proof that material suppliers can also follow the streamline of progress.

8 Conclusions and Discussion

The present chapter discusses the evolution, current practice, and future challenges of adhesive bonding technology in aircraft structures. The use of adhesive bonding in aircraft structures was started more than 80 years ago. For about seven decades, it remained limited; however, in the last 10 years, mainly with the production of the Boeing 787, the Airbus A380, and the Airbus A350, the use of adhesive bonding has been proportional to the use of CFRPs.

Adhesive materials, and mainly the epoxies, have evolved considerably regarding their mechanical properties. However, there is still a need for adhesives with higher Young's modulus, tensile strength, and mode I fracture toughness. Efforts in this direction comprise the improvement of chemical formulation as well as the reinforcement with nanofillers. Early experimental results on carbon nanotube-reinforced epoxy adhesives are very promising as they show a considerable increase of mechanical properties, the electrical and thermal conductivities. Very interesting developments are expected in this area in the forthcoming years.

The efficiency of the existing destructive characterization methods for bonded joints needs to be improved. A variety of mechanical tests exist; however, most of them have been developed for different materials or different material systems. At the same time, there is no complete study in the literature on the appropriateness of these tests. The standards for the widely used fracture toughness need to be modified to account for bonded joints and especially for composite adherends, while new, faster, and less expensive tests such as the centrifuge tests should be considered for future use. On the other hand, NDT methods play an important role on the quality assessment of bonded joints during manufacturing and maintenance. Yet, the existing NDT methods cannot characterize sufficiently neither the surface of the adherends before bonding nor the bondline after bonding. It is, therefore, of crucial importance for the ENDT methods currently under development to go into industrial use. Alongside independent development of destructive and nondestructive characterization methods, progress should be made on the correlation between them and the development of links between NDT methods and predictive models.

Certification of adhesive bonding will be a very important step in the evolution of aircraft structures because it will enable the realization of full cost and weight savings from the extensive use of CFRP materials in future aircrafts. To achieve this goal, investigation is performed toward the development of ENDT methods capable to ensure strength of each joint and the development of design features that will prevent the crack in the bondline from reaching a critical size. However, to be practical besides the technical progress, a validation of every aspect of aerocomposites' manufacturing cycle is required because each affects the joint performance. This full validation process involves the designers, the manufacturing equipment, its users, and also the material suppliers.

References

Adams RD, Drinkwater BW (1997) Nondestructive testing of adhesively-bonded joints. NDT & E Int 30:93–98

BOPACS (2017) (Boltless assembling Of Primary Aerospace Composite Structures) Project's Public Final Report, Project ID: 314180, Funded under: FP7 TRANSPORT

ComBoNDT (2015) (ComBoNDT—Quality assurance concepts for adhesive bonding of aircraft composite structures by advanced NDT) Grant Agreement No. 636494, Funder under H2020

da Silva LFM, Dillard DA, Blackman BRK, Adams RD (eds) (2012) Testing of adhesive joints. Wiley-VCH Verlag & Co, Weinheim

Davis GD (2011) Surface treatments of selected materials. In: da Silva LFM, Öchsner A, Adams RD (eds) Handbook of adhesion technology. Springer, Berlin. https://doi.org/10.1007/978-3-642-01169-6_8

ENCOMB (2014) (Extended Non-Destructive Testing of Composite Bonds) Project Final Report—Project ID: 266226. Funded under: FP7-TRANSPORT

Floros IS, Tserpes KI, Löbel T (2015) Mode-I, mode-II and mixed-mode I+ II fracture behavior of composite bonded joints: experimental characterization and numerical simulation. Compos Part B 78:459–468

Higgins A (2000) Adhesive bonding of aircraft structures. Int J Adhes Adhes 20:367–376

Katsiropoulos CV, Chamos AN, Tserpes KI, Pantelakis SG (2012) Fracture toughness and shear behavior of composite bonded joints based on a novel aerospace adhesive. Compos Part B 43:240–248

Markatos DN, Tserpes KI, Rau E, Markus S, Ehrhart B, Pantelakis S (2013) The effects of manufacturing-induced and in-service related bonding quality reduction on the mode-I fracture toughness of composite bonded joints for aeronautical applications. Compos Part B 45:556–564

Markatos DN, Tserpes KI, Rau E, Brune K, Pantelakis S (2014) Degradation of mode-I fracture toughness of CFRP bonded joints due to release agent and moisture pre-bond contamination. J Adhes 90:156–117

Michaloudaki M (2005) An approach to quality assurance of structural adhesive joints. Ph.D. Dissertation. Technical University of Munich

MOJO (2011) (Modular Joints for Aircraft Composite Structures) Final Report, Project ID: 30871, Funded under: FP6-AEROSPACE

Moutsompegka E, Tserpes KI, Polydoropoulou P, Tornow C, Schlag M, Brune K, Mayer B, Pantelakis S (2017) Experimental study of the effect of pre-bond contamination with de-icing fluid and ageing on the fracture toughness of composite bonded joints. Fatigue Fract Eng Mater Struct 40(10):1581–1591

Pantelakis S, Tserpes KI (2014) Adhesive bonding of composite aircraft structures: challenges and recent developments. Sci China Phys Mech Astron 57:2–11

Petrie E (1999) Handbook of adhesives and sealants. McGraw Hill Professional, New York

Tserpes KI, Koumpias AS (2015) A numerical methodology for optimizing the geometry of composite structural parts with regard to strength. Compos Part B 68:176–184

Tserpes KI, Pantelakis S, Kappatos V (2011a) The effect of imperfect bonding on the pull-out behavior of non-crimp fabric Pi-shaped joints. Comput Mater Sci 50:1372–1380

Tserpes KI, Ruzek R, Mezihorak R, Labeas GN, Pantelakis SG (2011b) The structural integrity of a novel composite adhesively bonded flap-track beam. Compos Struct 93:2049–2059

Tserpes KI, Ruzek R, Pantelakis S (2012) Strength of Pi shaped non-crimp fabric adhesively bonded joints. Plast Rubber Compos 41:100–106

Tserpes KI, Markatos DN, Brune K, Hoffmann M, Rau E, Pantelakis S (2014) A detailed experimental study of the effects of pre-bond contamination with a hydraulic fluid, thermal degradation, and poor curing on fracture toughness of composite bonded joints. J Adhes Sci Technol 28:1865–1880

Tserpes KI, Peikert G, Floros IS (2016) Crack stopping in composite adhesively bonded joints through corrugation. Theor Appl Fract Mech 83:152–157

Wachinger G, Thum C, Llopart L, Maier A, Wehlan H, Stöven T (2009) New trends in CFRP treatment and surface monitoring for automated structural adhesive bonding. In: International conference on composite materials ICCM17, Edinburgh
Wernik JM, Meguid SA (2014) On the mechanical characterization of carbon nanotube reinforced epoxy adhesives. Mater Des 59:19–32
Zielecki W, Kubit A, Trzepieciński T, Narkiewicz U, Czech Z (2017) Impact of multiwall carbon nanotubes on the fatigue strength of adhesive joints. Int J Adhes 73:16–21

Chapter 13
Bonded Repair of Composite Structures

Georgios Kanterakis, Roland Chemama, and Konstantinos Kitsianos

1 Introduction

Materials and structures employed in the aerospace industry and elsewhere may periodically require repair for in-service damage. Rapid inspection and repair of the aircraft are important for military and commercial applications in order to decrease downtime. Early detection of the damage and application of the suitable repair lead to significant extension of fatigue life and reduction of costs (Baker and Jones 1988; Marioli-Riga et al. 1997; Tsamasphyros et al. 1999, 2003; Armstrong et al. 2005; Christopoulos et al. 2016; SAE AIR5144 2017).

The use of composite materials is continuously increasing in aero structures and other components in aircraft like Boeing 787 and Airbus 350 (see figure below), which leads to the conclusion that the MRO (Maintenance–Repair-Overhaul) market for structural composite repairs is expected to have significant growth in the future, as shown in Fig. 13.1.

The next 20 years will see introduction of aircraft having more than 55% of composite structures.

Figure 13.2 is absolutely enlightening concerning the "enemies" surrounding each aircraft during servicing, with a proven potential of causing damage to its structure (fuselage, wings, etc.). "Aircraft feel much more comfortable when flying, as there are many threats for their structural integrity when on ground." Of course, this does not take into consideration dangers in the air, like bird strikes, lightning, and hail, as shown in Fig. 13.3.

All repairs should be conducted per the specific instructions outlined in the structural repair manual (SRM) for the aircrafts. For military aircraft, these have the form of technical orders (TOs), technical manuals, or equivalent. These manuals are pre-

G. Kanterakis (✉) · R. Chemama · K. Kitsianos
GMI Aero, Paris, France
e-mail: george@gmi-aero.com

S. Pantelakis, K. Tserpes (eds.), *Revolutionizing Aircraft Materials and Processes*, https://doi.org/10.1007/978-3-030-35346-9_13

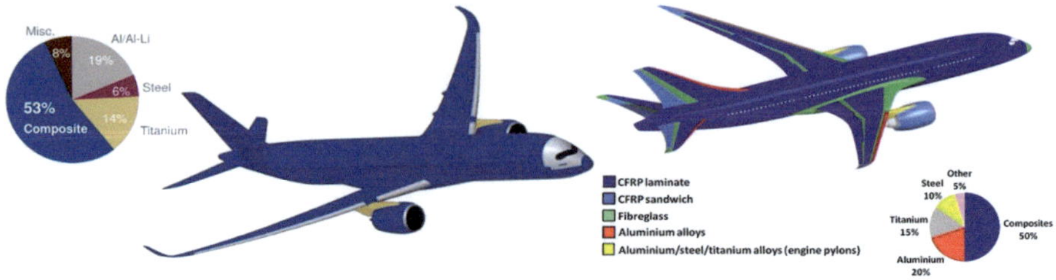

Fig. 13.1 Use of composite materials in modern aircraft

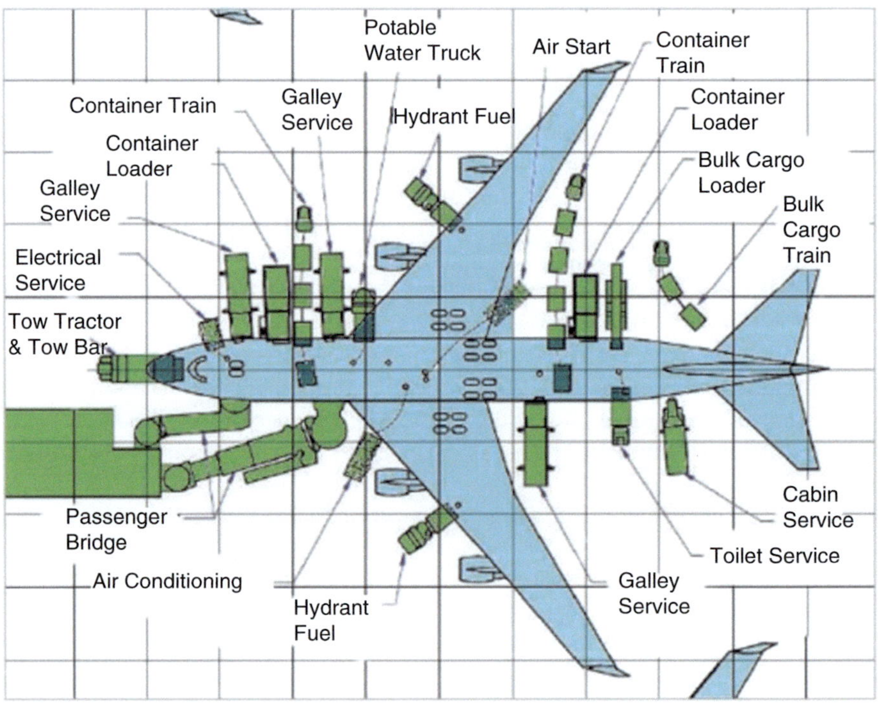

Fig. 13.2 Usual "ground threats" to aircraft structural integrity

pared by aircraft/helicopter manufacturer while being accepted and certified by the European Aviation Safety Agency (EASA), or equivalent authority, for commercial aircrafts. If the damage falls outside the allowable repair criteria, then either the part must be replaced or a specific repair must be designed for that particular component and damage level by an engineer specifically trained and qualified to design com-

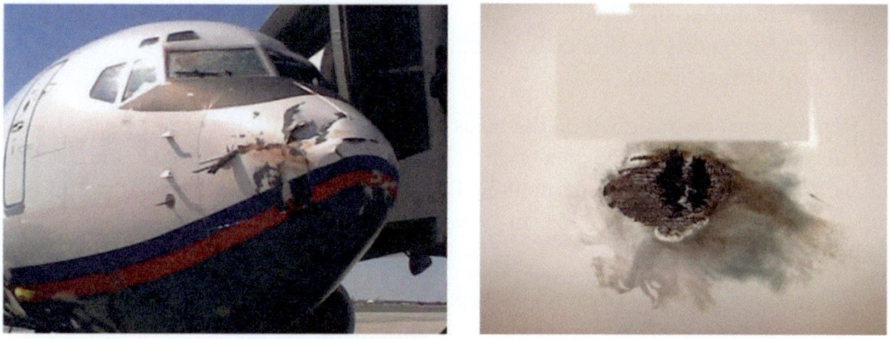

Fig. 13.3 Usual “flight threats” to aircraft structural integrity

posite repairs. This type of engineered repair design is usually performed in close cooperation with the aircraft manufacturer.

There are three main classifications of repairs defined as follows:

- Temporary preservation: Prevention of the entrance of foreign material into the defect or damaged area until the damage can be repaired. It is usually time or flight cycle limited. It should be performed as soon as the damage is discovered and removed once permanent repair is applied.
- Semipermanent repair: Restoration of a structure to fully ensure that the repair remains effective. This type of repair is often made at a field location and generally will be removed at the next major overhaul.
- Permanent repair: It restores the structural integrity of a component for the life of the aircraft. A permanent repair will replace a time or flight cycle limited repair, before the time or flight cycle limit expires.

The main purpose of the repair is to restore the initial structural integrity of the damaged part as follows:

- Capability in terms of strength, stiffness, and functional performance
- Safety of flight parameters
- Service life expectations
- Cosmetic and overall aesthetic appearance

The repair should return the structure to original load bearing capability and appearance. Before the repair process, the structural makeup of the component must be known, and the appropriate design criteria should be selected. The repair is in reality a joint, where a load is transferred from the parent material into and out of the “patch.” The very basic principles of composite repairs (by composite patching) include the following steps:

- Inspect to assess damage (extent and degree).
- Remove damaged material.
- Treat contaminated material.
- Prepare repair area.

- Complete composite repair.
- Inspect for quality assurance (e.g., delaminations, inclusions, etc.).
- Restore surface finish.

Repairs must be designed on the basis that the repair efficiency can be predicted, and they should be designed conservatively with respect to the various failure modes to include the surrounding structure. The critical phases for bonded composite repairs on composite structures, in light of their durability and efficiency, are as follows:

- Selection of adhesive system and material for patch repair
- Design of composite patch including environmental effects
- Surface preparation and bonding process

The older composite repair concept uses a pre-cured patch, which is cured before being bonded to the structure, while in the more recent co-curing approach, the raw composite plies are laid, cured, and bonded to the structure in a single process. The co-cured concept eliminates the need for a lay-up and provides better geometrical fit to the substrate. Table 13.1 presents a comparison of the pros and cons of pre-cured vs. co-cured composite material repairs.

The repair process and the necessary equipment needed for the repair of a damaged part always depend on the size of damage, the kind of damage, the available materials, and the accessibility and removability of the damaged part and on the structure of part (monolithic or sandwich). Aircraft parts are mainly divided in two groups, as presented in Fig. 13.4:

- Primary structure which contributes to carrying flight, ground, and pressurization loads
- Secondary structure, which carries only air or inertial loads

In older aircraft, most of composite parts are classified into secondary structure. These parts are most frequently repaired with composite patch technology. However, considering the rapidly growing volume of composite materials on primary structures of modern aircraft (A350, B787—see Fig. 13.1) and the increasing need for repairs of the older ones, innovative solutions are required to ensure the appropriate

Table 13.1 Pre-cured vs. co-cured patches

Co-cured patches	Pre-cured patches
Appropriate for complicated shapes	Appropriate for relatively flat surfaces
Diminished disbond occurrence since repair is flexible	More prone to disbonds
No need for special tools for patch fabrication	Need special tools which replicate repair area
Fast process, both prepreg and adhesive are cured together	Two-stage process—curing of patch and patch bonding
Possibility for certain degraded mechanical properties, if process not accurately followed	Superior mechanical properties

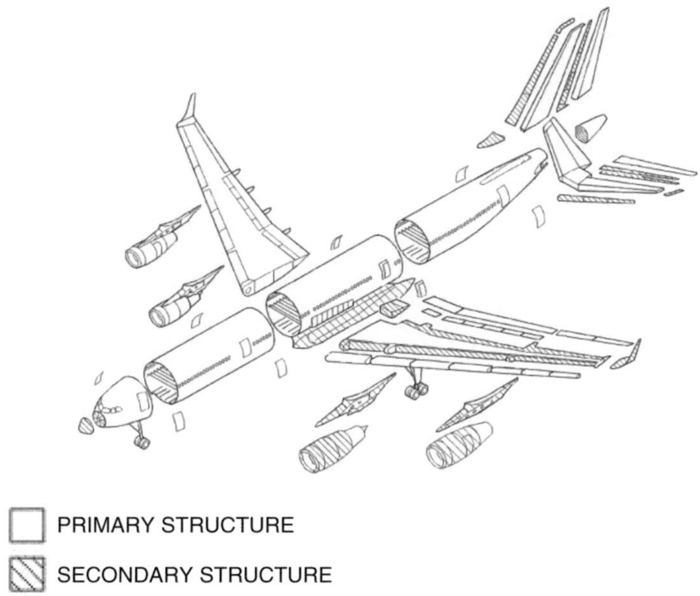

Fig. 13.4 Example of primary and secondary structure classification

level and reliability of application. As an alternative, these parts may be repaired using bolted structural repair, which, however, has important drawbacks.

2 Allowable Damage and Repair Classification

2.1 General

Allowable damage is defined as minor damage which does not affect the structural integrity or decrease the function of the component. It is not necessary to make a structural repair for this type of damage during the aircraft life. The parameters of allowable damage for specific components are given in the related chapter. Any parameters refer to the parent material but do not include the surface finish.

Some groups of damage do not affect the structural integrity or decrease the function of a component in normal operations; however, the damage could decrease the life of the component. This type of damage must be permanently repaired within a specified time or flight cycle limit. A temporary repair may also be required. Any parameters for damage with a time or flight cycle limitation will be given in the related chapter of aircraft manuals. Some groups of damage affect the structural integrity and the life of a component; damage of this type must be repaired immediately. Either a temporary or permanent repair will be specified in the related chapter.

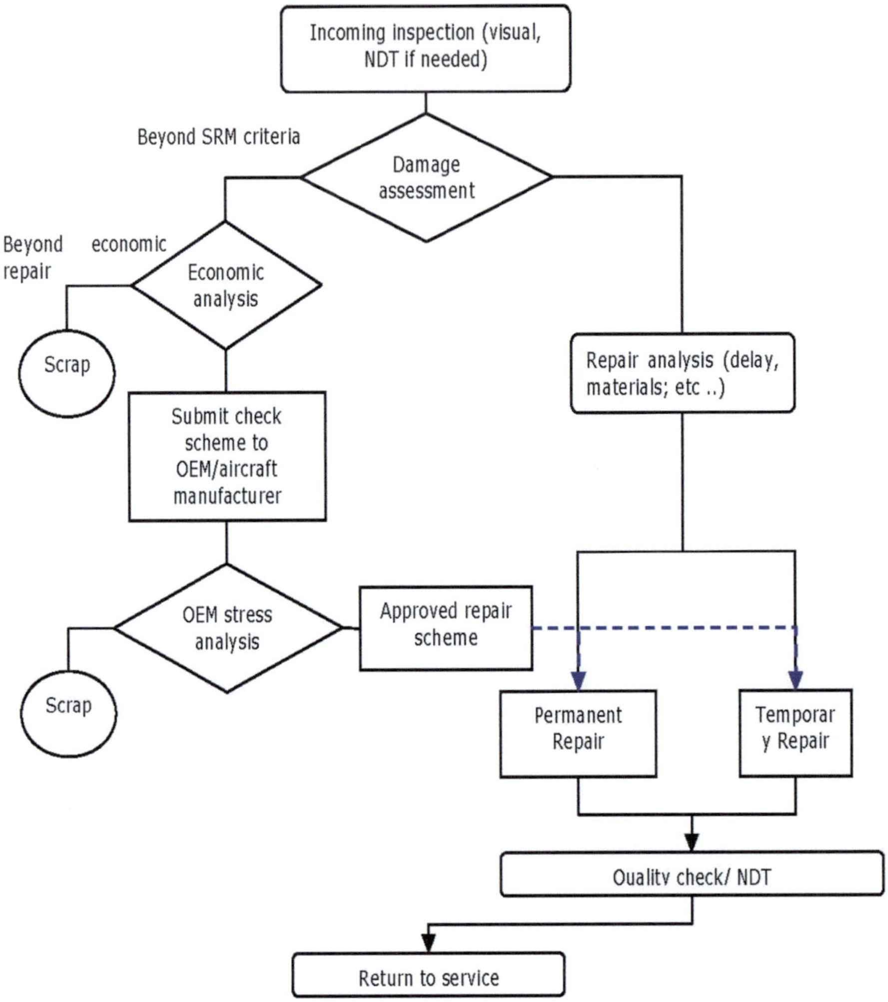

Fig. 13.5 Main steps to be followed for the inspection repair of damage

To repair damage, the steps usually followed are shown in the inspection/repair chart (Fig. 13.5).

2.2 *Common Types of Damage*

Non-Perforating Damage: Monolithic or Sandwich Structure

- Abrasion
- Scratches, gouges, and nicks

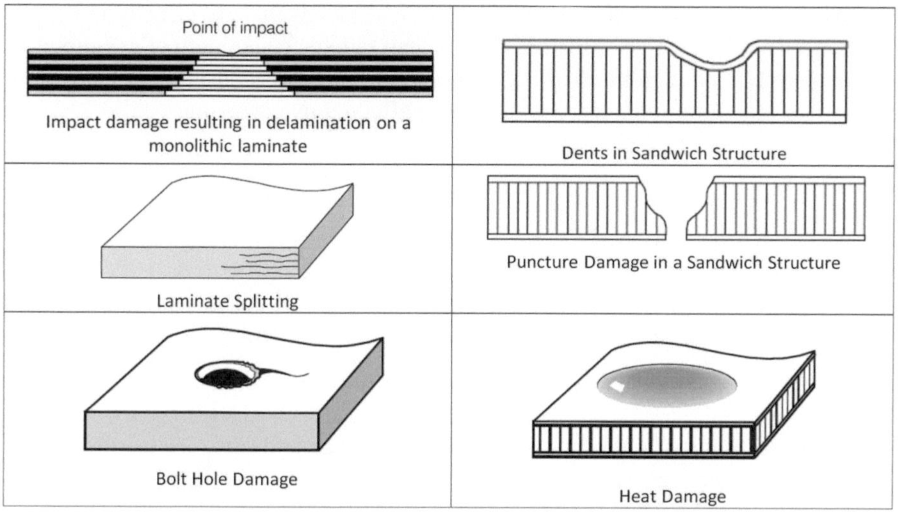

Fig. 13.6 Typical types of damage

- Dents
- Debonding: On monolithic parts, debonding can occur on the bond line with the surrounding structure, for example, ribs to skin interface. On sandwich structure, debonding can occur between the honeycomb core and the inner and/or outer skins. Debonding can be the result of an impact or a degradation of the bond line due to contamination by water or other fluids.
- Delamination: Following an impact, delamination can occur between the plies on monolithic and sandwich structures (Fig. 13.6).

Perforating Damage

- On monolithic structures, perforating damage will usually result in delamination around the perforation and damage to the structure underneath.
- On sandwich structures, perforation can affect one skin or both skins and the honeycomb. Debonding or delamination of the skins around the perforation usually occurs.
- If the perforation is left unprotected for a period of time, contamination (mainly by water) could occur.

Other Types of Damage

- Contamination: Water, skydrol, or other fluids can contaminate composite structures, especially sandwich parts, and reduce their strength characteristics.
- Heat effect: Overheating of composite structure can occur as a result of a lightning strike, engine fire, overcured repair, or mishandling. This would result in a local degradation of the structure and is usually associated with burn marks on the paint.

- Erosion: The leading edge of components and panels that are directly in the airflow may be affected by erosion. Erosion can reduce the strength of a component and permit fluid ingress. Erosion within certain limits is considered as abrasion.
- Corrosion: Aluminum honeycomb can be affected by corrosion.

2.3 Damage Inspection: Evaluation

The location of damage must be recorded exactly in all cases, using appropriate NDT techniques. More specifically, concerning the following:

- Abrasions: Examine the surface and determine the area and depth of the abrasion.
- Scratches, gouges, and nicks: Examine the surface and measure the length and depth of the damage.
- Delamination: Measure the area and depth of the delamination.
- Debonding: Measure the debonded area. On sandwich panels, inspect both skins.
- Contamination: Determine the extent of the contamination using an NDT method suitable for the type of contaminant. Cut the structure open if necessary.

The damage appraisal and the information given in the relevant aircraft manual are used to determine if the damage is either of the following:

- Allowable
- Repairable

It should be noted that the area to be considered as "damaged" may be larger than the actual area of damage, according to manufacturer's guidelines. For example, when two or more areas of damage are close together, the distance between the adjacent areas must not be less than a given value *X* (refer to Fig. 13.7). If the distance is less than *X*, then the areas must be considered as one damage. For the value of dimension *X*, refer to the relevant chapter. The overlapping of adjacent repair plies is not permitted. A minimum distance *Y* is necessary between repairs (refer to Fig. 13.8). Where distance *Y* cannot be achieved, the concerned adjacent areas must be considered as one. If distance *Y* is not given in the relevant chapter, consider it as 5 mm (0.2 in.)

In addition, the effect that damage can have on a component depends on the location of the damage and the size of the damage. Parts are divided into zones. These zones have differing structural importance, as for example presented in Fig. 13.9. These zones consider the following:

- Local stress levels
- Reserve factors, determined from structure testing
- Type of design and geometry of the concerned component

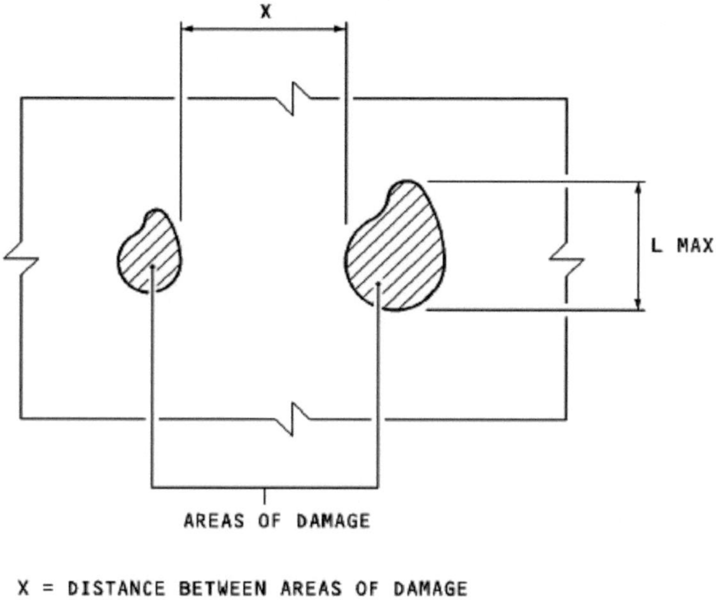

Fig. 13.7 Consideration of damage proximity

Finally, for some components, a *maximum number of damages/repairs* are defined for the life of the component in the SRM. The maximum number of damages which can be sustained and repaired is dependent on the damage size and the zone in which the damage occurred. This will vary from component to component and zone to zone, within that component. If this number is exceeded, contact the aircraft manufacturer.

3 Typical Repair Materials and Equipment

3.1 Composite Repair and Vacuum Bagging Materials

This chapter contains general information about repair materials, auxiliary materials, tools, and equipment which are necessary to make repairs on composite components. Table 13.2 lists common materials used on composite patch repairs.

In order for the bonding/curing to take place, a special vacuum bagging arrangement has to be prepared. A typical vacuum bagging lay-up for the co-curing of bonded composite patches, including examples of required materials, is presented in Fig. 13.10.

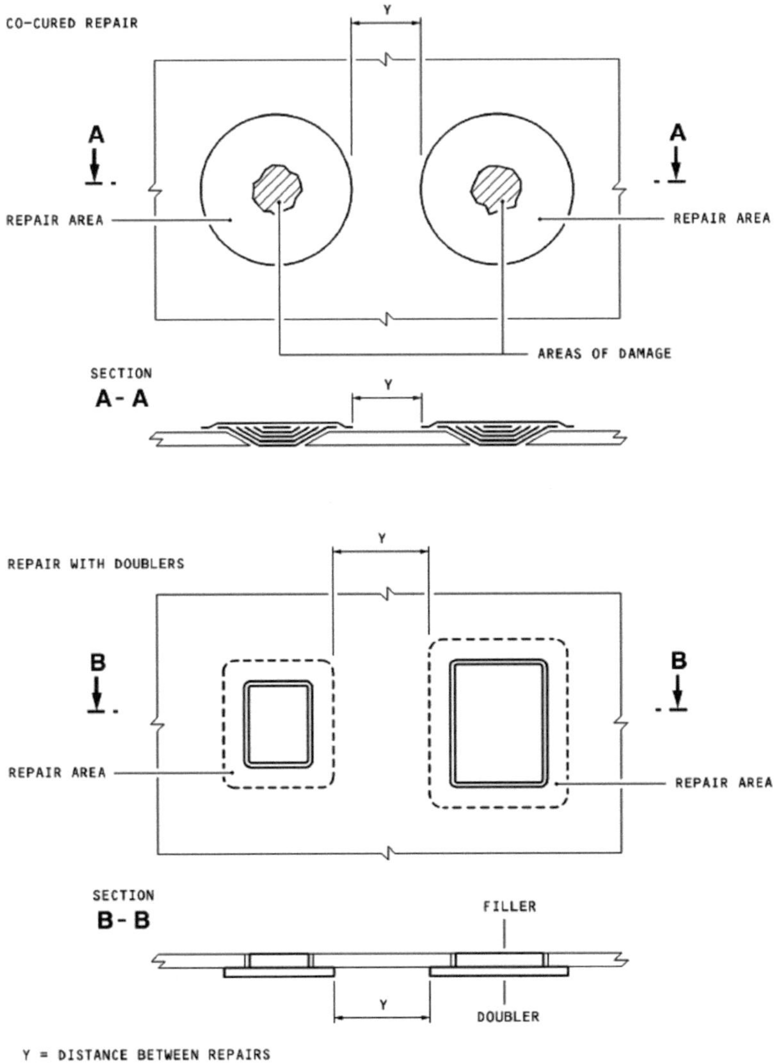

Fig. 13.8 Minimum distance between repairs

3.2 Bonding Consoles and Heating Elements

In order to understand the particular specifications of the thermal problem, it is important to establish the nature of the parts subject to this type of repair. If we try to categorize the parts, according to their thermal reaction for the bonding heating phase implementation, we will consider the following:

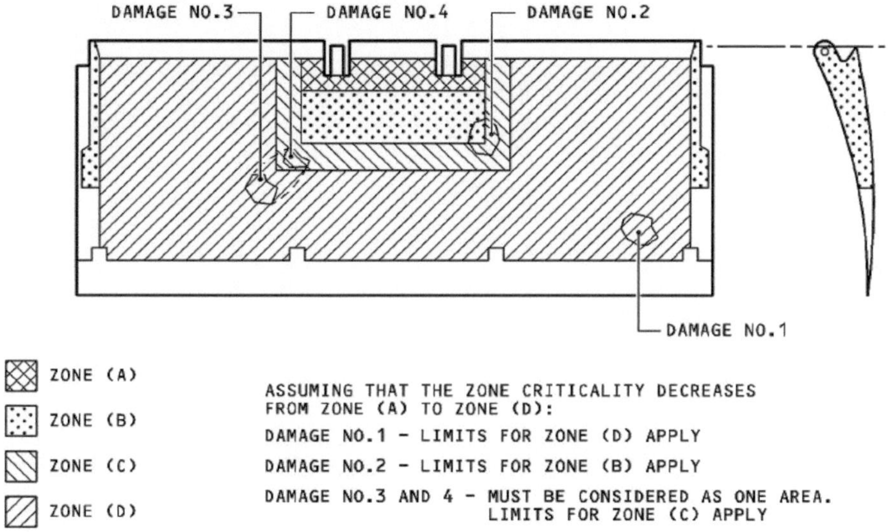

Fig. 13.9 Repair zoning classifications

- Regular and noncomplex reinforced structures
- Thick and/or complex structures with reinforcement

In the first case, the thermal problem is simplified by the regularity of the structure and the lack of heat sinks, in general. For small patches on thin carbon structures, the implementation techniques can be adapted to get suboptimum results. However, for new thick structures, metallic structures repaired by carbon, and future hybrid materials, full parameter control is required. Therefore, the difficulties of achieving an even temperature over the whole surface of the adhesive are added: asymmetries, reinforcements, variable composite or underlying honeycomb thickness, presence of fasteners or other metallic elements, massive areas that accumulate the heat, borders of the part and interfaces with the air that is a perfect insulator, etc. Finally, the anisotropic properties of the carbon structure may lead to unequal diffusion of heat on the surface and in depth.

To get a better understanding of the importance and the complexity of the heat transfer problem encountered in composite repairs, especially when dissimilar structure (i.e., heat sinks) is present, we will mention below the main thermal routes involved in a typical bonded composite repair case while providing some visualization in thermal transfer issues. More specifically, the four (4) thermal transfer routes, which interact when bonded composite repairs are performed, as can be seen in the left part of Fig. 13.11, are the following:

- Thermal Route No. 1: Transfer of thermal energy (*heating through conduction*), from the heating blanket to the repair area, regulated by the hot bonder (e.g., ANITA hot bonder developed by GMI) by means of electrical power, in order to reach the required temperature at each time step.

Table 13.2 Typical composite repair materials

Material	Class	Mono/sand	Shape	Thickness	Kind of part	Part raw materials				Kind of repair					Repair patch raw materials			
						Resin	Heat class	Fiber	Fiber feature	Hot	Wet lay up	Injection	Bonding	Riveting	Resin	Heat class	Fiber	Fiber feature
Composite	1	Monolithic	Simple	Thin														
				Thick														
			Complex	Thin	Elevator	Epoxy	180 °C	Carbon	AGP193-PW8SS2S									
				Thick														
		Sandwich	Simple	Thin														
				Thick														
			Complex	Thin	Elevator	Epoxy	120 °C	Carbon	Hexcel ref. G814	Yes	Yes	Yes	No	No	EA9396	170 °C	Carbon	Hexcel-dry carbon fabric ref. G814
																120 °C	Carbon	Hexcel-carbon prepreg ref. G814
				Thin	Aileron	Epoxy	180 °C	Carbon		Yes	Yes	Yes	No	No	EA9390	177 °C	Carbon	Hexcel-dry carbon fabric style 3K-70-PW (BMS 9-8)
															F263	177 °C	Carbon	Hexcel-dry carbon fabric style 3K-70-PW (BMS 8-212 or 8-256)
				Thick														
	2	Monolithic	Simple	Thin	Duct	Phenolic	80 °C	Glass	250/38%/644	No	Yes	No	No	No	Epocast54		Glass	Hexcel-glass dry fabric style 1581
					Firewall panel	Phenolic	180 °C	Carbon	200/40%/G803	No	No	No	No	Yes	RTV106	260 °C		Titanium plate or carbon phenolic plate
					Outer barrel	Epoxy	180 °C	Carbon	AGP193 or AH370	Yes	Yes	No	No	No	EA9390	177 °C	Carbon	Dry carbon fabric AH370
				Thick														
			Complex	Thin	Duct	Phenolic	80 °C	Glass	250/38%/644	No	Yes	No	No	No	Epocast54		Glass	Hexcel-glass dry fabric style 1581
					Outer barrel	Epoxy	180 °C	Carbon	8552S/35%/ACGP2280C	No	Yes (cosmetic)	No	No	Yes	N/A	180 °C	Carbon	Development in process
				Thick														

Composite	2	Sandwich	Simple	Thin	Radome	Epoxy	120 °C	Glass	Ref. 1581 (BMS 8-79)	Yes	Yes	Yes	No	No	EY-3804	66 °C	Glass	Glass dry fabric style 1581 or 7781 (BMS 9-3)
															F155	120 °C	Glass	Hexcel-glass prepreg 1581-F155-S-F69 (BMS 8-79)
						Epoxy	120 °C	Aramid	AEG EHA 250-33-50	Yes	Yes	Yes	No	No	LY564/ HY560		Glass	Hexcel-glass dry fabric style 1581
															–	120 °C	Aramid	AEG-aramid prepreg EHA 250-33-50
						Epoxy	120 °C	Aramid	Hexcel M14/54/788	Yes	No	Yes	No	No	–	–		
															–	120 °C	Aramid	Hexcel-aramid prepreg M14/54/788
						Epoxy	120 °C	Aramid	Hexcel ref 20914	No	Yes	Yes	No	No	LY5052/ HY5052	120 °C	Kevlar	Hexcel-aramid dry fabric style 20914
															–	–	–	
						Epoxy	180 °C	Quartz	Hexcel 1455/46%/581	Yes	Yes	Yes	No	No	LY5052/ HY5052	120 °C	Quartz	Hexcel-quartz dry fabric style 581
															–	120 °C	Quartz	
						Epoxy	120 °C	Quartz	Hexcel M14/48%/1237	Yes	Yes	Yes	No	No	LY5052/ HY5052	120 °C	Quartz	Hexcel-quartz dry fabric style 21237
															–	120 °C	Quartz	Hexcel-quartz prepreg M14/48%/1237
			Simple	Thick	Wing fairing	Epoxy	120 °C	Glass	Ref. 1581 (BMS 8-79)	Yes	Yes	Yes	No	No	EY-3804	66 °C	Glass	Glass dry fabric style 1581 or 7781 (BMS 9-3)
															F155	120 °C	Glass	Hexcel-glass prepreg 1581-F155-S-F69 (BMS 8-79)

(continued)

Table 13.2 (continued)

						Part raw materials				Kind of repair					Repair patch raw materials			
Material	Class	Mono/sand	Shape	Thickness	Kind of part	Resin	Heat class	Fiber	Fiber feature	Hot	Wet lay up	Injection	Bonding	Riveting	Resin	Heat class	Fiber	Fiber feature
Composite	2	Sandwich	Complex	Thin	Body fairing	Epoxy	120 °C	Aramid	Hexcel 1454/54%/914	Yes	Yes	Yes	No	No	EA9396	170 °C	Glass	Hexcel-glass dry fabric style 1581
															–	120 °C	Aramid	AEG-aramid prepreg EHA 250-33-50
						Epoxy	120 °C	Hybrid	Hexcel 913/54%/G973	Yes	Yes	Yes	No	No	EA9396	170 °C	Glass	Hexcel-glass dry fabric style 120
															–	120 °C	Aramid	Hexcel-hybrid prepreg 913/54%/G973
					S-duct fairing	Epoxy	180 °C	Aramid	Hexcel 1455/54%/914	Yes	Yes	Yes	No	No	LY5052/ HY5052	120 °C	Aramid	Hexcel-aramid dry fabric style 20914
																180 °C	Aramid	Hexcel-aramid prepreg 1455/54%/914
			Complex	Thick	Fan cowl	Epoxy	180 °C	Carbon	AGP193 or AH370	Yes	Yes	Yes	No	No	EA9390	177 °C	Carbon	Dry carbon fabric AH370
																177 °C	Carbon	Carbon prepreg style AGP 193 or AH370
					Body fairing	Epoxy	120 °C	Aramid	Hexcel 1454/54%/914	Yes	Yes	Yes	No	No	EA9390	170 °C	Glass	Hexcel-glass dry fabric style 1581
																120 °C	Aramid	AEG-aramid prepreg EHA 250-33-50
						Epoxy	120 °C	Hybrid	Hexcel 913/54%/G973	Yes	Yes	Yes	No	No	EA9396	170 °C	Glass	Hexcel-glass dry fabric style 120
															–	120 °C	Aramid	Hexcel-hybrid prepreg 913/54%/G973
						Epoxy	120 °C	Glass	AEG EHG 275-68-50	Yes	Yes	Yes	No	No	EA9396	170 °C	Glass	Hexcel-glass dry fabric style 1581
															–	120 °C	Glass	AEG-glass prepreg EHA 250-33-50

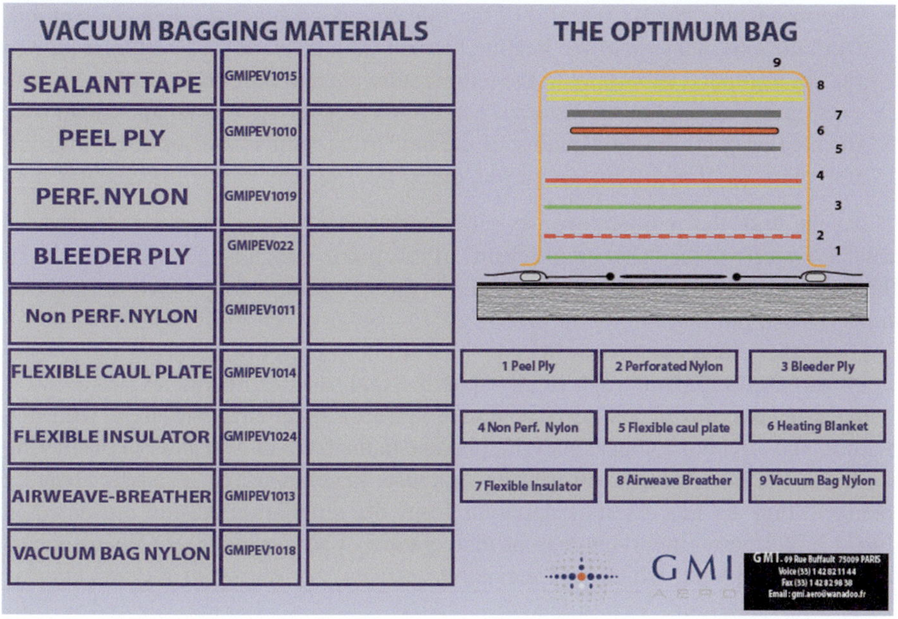

Fig. 13.10 Typical vacuum bagging lay-up for the co-curing of bonded composite patches

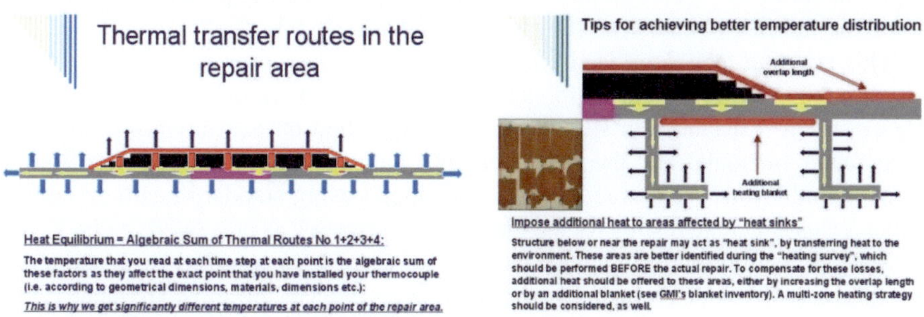

Fig. 13.11 Thermal transfer routes in repair area and effects of "heat sinks"—"tips" for achieving more homogeneous temperature field

- Thermal Route No. 2: Transfer of thermal energy (*heating through conduction*), from the repair to the rest of the structure, according to conduction coefficient (i.e., what kind of material is involved) and temperature. This way, the rest of the structure is heated at a temperature which is mainly related to its distance from the heat source and to the characteristics of the material.
- Thermal Route No. 3: Transfer of thermal energy (*cooling through convection*), from the repair area to the environment, according to the structural geometrical characteristics (e.g., presence of heat sinks), the temperature at each time step, and the "convection coefficient," related to the external conditions (i.e., if there is an air stream over the area or not, what is the room temperature, etc.).

- Thermal Route No. 4: Transfer of thermal energy (*cooling through convection*), from the external side of the heating blanket (i.e., through the "vacuum bag") to the environment, according to the temperature at each time step, the "convection coefficient" related to the external conditions (e.g., if there is an air stream over the area or not), and the "quality of the bag" (i.e., number of "airweave" insulation plies, level of vacuum achieved, etc.).

At the end, the temperature we get at every point of the repaired structure is related directly to the "heat equilibrium" of this point, which is the algebraic sum of Thermal Routes No. 1 + 2 + 3 + 4. *This is why we get significantly different temperatures at each point of the repair area.*

To further elaborate on the effect of "heat sinks" (i.e., geometrical dissimilarities) upon achieved temperature, a steady-state thermal transfer FE model is presented, calculating the spatial variation of temperature on a repaired structure with and without the effect of a "heat sink" (in this case in the form of a stringer), considering uniform heat flux directed towards the structure, as generated by a standard heating blanket (Fig. 13.12). As it is apparent from the simulation results presented in Fig. 13.12, a very important drop of temperature of several degrees Celsius occurs above the heat sink, which will inevitably lead to poor curing of the neighboring composite material.

The last 20 years have seen the introduction of bonding techniques for the repair of fiber-reinforced composites, and major repair organizations have acquired the expertise for bonding carbon fiber-reinforced structures using portable heating consoles and thermal blankets, as shown in Fig. 13.13.

This bonding technique is similar to the use of an oven or autoclave, applying vacuum pressure for consolidation and heating by conduction with the use of flexible electrical resistances in silicone mats, as specified in most civil aircraft structural repair manuals. For the improvement of the heating efficiency in different areas, a multi-zone technique has been developed using standard heating blankets which gives adequate results for a wide variety of repairs, as presented in Fig. 13.14.

Special heating blankets have been also developed, customized for heating different areas of the aircraft (e.g., radomes), according to the most frequent types,

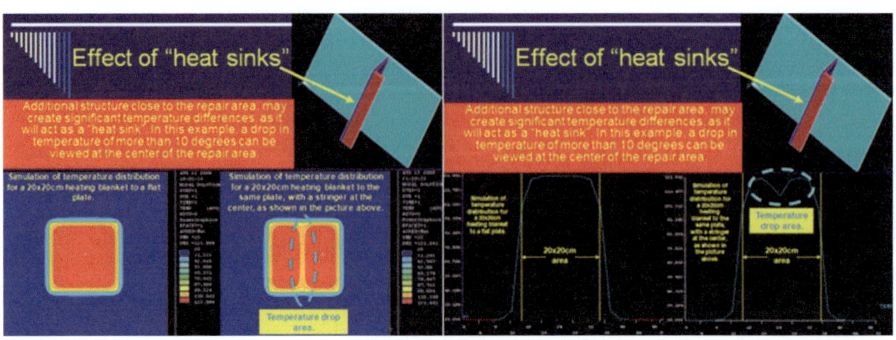

Fig. 13.12 Numerical simulation of the effects of "heat sinks" in repairs

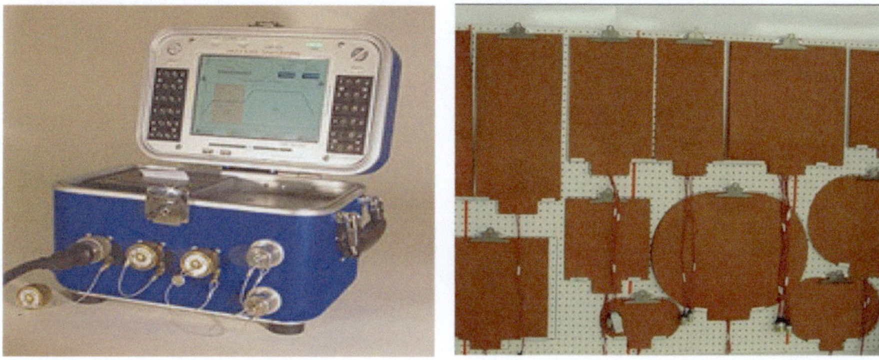

Fig. 13.13 Portable heating console (GMI Aero ANITA) and heating blankets used for curing of repairs

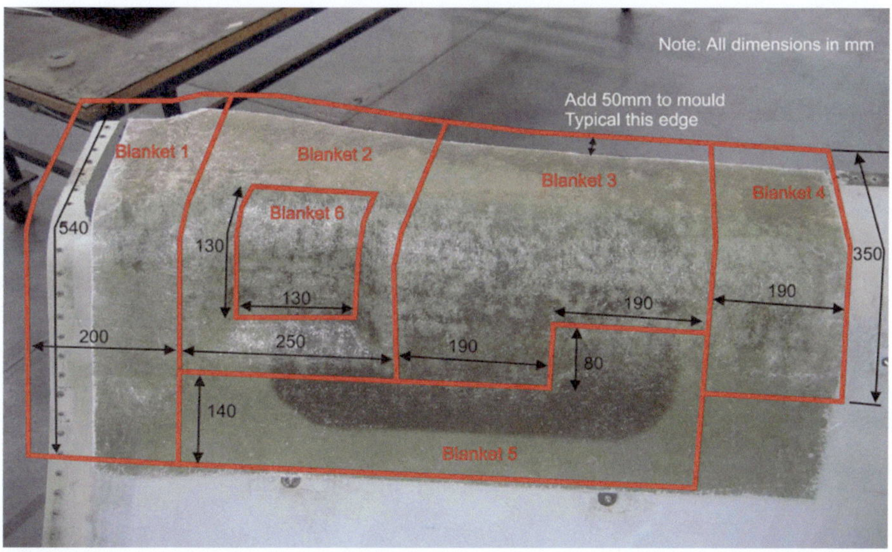

Fig. 13.14 Example of multi-zone heating technique

dimensions, and areas of damage. Moreover the so-called conformable or stretchable blankets have been developed for areas with abrupt geometrical changes. Using such blankets, improved heat conduction can be achieved, leading to better and faster repair (Fig. 13.15).

Other developed alternative heating solutions, to satisfy specific heating requirements, include guiding hot air on the repair area by means of foldable ovens and radiant panels, as can be seen in Fig. 13.16.

Both solutions have been proven adequate for small repair areas, while their application to the larger areas should be considered on a case-by-case basis, according to the results and the experience already accumulated by the end user (nonuni-

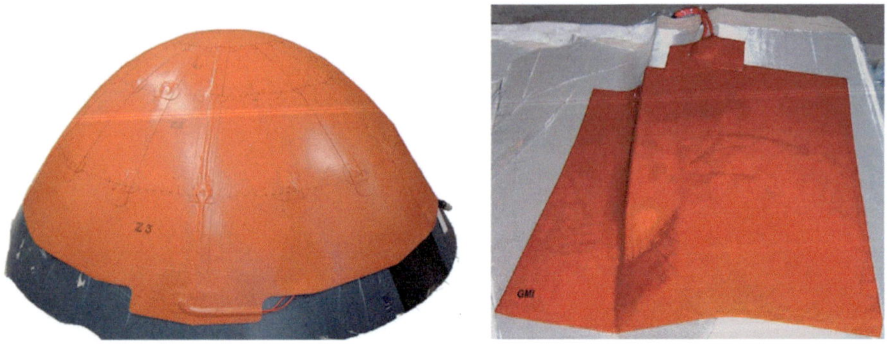

Fig. 13.15 Specially shaped (left) and "conformable" heating blankets

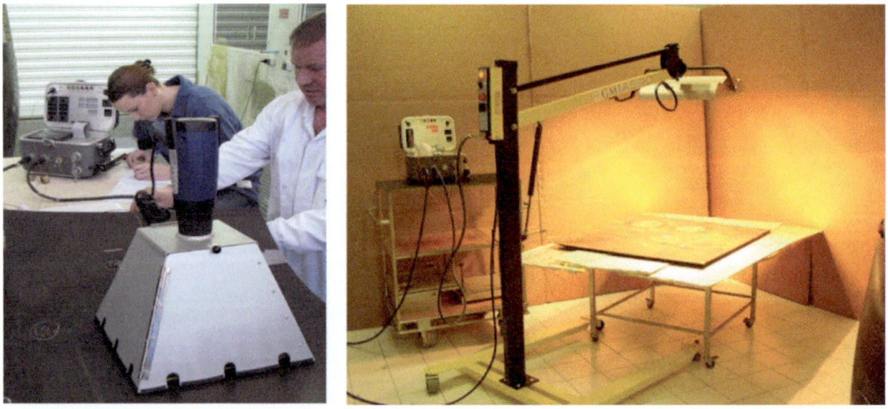

Fig. 13.16 "Hot air gun" with foldable oven (left) and IR panels (right) for heating areas where standard blankets could not be used

versal application of solutions, large energy requirements, important heat losses, flow of air restrictions, etc.).

3.3 Surface Preparation Toolkits

The surface preparation of the damaged area on a composite part is a lengthy, cumbersome, and delicate process. This is due to the need to achieve very specific and sometimes complex geometries (i.e., gradually increasing in diameter, stepping or scarfing of the damaged area, etc.) on a very strong and brittle material. Currently, surfaces are prepared for repair using high-performance drilling equipment, with diamond tooling, enabling delicate operations on the damaged structure. The majority of such equipment and associated methodology has been developed for the aeronautical industry.

Due to the hard fibers, the expensive tools wear out rapidly, causing variable surface quality. Therefore, the stringent requirements for successful repair are often difficult to meet. Furthermore, success in the use of the special tooling depends on the skill of the operator. This implies a large potential for "human error." The tooling often used for the surface preparation starts from simple sanders and extends to special equipment for drilling and scarfing. In the aeronautical sector, where the requirements for dimensions control of the processed area are very strict, specific toolings have been developed, as can be seen in Fig. 13.17.

In a similar way, standard machine tools, attachments, and cutters have been adapted to produce the desired results, as can be seen in Fig. 13.18.

The application methodology implies that there are two steps in the composite material machining:

- Opening of a hole by cutting the external composite material surface, in order to "clean" the structure from cuts/delamination or associated damage. At the same step, the internal "core" material potentially existing (honeycomb, foam, etc.) is simultaneously removed, as can be seen in Fig. 13.19. Especially for the honeycomb removal, special tooling has been developed, as shown in Fig. 13.20.

Fig. 13.17 Examples of "Leslie" series surface preparation toolings used in the aeronautical sector

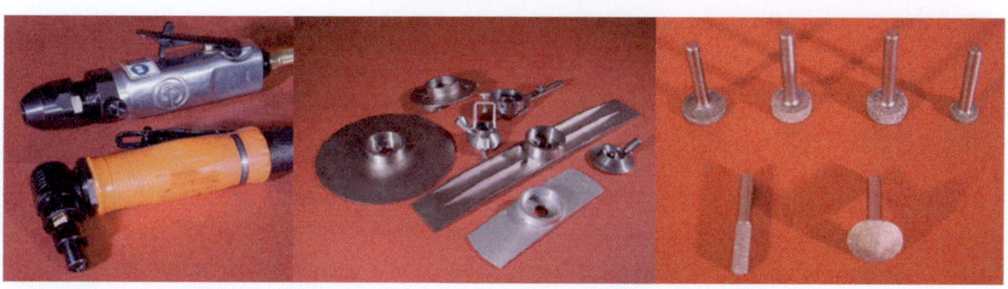

Fig. 13.18 Standard machine tools, attachments, and cutters

Fig. 13.19 Cutting of external composite surface and removal of core

Fig. 13.20 Honeycomb removal below panel

Fig. 13.21 Stepping—scarfing before application of the patch

- Stepping or scarfing of the external surface to be repaired, to achieve better bonding surface and reduce the shear stresses developed during operation, after the application of the patch, as shown in Fig. 13.21.

3.4 NDT Equipment

NDI (Non-Destructive Inspection) methods are employed in damage analysis and repair design of composite structures in three ways:

- Damage location
- Damage evaluation, i.e., type, size, shape, and internal position
- Post-repair quality assurance

The first and most important activity is to identify the damage, which is initially achieved by visual inspection. This localizes the damaged area and is then followed by a more sensitive NDI method, which maps the extent of any internal damage. Detailed NDI is particularly important when dealing with composites, because the majority of the damage is usually within the composite structure. A number of NDI techniques are currently available to identify damage in composite structures, as listed below:

- Tap test: It is based on the fact that the sound emitted when knocking on the structure changes when the thickness or material type changes or when porosity is present. It can also detect disbonds between the skin laminate and the main spar. There are three types of tap testing equipment: a manual tapping hammer, the "Woodpecker" portable bond tester, and the Computer-Aided Tap Tester (CATT) system. The automated tap methods have the advantage of producing a print of the damaged area, which is a useful and permanent record of the damage found. They work well for thin laminates, honeycomb structures, and other sandwich panels but are less effective on thicker parts.
- Ultrasonic NDI: UT can be carried out to investigate if the composite structure is damaged. It reveals flaws quickly and reliably and is the most common industrial NDI method for composite materials. Its main advantage is that it enables us to see beneath the surface and check for dry fiber or delamination. Existing portable U/S equipment, already used for the implementation of bonded composite repairs in the aeronautical sector, can be seen in Figs. 13.22 and 13.23.
- Infrared (IR) Thermography: Adhesive joints are critical points in a composite structure. For this reason, they are inspected particularly carefully. Infrared scanners are used to examine the structure throughout its length, measuring the identical points each time. The scanner can see through the laminate and check the adhesive joint. It records temperature differences in the adhesive, possibly identifying flaws, and takes a series of pictures. If there are any doubts, a point can be highlighted and later analyzed using electronic image processing.

Fig. 13.22 GMI Aero "ELISA" U/S NDT portable console, used for inspection of bonded composite repairs in the aeronautical sector

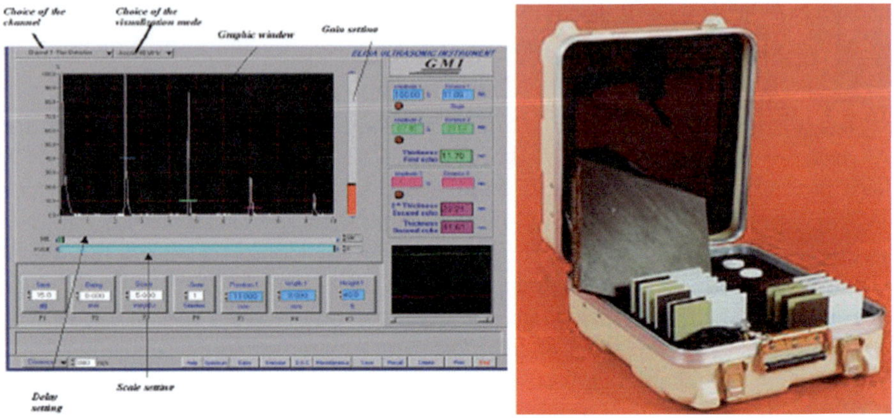

Fig. 13.23 Example of HMI (Human Machine Interface) (screen snapshot) and calibration specimens associated with "ELISA" U/S NDT portable console, used for inspection of bonded composite repairs in the aeronautical sector

4 Standard Composite Repair Procedures

This chapter contains typical procedures common for all wet lay-up or prepreg repairs to components made from epoxy resin reinforced with several layers of carbon fabrics. The most common construction is either a monolithic structure or a sandwich of two laminated skins and a honeycomb core. The repairs described are wet lay-up repairs requiring the rapid use of catalyzed or two-part resin materials, or prepregs. Specific allowable damage, repair limitations, and repair data can be found in the relevant section of the aircraft structural repair manuals, associated with each structural component.

4.1 Selection of Repair Method and Repair Materials

- Determine according to the relevant chapter of the SRM the design characteristics and classification of the damaged area.
- Determine the type and extend of the damage.
- Determine the allowable and repairable damage limits.
- Determine the method of repair.
- Determine the construction and used materials of the structure.
- Determine the required repair materials. Refer to the relevant section of the aircraft's structural repair manual for specific component-related temperature resistance requirements.

4.2 Removal of Water from Damaged Area

- Remove damaged skin plies to open up honeycomb area in the damaged area. Remove standing water using vacuum, low pressure, dry, oil-free compressed air, or a hot air gun.
- Sand the core to remove the adhesive. Removal of adhesive fillets on core is not required.

4.3 Remove Damage and Prepare for Repair

- Trim out the damaged laminate to a smooth shape with rounded corners or a circular or oval shape. Take care not to damage the undamaged plies, core, or surrounding material.
- When the core is also damaged, remove the core by trimming to the same outline as the skin.
- In area where contamination cannot be removed by cleaning or drying, remove the contaminated structure along with the other damage.
- When opposite inner skin is also damaged, trim out the damage to a smooth rounded shape.
- When core is removed from the inner surface, carefully smooth core down to adhesive film.
- Inspect cut out area to ensure that all damages has been removed.

4.4 Preparation of Damaged Areas

- Determine the number and material of the plies that have been cut. Determine the thickness of the laminate that has been removed. Mask off the area around the cleaned up damage allowing 25 mm overlap for each removed mm thickness, plus an additional 25 mm all around.
- Remove the paint finish using No. 280 or finer Scotch-Brite Abrasive or No. 150 or finer sandpaper in the masked off area.
- Abrade surfaces around repair using No. 280 or finer abrasive.
- Dry and clean the prepared area.

4.5 Preparation of Repair Plies

- The number of plies and types of materials used in the structure is stated in the aircraft's structural repair manual.

- From each type of material required, cut a piece of prepreg that is large enough for cutting the required number of plies for the repair patch. Use one ply of fabric in the patch for each damaged ply of the original laminate, plus extra repair plies as required for specific repairs.
- Place succeeding plies of the patch as described in the above steps and with orientation as detailed for the specific repair.
- Proceed to lay-up/bagging procedure.

4.6 Lay-Up/Bagging Procedure

- Place a layer of peel ply over the lay-up. Cut the peel ply so that edges extend 75 mm beyond the edge of the repair.
- Secure three (3) thermocouples (spaced evenly around repair) to the edge of the largest repair ply.
- Place for each two repair plies one layer of 120 g/m^2 glass fabric, with maximum of three (3) layers, over the repair as surface bleeder.
- Place a layer of non-perforated parting film (0.2 mm thick) over the lay-up so that its edges extend 50 mm beyond the edges of the bleeder.
- Place an optional metal plate manufactured from 0.4 mm thick aluminum over the glass bleeder. Make the aluminum plate slightly smaller than the bleeder.
- Place a heat blanket on the lay-up. The heat blanket must extend a minimum of 50 mm beyond the repair patch edges.
- Place on the lay-up one (1) ply of 120 g/m^2 glass fabric as breather cloth. The breather cloth must extend beyond the parting film and also must extend beyond the heat blanket and aluminum plate (if they are used). The breather cloth must also make contact with the surface bleeder cloth.
- Apply extruded sealing compound or sealing tape around the entire area approximately 5–15 cm outside the heating blanket edge.
- Secure the vacuum outlet to the surface breather cloth (outside of the repair ply area).
- Lay a piece of vacuum bag material over the entire repair area. Seal the edge with extruded sealing compound around the repair area.
- Evacuate the space under the vacuum bag and maintain a minimum of 74 kPa (20 in of mercury) vacuum during the entire cure cycle.
- Check the vacuum bag and ensure that there are no leaks (Fig. 13.24).

4.7 Curing of the Repair

- Heating blankets should be used to accelerate the cure. If these cannot be used, an infrared heat lamp may be alternatively used.
- The temperature on the repair surface must be monitored throughout the accelerated curing period using thermocouples placed at the edge of the repair.

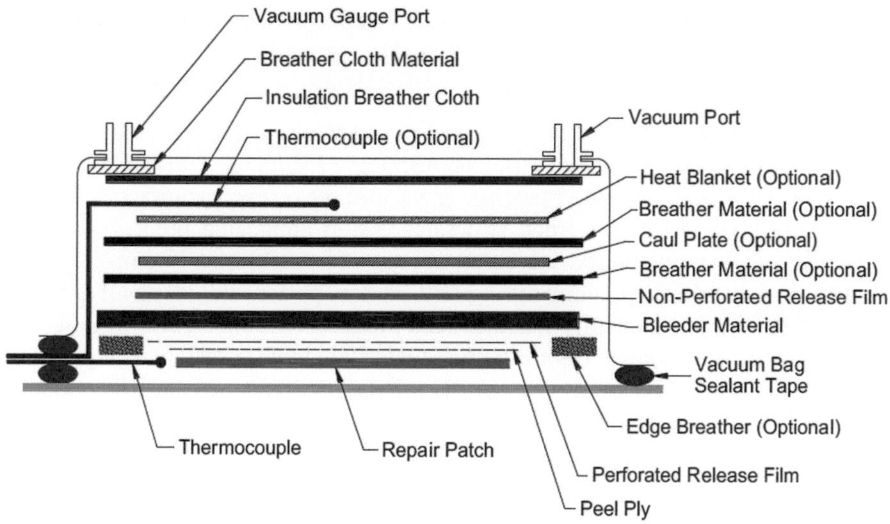

Fig. 13.24 Typical composite repair vacuum bagging

- The cure time does not include the time required for the mold and repair part to heat up to temperature. Cure time does not start until the repair part has reached the designated temperature.
- Heat the repair area with a heat up rate of max 3 °C/min at the temperature and for the time detailed in resin data sheet.
- Maintain vacuum of 74 kPa (20 in. of mercury) minimum during the entire cure cycle.
- Remove bagging and parting film after curing.

4.8 Post-repair Inspection and Finishing

- Inspect the completed repair. The patch should be free from pits, blisters, starved areas, and excess resin deposits.
- Lightly sand the edge of the topmost repair ply as necessary to blend the edge and produce a smooth surface finish. Ensure that when sanding the fibers are not damaged.
- Restore the surface finish in accordance with the aircraft maintenance manual.

5 Latest Composite Repair Innovations and R&I Areas

Research and innovation (R&I) activities are mainly focused in the field of repair and maintenance of composite structures, through the design of innovative tooling, equipment, and methodologies. In this chapter, a number of such innovative solutions and the corresponding challenges answered are analyzed.

5.1 Repairs Using OLGA for Patch Preparation (to Overcome Porosity Issues) and Secondary Bonding on Aircraft

As there are several issues to be addressed in order to perform bonded repairs on very thick structures, like the ones used on A350/B787, among efforts to solve them, appropriately shaped pre-cured patches of ~1 mm have been manufactured, to be secondarily bonded on concerned structures, using applicable prepreg material. However, as NDT discovered increased porosity of patches, an attempt to apply positive pressure during manufacturing will be now performed, using OLGA (see Fig. 13.25), in order to identify whether such procedure would significantly reduce patch porosity. Additional room for improvements could be traced, by using OLGA during patch bonding on aircraft, to optimize the patch preparation/bonding process using OLGA, considering actual aircraft materials.

5.2 Problems in Achieving Homogeneity During Repair Co-curing on Thick Geometrically Complex Structures

This is a well-known problem and has been treated in the past using multi-zone heating approach. However, as this process adds complexity and is time-consuming, a compromise has been identified, taking advantage of A350 structural periodicity. To this end, the **AdaptHEAT** adaptive heating blanket has been designed (see Fig. 13.26), tailored to A350 structural components, which could be applied using existing two-zone ANITA bonding consoles, greatly improving homogeneity. Additional T homogeneity improvements could be achieved, using variable insulation.

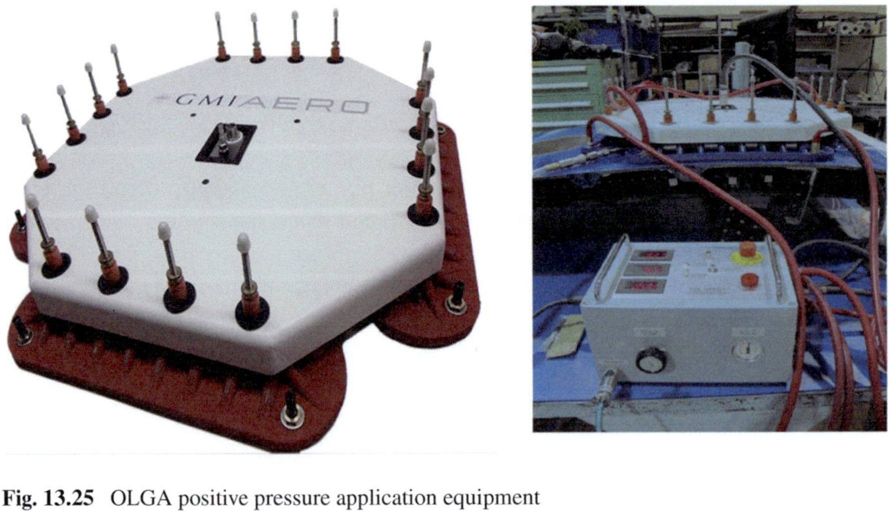

Fig. 13.25 OLGA positive pressure application equipment

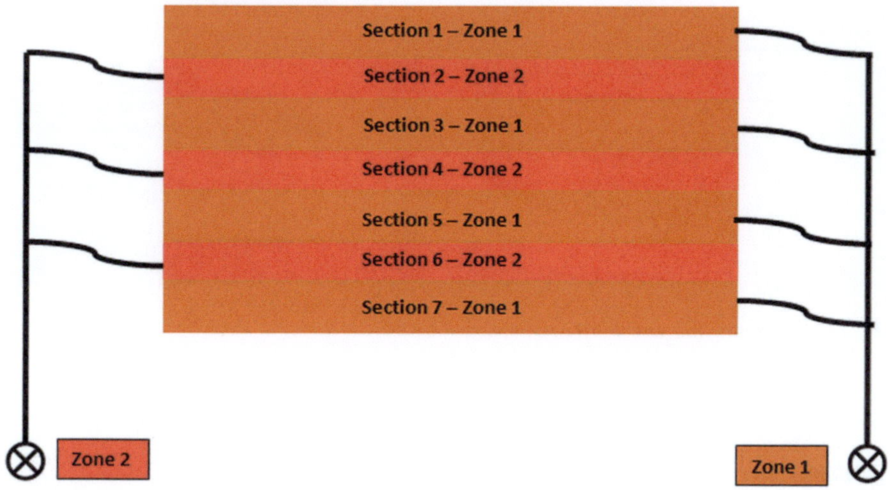

Fig. 13.26 AdaptHEAT operation principle

5.3 Increased Quality Control Requirements During Repair Operations on Complex Structures

As the repair areas get more and more complex, more accurate quality control procedures are necessary, to ensure compliance with temperature homogeneity requirements. To this end, a 12/24 Thermocouple wireless scanner has been developed, which is "plug and play" to existing bonding consoles, thus significantly improving sensitive area monitoring capabilities. Data is real-time recorded and stored by current ANITA versions to be subsequently electronically transmitted, as required. This concept could be simultaneously introduced with appropriately prepared sensing mats, enabling temperature measurements without marking the part (Fig. 13.27).

5.4 Development of Digital / Physical Twin of Composite Repair

In a similar manner to the abovementioned problematic, the following concept has been developed (Fig. 13.28) using dielectric curing sensors. When a safety critical structural repair is performed by an airline or an MRO, all repair data (temperatures, vacuum level, etc.) are recorded and real-time transmitted to the aircraft manufacturer. At the aircraft manufacturer's facilities, a setup using same materials is prepared and is simultaneous to the actual repair cured, using appropriately selected portion of transmitted real-time data (e.g., lagging thermocouple), so as to enable imminent destructive or nondestructive testing of produced material. In addition, within the setup prepared at the aircraft manufacturer's facilities,

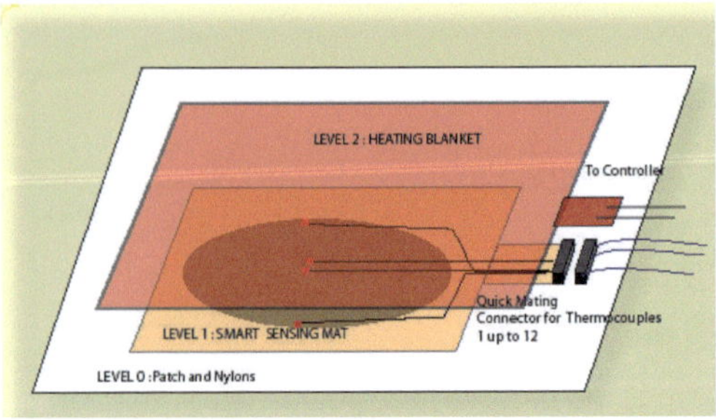

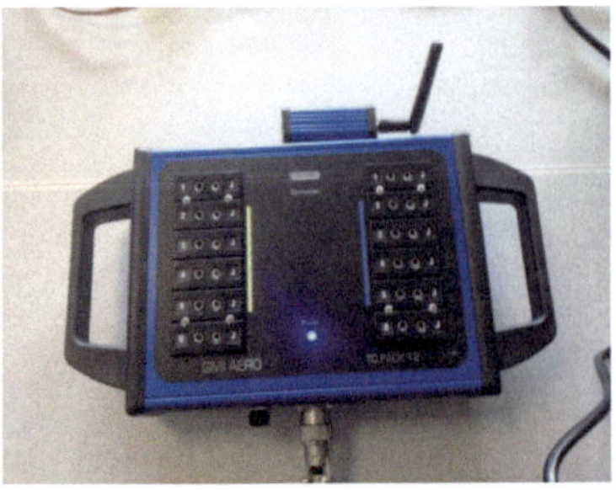

Fig. 13.27 Sending mat and thermocouple extender

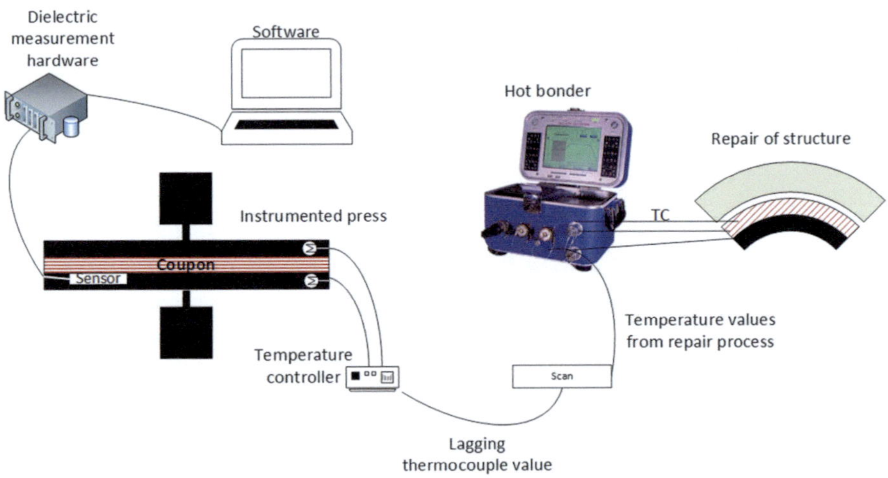

Fig. 13.28 Use of dielectric sensors for quality control

dielectric sensors for curing degree monitoring are included, thus providing real-time degree of cure data, equally valid for the remotely performed structural repair on the aircraft.

5.5 *Heating for Quick Repairs of Limited Dimensions*

Certain (limited) repairs could be applied overnight in aprons or hangars. Simple heating solutions (Fig. 13.29) assist in the implementation of such operations. The range of application and the corresponding bonding procedure (e.g., allowed T margins) need to be defined for the specific aircraft needs.

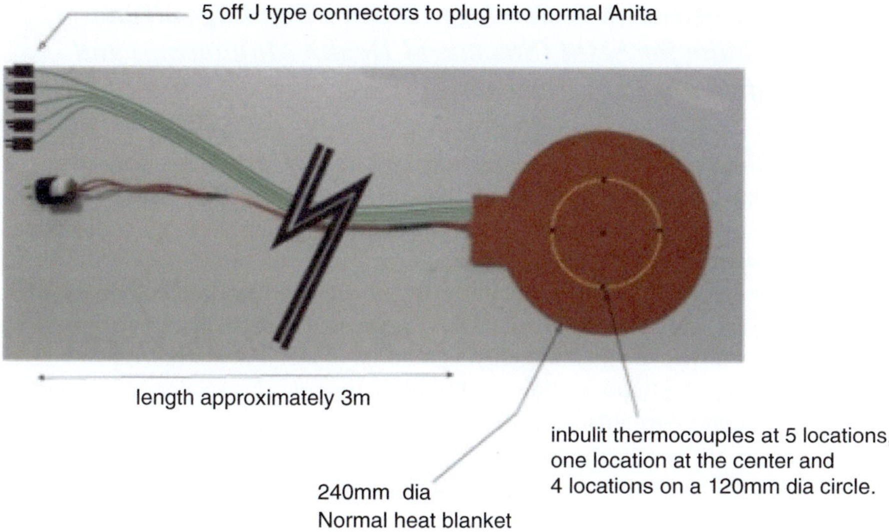

Fig. 13.29 Simple heating solutions for limited size repair

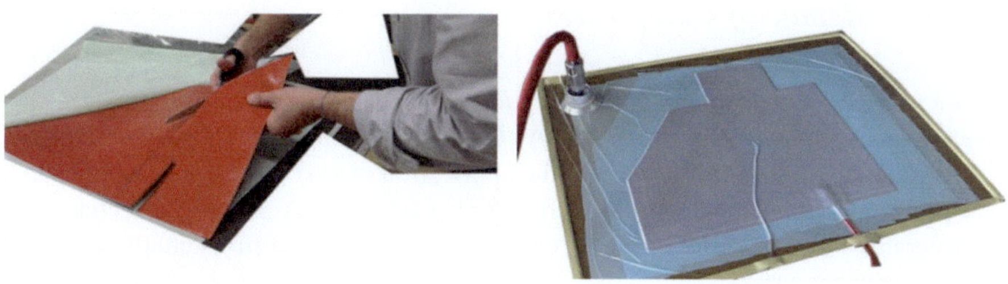

Fig. 13.30 CONDUCTOR cuttable heating blankets

5.6 *Consumable Heating Blankets*

In several cases, heating blankets of a very specific shape are required, while in others local increase of heating is necessary to achieve required homogeneity. Through CONDUCTOR (Fig. 13.30), GMI brings a very promising product that improves the quality of bonding operations and productivity by totally suppressing the lead time for heating element fabrication. The product appears as a standard silicone blanket integrating a mesh of specially coated fibers, which conduct electrical power in a distributed way over the whole surface of the blanket, thus gaining in homogeneity, while minimizing the risk to overheat areas outside repair zone. The special weaving of the fibers allows for cutting the blanket in different shapes whenever there is the need to follow contours that inhibit the use of a standard rectangular heating blanket. Power wires installed at predefined spots allow easy connection for power supply, through connectors that have been specially selected to be easily mounted with maximum safety.

5.7 *Magnetostrictive Technology: Ensure Bonding Adequacy/Durability for SHM (Structural Health Monitoring) and Certification Requirements*

The magnetostrictive sensing methodology is based on the coupled mechanical/magnetic behavior of certain materials, appropriately formed to become sensors (wires or ribbons) and then bonded on top of repair patches or other structures. Specially developed coils are used to excite and read magnetic response, which, thanks to a special software, reveals internal damage due to delaminations or other defects, through anomalies created in the magnetic response. The new proposed feature has to do with the methodology's potential to assist in aircraft/repair certification, providing continuous strain mapping of the area of interest, compared to spot measurements currently performed using strain gauges or other methods (e.g., A350 fuselage parts, extensive repairs). Figure 13.31 presents an experimental setup as well as a field instrument for magnetostrictive sensing.

5.8 *Out-of-Autoclave (OOA) Heating Solution for Production and Repair of Composites*

Advanced heating system and control mode for homogeneous high-temperature curing of large composite repairs—"ADVANCED" (Fig. 13.32) concerns the development of innovative solutions either for the application of very large composite repairs, to be performed outside autoclaves, or for OOA production. Achieving strict temperature tolerances (±5 at 180 °C or 225 °C) for repairs of several m^2 is challenging, but

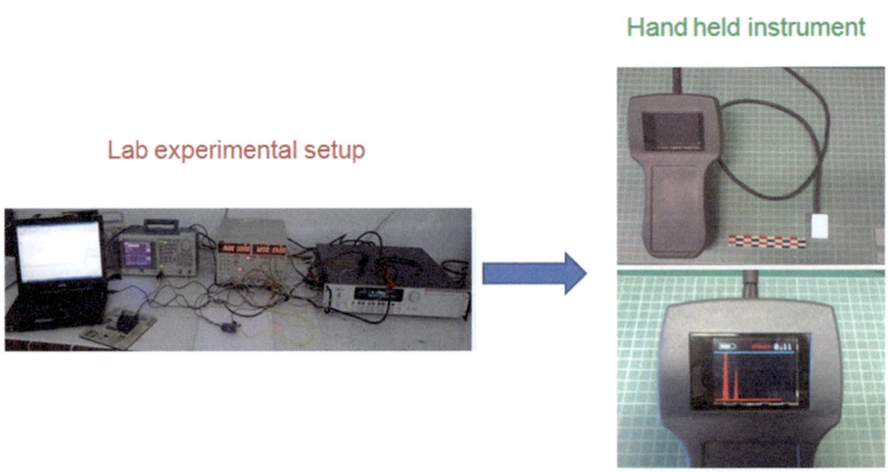

Fig. 13.31 Magnetostrictive sensing instruments

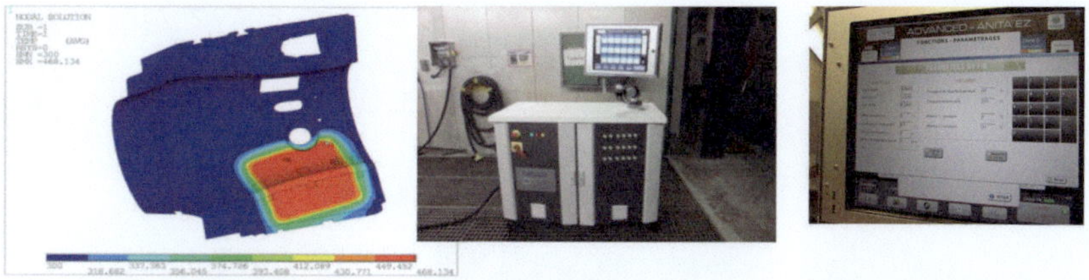

Example of thermal transfer simulation of A380 cowling Inner Fixed Structure to retrieve thermal signature of repair.

ADVANCED industrialized heating control and power supply unit delivered to Aircelle Le Havre.

Close-up of ADVANCED HMI unit, presenting maximum interoperability with ANITA.

Fig. 13.32 ADVANCED main features

expected benefits are significant: reduction of autoclave utilization induces direct cuts to overall repair cost and CO_2 footprint, as energy requirements for OOA curing are minimal, compared to autoclave. The main development steps include the following:

- 3D thermal FE simulation of full-repair case to retrieve "thermal signature" of the repair
- Customization of heating blankets to "thermal signature" of part
- Optimization of blankets, reducing number of control zones (nonuniform heating generation elements)
- Power supply and control unit development
- Development and validation at industrial environment of power supply and control unit (PS&CU) of 48 KW, capable of heating up to 18 heating zones at a target temperature tolerance of ±5 °C at 225 °C
- Simultaneous data acquisition from eighty (80) control and monitoring thermocouples, using innovative control algorithms with increased flexibility in defining control mode
- User-friendly HMI (similar to standard ANITA EZ), for immediate transition of operating personnel

5.9 Heating Pins: Heating Bolts

During drilling operations on composite laminate parts, damage can occur at the entrance/exit of the drill or internally. The *heating pins/bolts* have been developed to provide heat for curing the resin applied to repair these damages; compaction forces can be also mechanically generated. Inserted in the hole, the tool provides heat to raise the temperature at the periphery of the hole and in a localized area on the two in/out skins at a temperature of 80 °C approximately or at another pro-

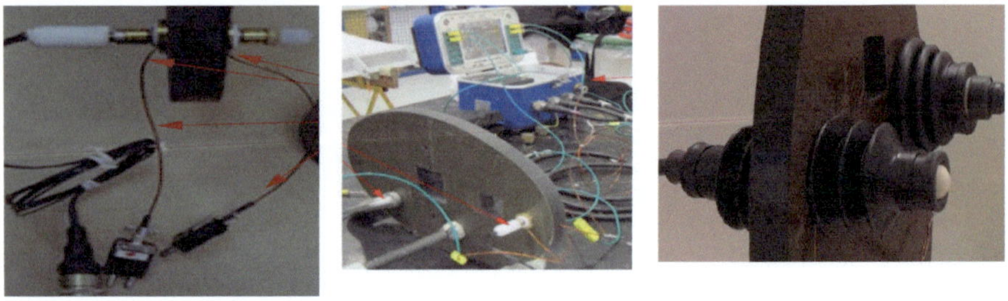

Temperature sensing on both skins by thermocouples.

Control oftwo heating devices of small diameter through ANITA.

Heating devices installed with appropriate insulation covers.

Fig. 13.33 Various heating pins and heating bolts

grammed value. To control the process, specific electrical and electronic interfaces with advanced programmable software have been developed. Diameters range from 3.2 to 19.1 mm, as presented in Fig. 13.33.

5.10 ANITA Remote Control Tablet (ARCT)

The continuous presence of operators in front of repairs to supervise curing cycles induces great cost of man-hours, without real production benefit. Remote control of hot bonders and supervision of running curing cycles would assist in increasing productivity of operators while facilitating their tasks. The ANITA Remote Control Tablet (ARCT) answers to this challenge by enabling the remote and wireless representation of ANITA EZ screen and sound alarms on an 8″ touch screen tablet. The ARCT presents a duplication of the screen and permits remote controlling of ANITA EZ. No process information is stored in the tablet. A portable, battery-operated, wireless loudspeaker, which is carried by the operator, reproduces all sound alarms. The system is fully "plug and play" and easy to use while being able to simultaneously control a maximum of four (4) ANITA EZ consoles, as presented in Fig. 13.34.

The system is now complemented by PANOPTES, a software permitting the remote supervision of up to eight (8) ANITA hot bonders. The overall architecture of the proposed solution is shown in Fig. 13.35.

6 Conclusions

As transport vehicle manufacturers are making great efforts to reduce the cost of production, they are investigating more cost-efficient manufacturing methods while developing more efficient vehicles, increasing the vehicle's performance, and reduc-

Duplication of ANITA EZ screen on an 8'' touch screen tablet (ARCT) in industrial environment.

Detail of the ARCT, showing the remote control screen. Up to 4 ANITA consoles could be simultaneously controlled.

Fig. 13.34 ANITA hot bonder remote control tablet

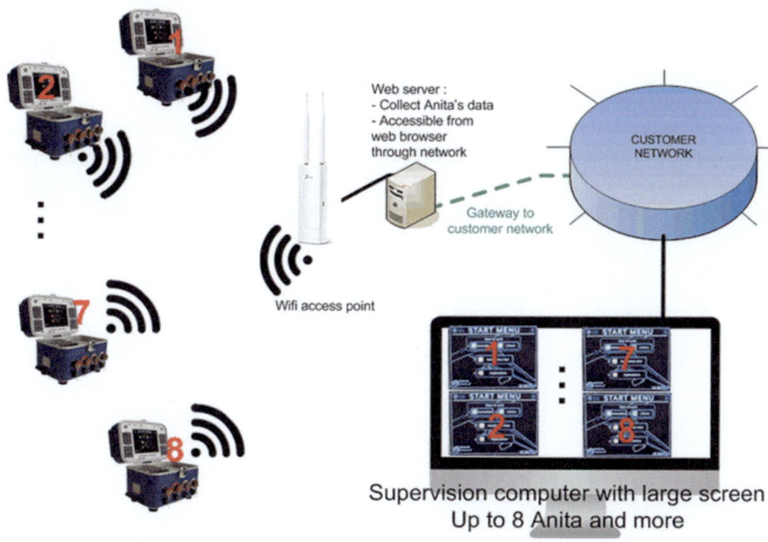

Fig. 13.35 PANOPTES network topology

ing its weight, fuel consumption, and emissions. As a result, the newly constructed aircraft of manufacturers like Airbus, Boeing, ATR, Bombardier, and Dassault have a structure made of 70–80% of carbon composite materials (CFRP). Even fuselage and other class I (i.e., safety critical) parts of the structure are now made in carbon. While composite structures become larger, more integrated, and more critical, guaranteeing the integrity of a major repair when the aircraft returns to service is a growing imperative. Research into methods of applying and inspecting (either online or off-line) repairs in situ could pave the way for even wider application of composites on aircraft while maintaining airworthiness and keeping costs at acceptable levels. Bonded composite repairs and associated equipment and application methodologies are thus expected to play a dominant role in aeronautics within coming decades.

References

Armstrong KB, Bevan G, Cole W (2005) Care and repair of advanced composites. In: SAE International, 2nd revised edition (30 Jun 2005)

Baker AA, Jones R (1988) Bonded repair of aircraft structures (Chapters 1,3,6&7). Martinus Nij Publishers, Dordrecht

Christopoulos A, Koulalis I, Chemama R, Hristoforou E, Kanterakis G, Tsamasphyros G, Kitsianos K (2016) Strain monitoring and damage detection of bonded composite structures, using magnetostrictive sensors—latest developments and applications. In: 6th EASN international conference, 18–21/10/2016, Porto, Portugal

Marioli-Riga ZP, Tsamasphyros GJ, Kanderakis GN (1997) Development of a method for A/C emergency repairs by composite patches. In: Paipetis SA, Gdoutos EE (eds) 1st Hellenic conference on composite materials, vol II, pp 143–156, Xanthi, Greece

SAE AIR5144 (2017) Heat application for thermosetting resin curing, Revised 2017-04

Tsamasphyros GJ, Kanderakis GN, Marioli-Riga ZP (1999) 3D FEA of debonding & thermal effects near the crack-tip of a metal struct. Rep. by Comp. Patch. In: Aravas M, Katsikadelis JT (eds) 3rd Nat. Cong. on Comput. Mechs, Univ. of Thessaly, vol II, pp 429–436, Volos, Greece

Tsamasphyros GJ, Furnarakis NK, Kanderakis GN, Marioli-Riga ZP (2003) Detection of patch debonding in composite repaired cracked metallic specimens, using optical fibers and sensors. In: SPIE optical metrology conference, 23–26 June 2003, Munich, Germany

Index

S. Pantelakis, K. Tserpes (eds.), *Revolutionizing Aircraft Materials and Processes*, https://doi.org/10.1007/978-3-030-35346-9

B

MIX
Papier aus verantwortungsvollen Quellen
Paper from responsible sources
FSC® C105338

If you have any concerns about our products,
you can contact us on
ProductSafety@springernature.com

In case Publisher is established outside the EU,
the EU authorized representative is:
Springer Nature Customer Service Center GmbH
Europaplatz 3, 69115 Heidelberg, Germany

Printed by Libri Plureos GmbH
in Hamburg, Germany